XXIII School-Seminar of Young Scientists and Specialists Under the Leadership of the Academician, Professor A.I. Leontiev "Problems of Heat and Mass Transfer and Gas Dynamics in Power Plants" (XXIII SSYSS 2021)

Journal of Physics: Conference Series Volume 2039

Yekaterinburg, Russia
24 – 28 May 2021

ISBN: 978-1-7138-5010-6
ISSN: 1742-6588

Printed from e-media with permission by:

Curran Associates, Inc.
57 Morehouse Lane
Red Hook, NY 12571

Some format issues inherent in the e-media version may also appear in this print version.

Institute of Physics
Dirac House, Temple Back
Bristol BS1 6BE UK

Phone: 44 1 17 929 7481
Fax: 44 1 17 920 0979

techtracking@iop.org

Additional copies of this publication are available from:

Curran Associates, Inc.
57 Morehouse Lane
Red Hook, NY 12571 USA
Phone: 845-758-0400
Fax: 845-758-2633
Email: curran@proceedings.com
Web: www.proceedings.com

TABLE OF CONTENTS

Preface

School-Seminar of Young Scientists and Specialists (SSYSS) under the leadership of the academician of RAS A.I. Leontiev "Problems of Heat and Mass Transfer and Gas Dynamics in Power Plants" is organized by National Committee for Heat and Mass Transfer, Russian Academy of Sciences. It has been held every two years since 1977. The School-Seminar brings together young researchers and leading scientists working in the field of heat and mass transfer, and provides an opportunity for open discussion and assessment of the results obtained.

In 2021, the Institute of Thermal Physics of the Ural Branch of RAS was the co-organizer of the School-Seminar, therefore XXIII SSYSS was held for the first time in the Urals near Ekaterinburg on May 24th - 28th 2021.

More than 150 reports were included in the scientific program, including about 30 lectures of leading Russian thermophysicists. Due to the current situation of COVID-19 epidemic, some lectures took place online.

Topics of the XXIII SSYSS:
1) Heat and mass transfer in single-phase convection
2) Metastable states and fluctuation phenomena
3) Heat and mass transfer in two-phase and multiphase media
4) Complex heat and mass transfer and porous media
5) Chemical transformations and heat and mass transfer
6) Applied problems of heat and mass transfer: intensification and efficiency
7) Numerical methods in the theory of heat and mass transfer

These topics were discussed, among other things, in relation to power engineering application and to gas turbines, rocket engines, space facilities, nuclear reactors, cryogenic devices, environment protection. In addition, fundamental issues of phase metastability, molecular dynamics modeling, fluctuation phenomena and heat transfer during phase transformations, surface phenomena and thermophysical properties were considered. The special attention was paid to environmental safety issues.

According to the tradition of the School-Seminar, the focus was on a detailed discussion of sectional poster presentations with a selection of the best reports in each section.

XXIII SSYSS Proceedings is a collection of scientific papers submitted to the School-Seminar. All the papers were subjected to peer-review by conference committee members and leading Russian scientists in the field of heat and mass transfer.

We would like to express gratitude to all members of SSYSS Committee and National Committee for Heat and Mass Transfer. Sincerest appreciation goes to Dmitry Olegovich Onishchenko for financial support, JSC "MPEI Publishing House" and personally Georgy Glebovich Yankov for organizing and supporting the process of preparing papers for publication. Special thanks to Andrey Vladimirovich Vinogradov, Director of the Institute of Thermal Physics UB RAS, as well as to all staff of this Institute for their hospitality and efficient work in organizing XXIII SSYSS. Finally, we want to thank all the authors and conference participants for the productive cooperation.

XXIII SSYSS Committee

Committee

Chairman
Acad. RAS **Alexander Ivanovich Leontiev**, Bauman Moscow State Technical University

Deputy Chairman
Corr. Member RAS **Alexander Nikolaevich Pavlenko**, Institute of Thermophysics SB RAS

Scientific Committee Members
Acad. RAS **Sergey Vladimirovich Alekseenko**, Institute of Thermophysics SB RAS
Acad. RAS **Oleg Mikhailovich Alifanov**, Moscow Aviation Institute
Prof. **Alexander Nikolaevich Arbekov**, Bauman Moscow State Technical University
Corr. Member RAS **Alexey Yurievich Varaksin**, Joint Institute for High Temperatures RAS
Prof. **Andrey Vladimirovich Vinogradov**, Institute of Thermal Physics UB RAS
Prof. **Yuri Fedorovich Gortyshev**, Kazan National State Technical University, Institute of Energy Efficient Technologies
Prof. **Alexander Igorevich Guryanov**, Rybinsk State Aviation Technological University
Prof. **Igor Vladimirovich Derevich**, Bauman Moscow State Technical University
Corr. Member RAS **Alexey Viktorovich Dedov**, National Research University "Moscow Power Engineering Institute"
Prof. **Alexander Sergeevich Dmitriev**, National Research University "Moscow Power Engineering Institute"
Prof. **Leonid Aleksandrovich Dombrovsky**, Joint Institute for High Temperatures RAS
Corr. Member RAS **Ivan Vladimirovich Egorov**, Central Aerohydrodynamic Institute

Prof. **Viktor Vladimirovich Yagov**, National Research University "Moscow Power Engineering Institute"
Prof. **Leonid Samoilovich Yanovsky**, Central Institute of Aviation Motors, Institute of Problems of Chemical Physics RAS
Prof. **Georgy Glebovich Yankov**, National Research University "Moscow Power Engineering Institute", JSC "MPEI Publishing House"

Scientific Secretary
Dr. **Natalia Vladimirovna Medvetskaya**, Joint Institute for High Temperatures RAS

Secretary
Rimma Sergeevna Gromadskaya, Joint Institute for High Temperatures RAS

Local Organizing Committee (Institute of Thermal Physics UB RAS)
Dr. **Valentina Nikolaevna Andbaeva**
Dr. **Konstantin Anatolievich Busov**
Dmitry Alekseevich Galkin
Svetlana Yurievna Elina
Dr. **Maksim Sergeevich Zakharov**
Dr. **Evgeny Vladimirovich Lipnyagov**
Dr. **Petr Evgenievich Mezentsev**
Andrey Sergeevich Tomin
Maria Nikolaevna Khotienkova

Peer review declaration

All papers published in this volume of Journal of Physics: Conference Series have been peer reviewed through processes administered by the Editors. Reviews were conducted by expert referees to the professional and scientific standards expected of a proceedings journal published by IOP Publishing.

- **Type of peer review: Single-blind**
- **Conference submission management system: –**
- **Number of submissions received: 42**
- **Number of submissions sent for review: 42**
- **Number of submissions accepted: 36**
- **Acceptance Rate (Number of Submissions Accepted / Number of Submissions Received X 100): 85**
- **Average number of reviews per paper: 2**
- **Total number of reviewers involved: 28**
- **Any additional info on review process: –**
- **Contact person for queries:**
  Name : Dr. Maksim Zakharov
  Affiliation: Institute of Thermal Physics, Ural Branch, Russian Academy of Sciences
  Email : maksim.s.zakharov@gmail.com

Investigation of aerodynamics and heat transfer of the modular jet recuperator

P D Alekseev and Yu L Leukhin

Northern (Arctic) Federal University named after M.V. Lomonosov
Russia, 163002, Arkhangelsk, Severnaya Dvina Emb. 17

p.alekseev@narfu.ru

Abstract. A study of the aerodynamics and heat transfer of a jet modular recuperator with a change in its geometric characteristics has been carried out. The influence of the in-line and staggered arrangement of the blowing holes, as well as the diameter of the perforated pipe is considered. In all considered variants, the number of holes, their diameter and gas flow rate through the recuperator remained unchanged. Numerical modeling of the problem was carried out in a three-dimensional setting using the ANSYS Fluent 15.0 software package. It was found that with the in-line arrangement of the blowing holes, secondary flows are formed between their longitudinal rows in the form of swirling jets of opposite rotation directed towards the outlet section of the recuperative device, through which the main part of the heated air flows out. With the staggered arrangement of the blowing holes, the formation of spiral vortices is disturbed, the air flow is carried out along the entire cross section of the annular channel, increasing the drift effect of the flow on the impact jets, which leads to a decrease in the intensity of heat transfer and its uniformity along the length of the working surface. An increase in the diameter of the inner perforated pipe leads to a decrease in the drift effect of the cocurrent flow on the jets, an increase in the distribution uniformity of the heat flux along the length of the heat transfer surface, and an increase in the heat transfer coefficient.

1. Introduction

Modern trends in recuperator engineering, taking into account the most efficient use of production areas, require the creation of compact recuperators with high thermal efficiency and operational reliability. To a large extent, these requirements are met by modular recuperators, which are easy to operate, can be easily built into vertical and horizontal gas ducts, and have the ability to get any required heat transfer surface by selecting the required number of modules [1]. One of the ways to intensify convective heat transfer in recuperative devices is the jet leakage of the heat transfer gas onto the surface. A sharp intensification of heat transfer is achieved by a short length of the jet path due to significant turbulence of the near-wall gas layer at the blown surface [2]. The jet modular recuperator shown in figure 1 works as follows: cold air is fed into the inner tube 1, perforated with blowing holes, from which, in the form of an impact jets system, it is supplied to the inner surface of the heat transfer tube 2 and heats up. The module itself is installed in the horizontal gas duct 3 and is washed from the outside by the flue gases escaping from the power engineering installation.

The purpose of this work is to study the aerodynamics and heat transfer of a modular jet recuperator in order to develop a design with a higher thermal efficiency.

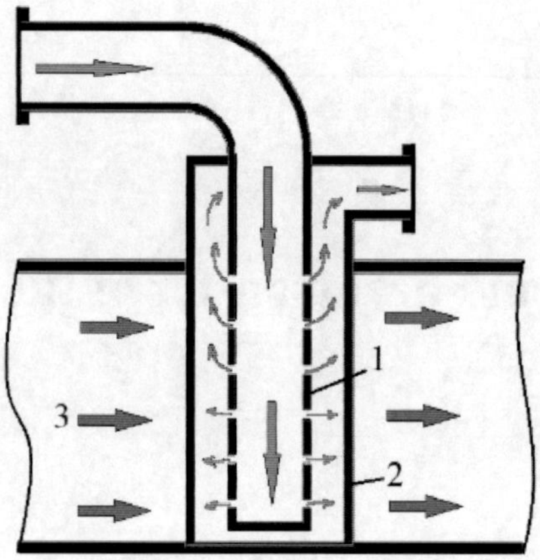

Figure 1. Scheme of a jet modular recuperator: 1 - internal perforated pipe; 2 - outer pipe; 3 - gas duct.

2. Research method

Numerical modeling of aerodynamics and heat transfer in a jet modular recuperator was carried out with a model, the geometrical dimensions of which corresponded to a real recuperator. The working length of the outer (heat transfer) pipe was 1389 mm, and its inner diameter d_2 was 100.5 mm. The outer diameter of the perforated insert was d_1 = 60.5 mm, its wall thickness was 5 mm, and the diameter of the blow holes was d_0 = 3.5 mm. The number of holes along the perimeter of the cross section was 5, and the number of rows was 33. The holes along the length had in – line arrangement. The temperatures of the air supplied to the device and of the outer pipe surface were 19.8°C and 100°C respectively, the air flow rate was 0.0055 $m^3 s^{-1}$. The air flow rate and temperatures were chosen to verify the calculation results according to the experimental data obtained on the experimental model of the jet modular recuperator with similar geometric dimensions and boundary conditions [1]. In this work, we additionally investigated the options for the staggered arrangement of the blowing holes, as well as with an increased outer diameter of the perforated pipe up to 75.5 mm. In all considered variants, the number of holes and gas flow rate (hence, the velocity of the jet outflow) remained unchanged.

Mesh models were built in the ANSYS ICEM CFD software module, implemented by the boundary correction method (Octree) and are unstructured tetrahedral meshes with a maximum dimension of 8 million cells (a fragment of the grid for one of the jets is shown in figure 2). Calculations performed on grids of higher dimension did not lead to a noticeable change in the calculation results.

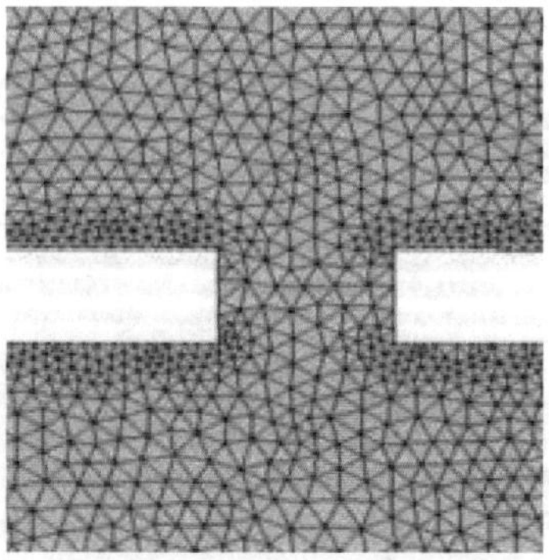

Figure 2. Fragment of the mesh model near the blow hole.

The study of the problem was carried out in a three-dimensional setting using the ANSYS Fluent 15.0 software package. The flow was described through Navier-Stokes equations of continuity and energy that are Reynolds-averaged. Reynolds equations closure is performed using both: a two-parameter turbulence model SST (Shear Stress Transport) *k-ω* streamlines curvature correction. To solve the problem, the following boundary conditions were used: 19.8 ° C temperature and 0.0055 m^3s^{-1} flow rate of air supplied to the inner perforated pipe of the device; 100 ° C outer pipe surface temperature; the air pressure at the outlet from the device was equal to atmospheric. The outflow velocity of the jets and the Reynolds number, determined from this velocity and the diameter of the blowing holes, were equal to 34.6 m s^{-1} and $8.06 \cdot 10^3$, respectively. The air flow rate and temperature were chosen to verify the calculation results according to the experimental data obtained on the experimental model of the jet modular recuperator with similar geometric dimensions and boundary conditions [1].

3. Research findings

The modular recuperator aerodynamic features of the type under consideration are determined by the conditions of impact jets interaction, flowing from the inner perforated pipe with a concave cylindrical surface while one-sided flow diversion from an annular channel. Due to the high resistance of the blowing holes, the gas flow rates through them differ by a maximum of 7%, and a higher flow rate is observed through the holes located at the blind end. The jets coming out of this holes, hit the surface of the outer tube forming a set of unsteady vortex structures. The combination of these structures leads to the formation of secondary flows in the form of spiral vortices rotating in opposite directions. Figure 3 shows the streamlines projections onto the channel cross-section, and figure 4 shows the distribution of secondary vortex flows along the length of the annular channel along one row of holes. With the in-line arrangement of the blowing holes, two swirling jets of opposite rotation are formed between their longitudinal rows, directed towards the outlet section of the recuperative device. As the vortices move, their intensity increases due to the receipt of additional energy and mass from the jets located closer to the outlet. Through the formed secondary vortices, the main part of the heated air flows out of the module.

With an increase in the air flow rate to the outlet section of the annular channel and the intensity of the secondary vortex motion, the drift effect of the cocurrent flow on the impact jets increases. Figure 5 shows the change in temperature along the annular channel length (the flow out of the channel occurs on the left). Along the lower part of the cross - section, which coincides with the blowing holes axes, it can be seen that as the outlet is approached, the drift effect of the cocurrent flow gradually increases, and in the upper part, located between the rows of jets, the heated air transfer by secondary vortices to the surface of the inner pipe is observed. Due to the air flow rate increase and the constrained jets arrangement, not only the increase in the speed of the rotational movement occurs, but also their spinning.

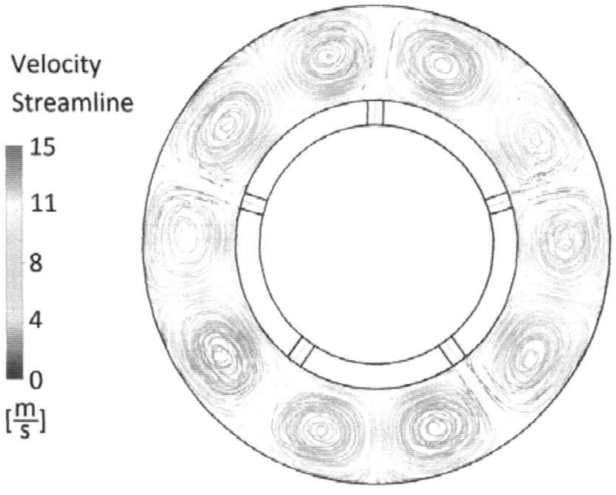

Figure 3. Streamline projections on the cross-sections of the annular channel.

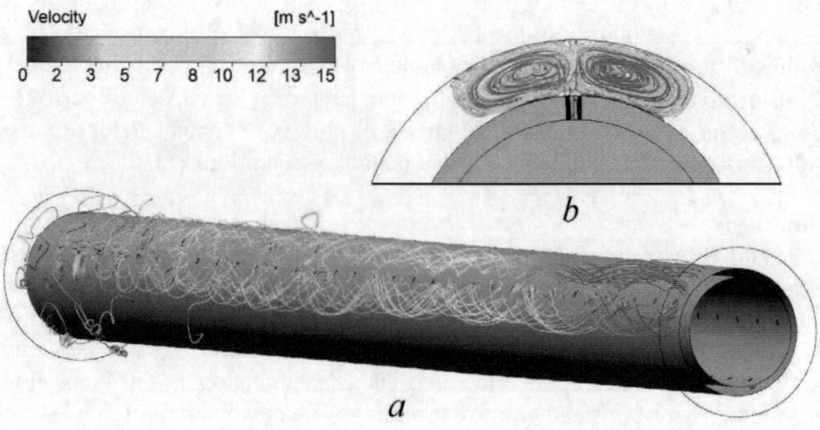

Figure 4. Distribution of secondary vortex flows along the length of the annular channel.

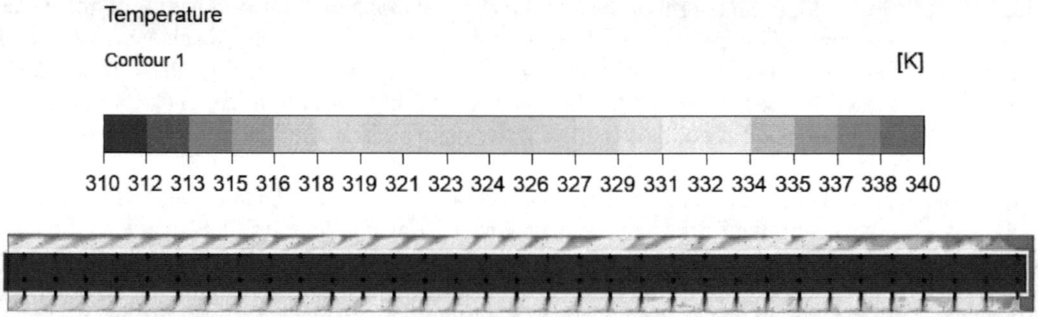

Figure 5. Temperature distribution along the length of the annular channel.

The nature of the distribution of the heat flux density represented in figure 6 shows that the most intense local heat transfer on the surface of the outer (heat-transferring) pipe is observed on its half, located closer to the blind end. As the flow moves towards the outlet, the jet action influence to the surface weakens. The maximum values of the heat transfer coefficient along the generatrix of the outer tube, located opposite the blowing holes, decrease towards the outlet section by about 40% (figure 7 line 1), and its minimum values on the same generatrix (between the jet axes), on the contrary, to the exit increase by about 30%. It should be noted that there is a significant non-uniformity of heat transfer along this typical line. On average, the maximum values exceed the minimum ones by 3.6 times.

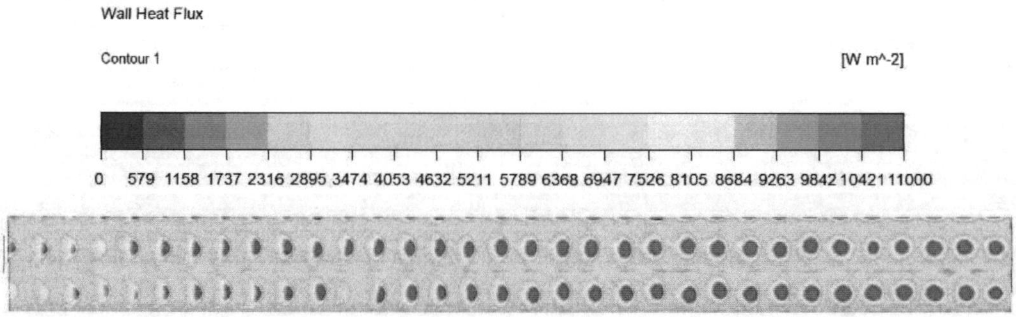

Figure 6. Distribution of the heat flux density on the surface of the outer pipe with the in-line arrangement of the blowing holes.

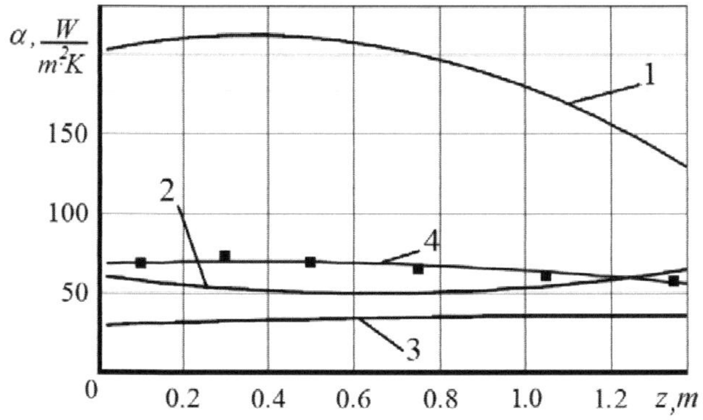

Figure 7. Distribution of heat transfer coefficients along the length of the outer pipe: line 1 and line 2 - maximum and minimum values along the generatrix located opposite the holes; line 3 - on the generatrix located between the rows of holes; line 4 - surface averages. Points - experimental data.

Line 3 shows the change along the length of the heat flux density along the generatrix of the outer surface located between the rows of blow holes. Heat transfer in this area is the lowest, despite the fact that it slightly increases towards the exit under the influence of secondary vortices. Moreover, it is 5.7 times less than the average maximum value (line 1).

The change along the length of the surface-averaged heat transfer coefficient (line 4) and its comparison with the experimental data show satisfactory agreement.

With the staggered arrangement of the same number of blowing holes, the conditions for the interaction of the jets with the surface of the outer tube and the formation of secondary vortices change. The absence of "corridors" does not allow the formation of spiral vortices directed from the blind end to the exit. Air flow is carried out over the entire cross-section of the annular channel, increasing the drift effect on the jets. The distribution of the heat flux density over the surface of the outer pipe with a staggered arrangement of the blowing holes is even more uneven in length (figure 8) than with a in- line one. In this case, the heat transfer coefficient averaged over the surface is 9.7% less.

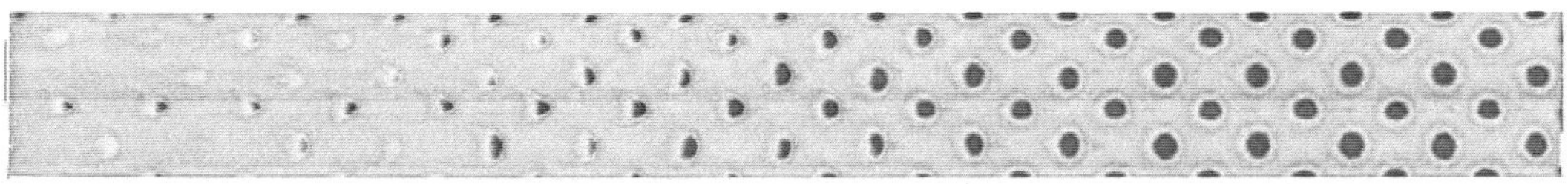

Figure 8. Distribution of the heat flux density over the surface of the outer tube with a staggered arrangement of the blowing holes.

With an increase in the diameter of the inner perforated pipe from 60.5 mm to 75.5 mm, the relative distance to the heat transfer surface d_h / d_0 decreases by 37.5% ($d_h = d_2 - d_1$ is the hydraulic diameter of the annular channel). As a result, the drift effect of the cocurrent flow on the jets is reduced, and the uniformity of the heat flow distribution along the length of the heat transfer surface is increased (figure 9). The heat transfer coefficient averaged over the surface increases by 5%, while the resistance of the module increases by 10%.

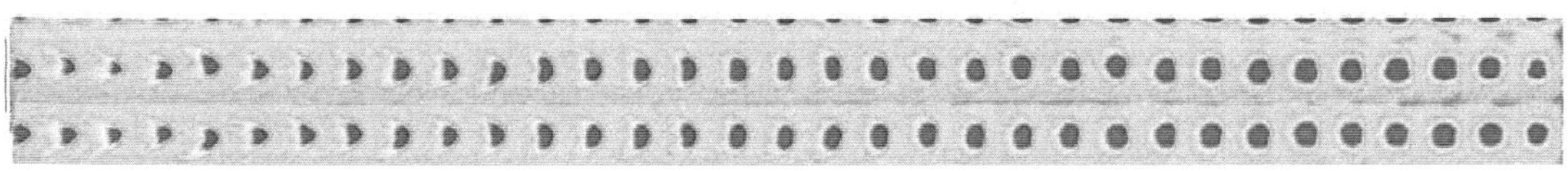

Figure 9. Distribution of the heat flux density over the heat transfer surface with a perforated pipe diameter d_1 = 75.5 mm.

It should be noted that in all the variants considered in the work, there is a significant non-uniformity of heat transfer at the working surface, due to the local impact of jets on it.

4. Conclusions

With the in-line arrangement of the blowing holes, secondary flows are formed between their longitudinal rows in the form of swirling jets of opposite rotation, directed towards the outlet section of the recuperative device, through which the main part of the heated air flows out.

The staggered arrangement of the holes disrupts the formation of spiral vortices. Therefore, the air flow is carried out along the entire cross section of the annular channel, increasing the drift effect of the flow on the impact jets, which leads to a decrease in the intensity of heat transfer and its uniformity along the working surface length.

An increase in the diameter of the inner perforated pipe leads to a decrease in the drift effect of the cocurrent flow on the jets, an increase in the distribution uniformity of the heat flux along the length of the heat transfer surface, and an increase in the heat transfer coefficient.

References

[1] Saburov E, Ostashev S, Orekhov A, Leukhin Y, Nikiforov V, Kreinin E and Shaginyan I 1988 *J. Industrial energy.* **6** 33
[2] Dyban E and Mazur A 1982 *Convective heat transfer with jet cooling of bodies.* (Kiev: NAUKOVA DUMKA)
[3] ANSYS 15 - User's guide

On the dynamic effects of evaporation and boiling of drops on inhomogeneous graphene substrates for effective thermal management of energy devices

D D Babenko, A S Dmitriev and I A Mikhailova

Thermo Nano Laboratory, Low Temperature Department,
National Research University «Moscow Power Engineering Institute»
Russia, 111250, Moscow, Krasnokazarmennaya, 14

rainbowdm1802@gmail.com

Abstract: This research examines thermal performance of metal-graphene nanocomposites for thermal management. Graphene nanoflakes and graphite microparticles were compressed with cooper meshes under different pressures. The prepared samples were characterized using optical microscopy. The fabricated samples of cooper-graphene composites were studied to determine the droplet evaporation rates and contact angles at different temperatures including film boiling.

1. Introduction

Due to constant growth of thermal design, power of modern electronic devices there is growing interest in new thermal interfaces materials.[1-4] One of the main goals of the work is to study the properties of thermal interface materials for intensifying heat transfer and increasing the efficiency of cooling systems in microelectronics and high-current electronics. The physicochemical and thermohydrodynamic properties of such surfaces are important for more efficient heat dissipation in cooling systems using the drop cooling method. In such systems, the intensity of evaporation of the working liquid from the surface, as well as the conditions for the transition from the bubble mode of boiling to the film mode, are extremely important. The structure of the tablet has a strong effect on the evaporation process. Moreover there was investigated that pressing pressure on graphene composite heavily impacts on evaporation processes and on the destruction of the surface at high temperatures. The structure with a grid on the surface was chosen to check an impact on evaporation time and contact angles at wide range of temperatures.

Previously, surface modification using various polymer networks to study wettability and evaporation rates has already been used [5]. However, we are using metal meshes with graphene flakes here for the first time to study contact angles and evaporation rates of droplets.

In this work, we investigated the processes of evaporation and boiling, up to the Leidenfrost temperature, of liquid droplets (mainly water) on the surface of inhomogeneous graphene and graphite substrates with metal meshes. The effect of an inhomogeneous surface on the evaporation rate of droplets, as well as the behavior of contact angles depending on the temperature and parameters of the substrate surface, is shown.

2. Materials and method

Graphene nanoflakes with an average size of 10-30 μm (thickness - 1.5-3.5 nm) were received from Graphene Institute Ltd (Russia). Chemically clean graphite microparticles ~300 μm were used. As a mesh we used cooper M1(Cu>99.6%) mesh with cell side size 0.3 mm and wire diameter 0.1 mm supplied by TDMC, Russia. Samples were pressed in hydraulic press under 180-300 bar pressure in inch diameter tablet. We prepared 4 samples in total: a) graphene nanoflakes under 300 bar; b) graphene nanoflakes with cooper mesh under 300 bar; c) graphene nanoflakes with cooper mesh under 180 bar; d) graphite microparticles with cooper mesh under 300 bar (figure 1).

The technique for measuring the contact angle and the evaporation rate is similar to that presented in our works [6,7].

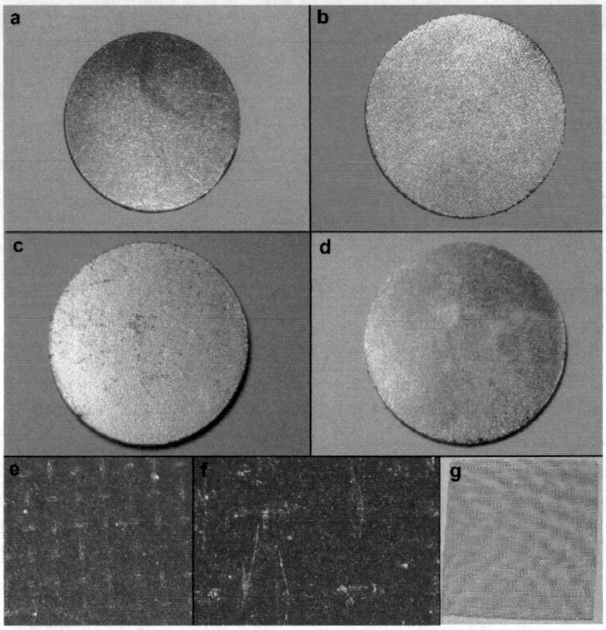

Figure 1. Photos and microscopy of samples. a) graphene nanoflakes p=300 bar; b) graphene nanoflakes with cooper mesh p=300 bar; c) graphene nanoflakes with cooper mesh p=180 bar; d) graphite microparticles with cooper mesh p=300 bar; e) sample B optical microcopy x4 magnification; f) sample B optical microcopy x10 magnification; g) sample of cooper mesh 03x01.

As we see from microscopy metal wires are not cylindric so we consider the surface as flat and somewhere graphene covers metal wires as we see in 10x magnification. For all types of samples, contact angles and evaporation rates were studied. Kruss EasyDrop was used as a system for measurement the contact angle and evaporation rates of drops on the composites. Also to measuring system was added a special heater with temperature control realized by PID controller for making a research in wide range of temperatures. Evaporation rate and contact angles were measured for all 4 samples. For each sample and each temperature was made from minimum 3 measurements at temperatures 30-40 °C and minimum 5 measurements for 50 °C and more. Distilled water was used as a working liquid.

3. Contact angles for samples nanocomposites and evaporation rates

Wettability makes a heavy impact on heat exchange between operating fluid and working surface. Various materials can be used for this purpose and more various methods can be used for surface modification [4-5]. In our work we made metal-graphene composite with regular structure made by webbed wires of cooper mesh (figure 1 e,f).

As a result, we can see that adding cooper mesh in graphene tablet leads to increase of contact angle from several degrees till 20 degrees under different surface temperatures (figure 2). But from another hand we see that on the sample C contact angle is dropping relatively to sample A. For proper description we need further researches, but preliminarily we explain it with porosity of graphene samples made with relatively low pressure. Also pressure heavily impacts on surface durability on high temperatures which described in more details in next section.

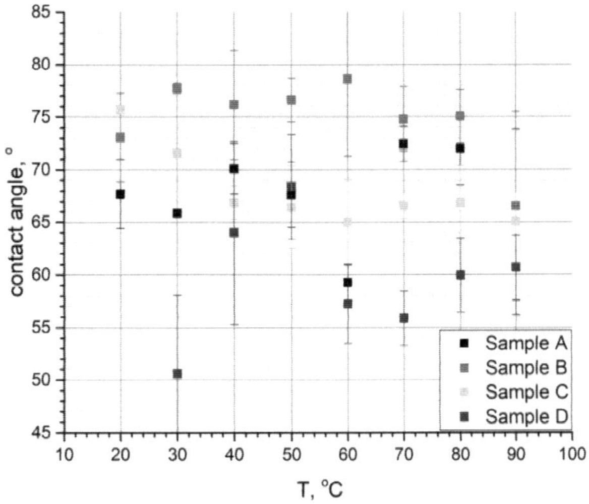

Figure 2. Contact angles at the start of the evaporation of water droplets for nanocomposites.

Relating to evaporation rates we see relatively the same times on 40-90 °C. But closer to boiliIing point of water we see great difference between samples. As we see from figure 3 adding a mesh to the surfaces decreasing the evaporation time on both ranges from 100 °C to Leidenfrost point and over Leidenfrost point. On lower pressure sample we see even more rapid evaporation in 100 °C to Leidenfrost point but over it while film boiling evaporation rate is lower than on sample B. Even more Leidenfrost point is higher approximately on 10 °C. On graphite sample we see medium situation relatively to sample B & C on 100 °C to Leidenfrost point range, but after it there is very rapid difference in evaporation rate.

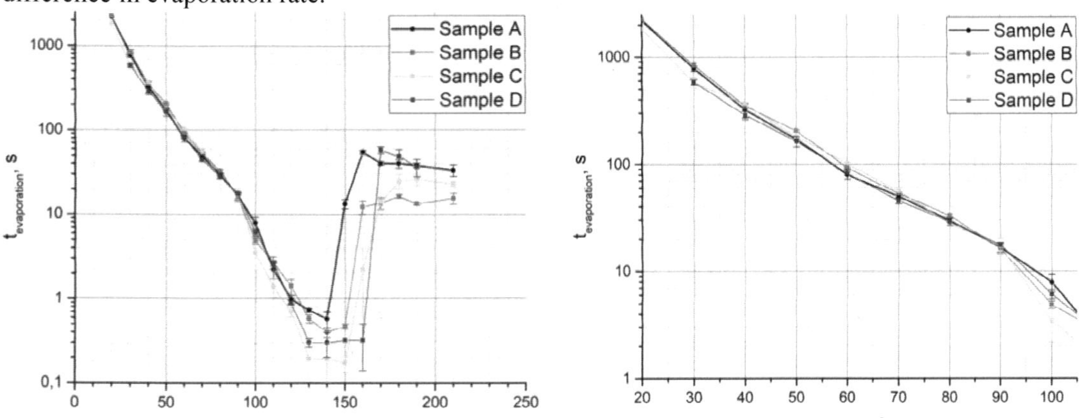

Figure 3. Evaporation time of water droplets for nanocomposites at different surface temperatures.

4. Surface destruction

Previously we described surface destruction of surface of graphene tablets [6]. In this work we made 3 samples under 300 bar pressure. For this samples there is no visual destruction even in optical microscope with 40x magnification. But we investigated that no matter how you clean the sample before the droplet evaporation, on temperatures higher than Leidenfrost point, levitating droplet finds some particles and we see an evaporation of colloid solution (water + graphene nanoflakes).

Figure 4. Levitating droplet of water with graphene nanoflakes from surface.

Cleaning the surface after each droplet didn't helped as well. This particles are not visible on the surface and we observed them only closer to the end of the evaporation process because color of water changes from transparent to black during the process and leaves batch of graphene nanoflakes at the end (figure 4).

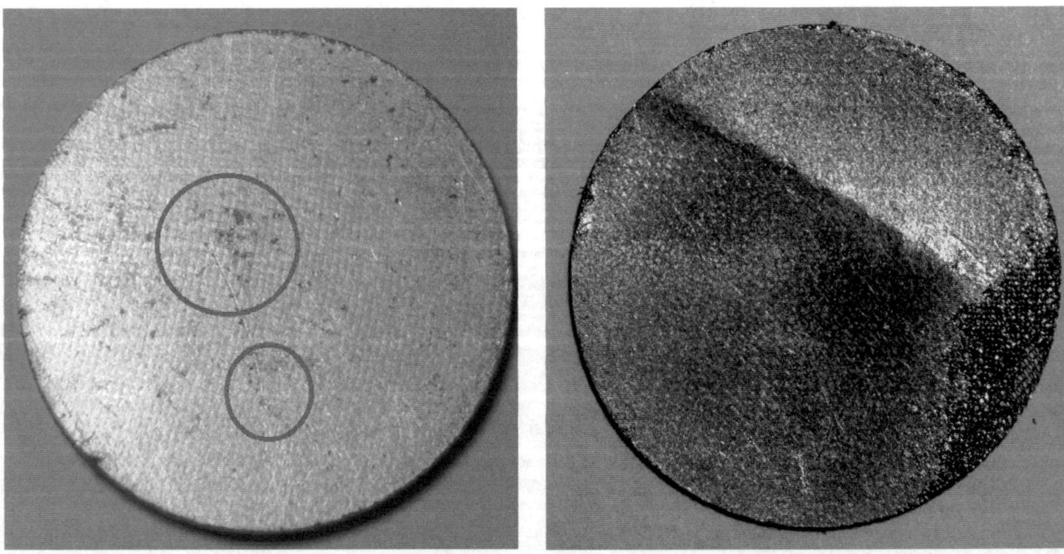

Figure 5. Photos of samples exposed to destruction sample C (left), sample D (right).

For the sample C we observed nearly the same situation that in [6]. The only difference that size of partial destruction is limited by the metal cell dimensions (figure 5). Also graphene detachment from surface of the wire remains possible. So in case of durable using we can get a rough surface that radically differs from initial.

For the sample D we observed a bit different situation. If for the graphene samples 300 bar is enough pressure to obtain stable surface, graphite has lower adhesion force between particles, and after all measurements there was significant surface destruction at the edge of the sample (figure 5). Wires of the mesh was separated from the graphite at the edge and inner layers of graphite was more subjected to destruction, so when the droplet moved on the edge of the sample character of evaporation radically changed and early on we limited surface of the sample, and later made the 2nd same sample for measurements on high temperatures. To confirm accuracy of results we compared evaporation rate on both samples and didn't observe significant differences. That's why we didn't make any other samples containing graphite as surface destruction is too heavy for using. To prevent this destruction there is a possibility of sintering processes while manufacturing or adding some polymers to keep the sample intact.

5. Results and discussions

In this work, we were the first to use metal meshes with graphene flakes to study the contact angles and evaporation rates of droplets. In our opinion, such inhomogeneous structures are very promising for active thermal interface materials with drop cooling of surfaces. This is especially important when the surface temperature is close to or higher than the Leidenfrost temperature. In this situation, thermal stability of the surface is required as a thermal interface material.

The structure with a grid on the surface reduces the evaporation time in the temperature range between the boiling point of the working fluid and the Leidenfrost temperature, moreover, for a sample with a lower pressing pressure, as a consequence, with a higher porosity, by more than 4 times. On a sample with graphite and a grid, it was also possible to achieve improvements in the rate of evaporation, as well as the values of contact angles are smaller than on samples with graphene, but graphite samples are also less resistant to water, as well as sample C (see Fig. 5) that they become nearly impossible to use in real arrangements. Moreover very high evaporation times while film boiling could be imperfect on the working temperatures close to Leidenfrost point. Using structures with high porosity allows us to increase Leidenfrost point and also adding cooper mesh to the surface leads to the same result, but we should admit that evaporation time after Leidenfrost point on the sample with high porosity is higher than on the same surface pressed under higher pressure with less pores. Destruction of samples with pressed graphene nanoflakes is need to be investigated, but as we can see 300 bar is enough pressure to evade much surface destruction.

Acknowledgments

The work is supported by the Russian Science Foundation (Grant No 17-19-01757).

References

[1] Alexander S. Dmitriev 2019 Hybrid Graphene Nanocomposites: Thermal Interface Materials and Functional Energy Materials, Graphene Production and Application, Sadia Ameen, M. Shaheer Akhtar and Hyung-Shik Shin, *IntechOpen*

[2] Hanssona J, Nilssona T, Ye L and Liu J 2017 *Int. Mat. Rev.* No **63** p 22

[3] Zhang P and Lv P 2015 *Energy* **82** p 1068

[4] Vorobyev A and Guo C 2013 *Laser Photonics Rev.* **7** no 3 385–407

[5] Kurusu R and Demarquette N 2018 *Int. Mat. Rev.* **64** no 5 p 249

[6] Babenko D et al 2018 *IOP Conf. Ser.: Mater. Sci. Eng.* **447** 012044

[7] Babenko D et al 2020 *J. Phys.: Conf. Ser.* **1565** 012010

Numerical simulation of a separated flow in a ribbed channel by the RANS and LES methods

A V Barsukov, V V Terekhov and V I Terekhov

Kutateladze Institute of Thermophysics SB RAS, Novosibirsk, Russia

andreybarsukov96@gmail.com

Abstract. The results of numerical simulation of a turbulent flow in a flat channel in the presence of vortex generators in the form of periodic solid ribs by the RANS and LES method are presented. The Reynolds number calculated by the rib height and the average flow rate is Re = 12600. The influence of the distance between the ribs on the flow structure is investigated. The boundaries of different types of roughness and their influence on the heat transfer intensity are shown.

Introduction

The study of complex separated flows is grounded by a large number of practical applications. One of the most important tasks in this area is to improve the efficiency of heat exchange equipment [1]. To control separated flow parameters, affecting heat transfer directly, various methods are used, which can be conditionally divided into the active and passive ones. Oscillating ribs [2], flow injection [3] or suction [4] can be used as the active methods. These methods allow a smooth control of separated flow parameters; however, the use of such methods is difficult. Passive control methods are preferred due to their simpler implementation. These methods include the presence of transverse ribs in front of the backward-facing step, which increases heat transfer on the lower wall of the channel at the point where the flow attaches. At that, an increase in heat transfer can reach 30%. In addition, tabs were used as vortex generators in [5, 6]. They allowed the achievement of a significant increase in heat transfer in the area located immediately behind the step, and the hydraulic resistance increased insignificantly.

In addition, transverse ribs can be also used as vortex generators. At that, depending on the distance between them, they can form a k- or d-type roughness. So the authors of [7] studied the flow structure at different distances between the adjacent ribs. A transition between two types of roughness was shown. But the question about the effect of a distance between the ribs on heat transfer was not considered.

This work deals with numerical simulation of a separated flow in a flat channel with a backward-facing step in the presence of periodic ribs. The effect of the inter-rib distance on the type of roughness and intensity of heat transfer is analyzed.

Computation details

The paper presents the results of numerical simulation of a separated flow in a flat channel with periodic ribs. The Reynolds number, calculated from the rib height and the average flow rate, was Re = 12600. The distance between the ribs was presented by variable parameter $w/x = 1 \div 21$. A constant heat flux was set on all walls of the channel, except the upper one, which was thermally insulated. At the inlet to the computational domain, a fully developed velocity profile and a constant temperature of the medium were set. The computational domain is presented in figure 1.

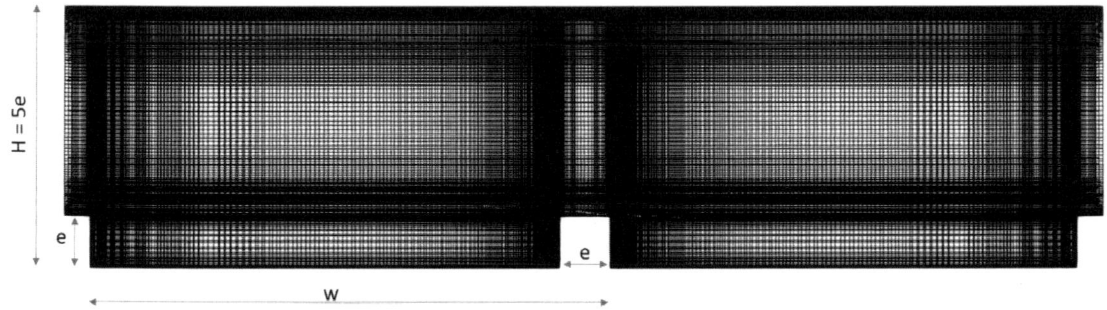

Figure 1. The scheme of computation domain.

Numerical simulation was performed using the RANS method. Integration of all equations included in the mathematical model was carried out using the control volume method of the second order of accuracy over the space in a stationary (iterative) setting using the cyclic TemperatureFoam solver from the OpenFOAM package.

A rectangular grid, condensed to the channel walls, was build for this problem. The criteria for finding the optimal number of cells were based on the constancy of solution with using more number of cells and $y^+ < 1$ (figure 2). In preliminary calculations, the results obtained were compared with numerical calculations and experiment [8].

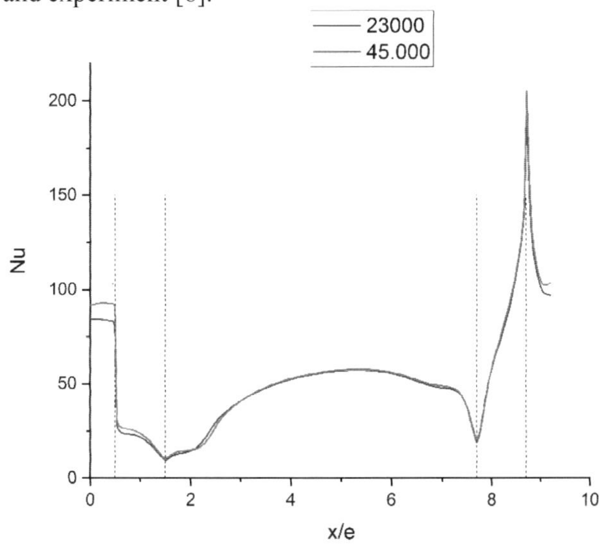

Figure 2. Distribution of the Nusselt number in a rectangular channel for various grids.

Result and discussion

Data on the dynamics and heat transfer of a turbulent flow in a flat channel with periodic ribs are obtained. Thus, figure 3 shows the flow structure for w/e = 2, 7.7, and 9.7. According to figure 3b, it can be seen that when the distance between the ribs is equal to two rib heights, the so-called d- type roughness is implemented. In this case, the presence of ribs does not affect the main flow. With a larger distance between the ribs (figure 3a), roughness of the k-type is implemented. At that, the ribs almost do not affect each other and the flow structure for a flat channel with a backward-facing step is formed. Accordingly, in the intermediate version, a transitional type of roughness is implemented, which is shown in figure 3c.

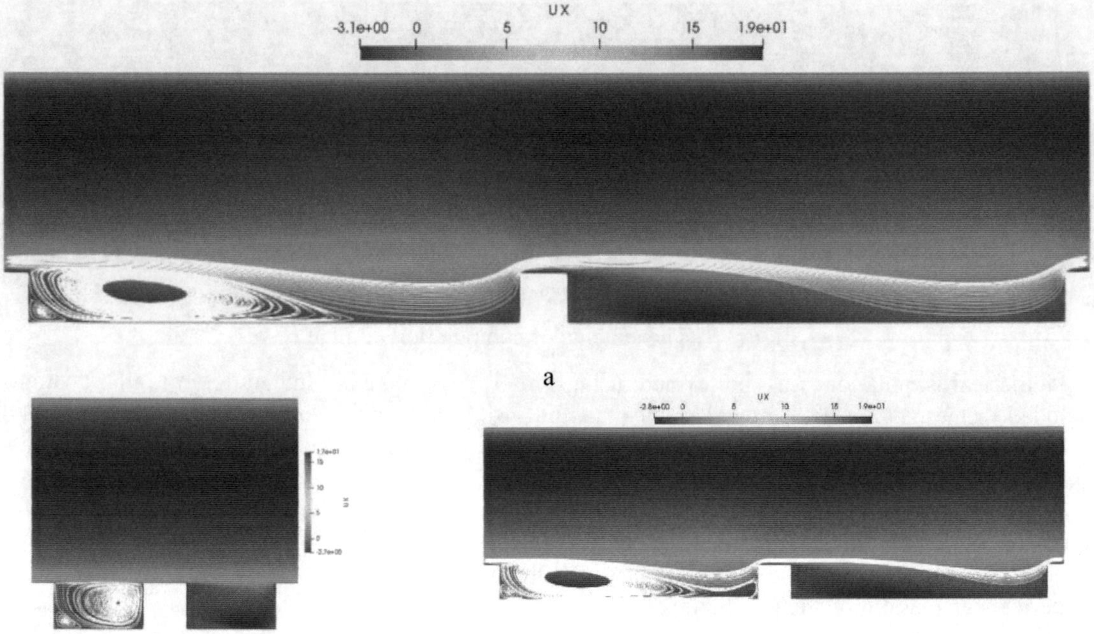

a b c

Figure 3. Flow structure for w/e: a) 9.7, b) 2, c) 7.7.

To study the effect of a transverse rib on the flow, a calculation by the large eddy method (LES) was carried. Visualization of the evolution of the vortex flow structure using the Q – criterion is presented in figure 4. It can be seen that vortices originate at the rib (figure 4a) and develop further along the channel (figure 4b). The rib leads to formation of large-scale vortices that act on the mixing layer, which leads to more intense heat transfer.

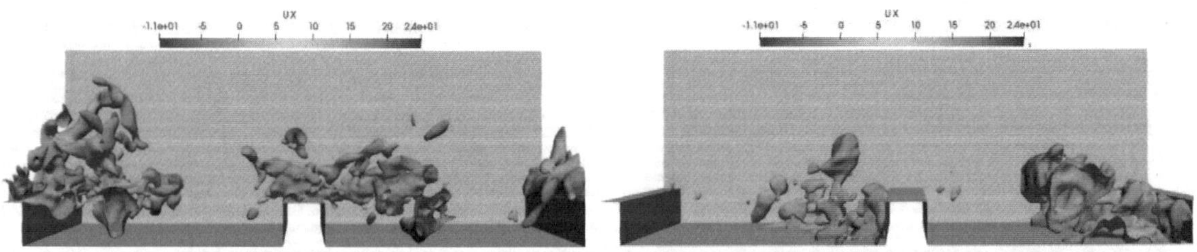

a b

Figure 4. Vortex visualization using the Q –criterion: a) initial moment, b) following moment.

The distribution of Nusselt number for the channel wall located between the ribs is presented in figure 5. It can be seen from the figure that with a roughness of type *d* or transition type, heat transfer increases significantly. When the point of separated flow attachment appears, a secondary vortex is formed near the next rib, which enhances heat transfer. This can be seen from the peak located at the end of a gap between the ribs.

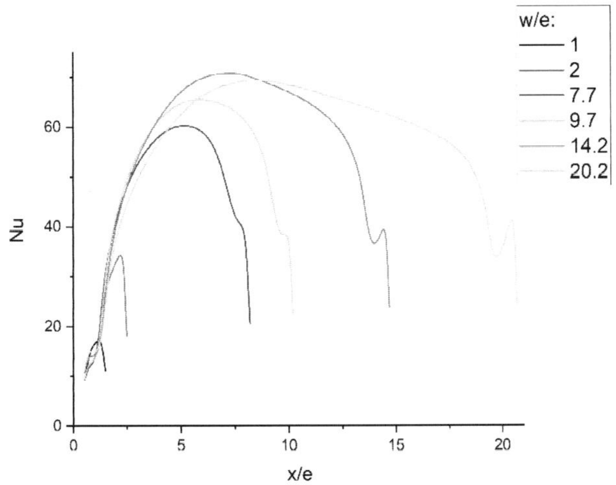

Figure 5. Nusselt number distribution.

Figure 6 shows the maximum value of the Nusselt number for various distances between the ribs. It follows from the figure that at w/e = 10 the maximum Nusselt number almost stops changing. At such large distances, the ribs stop influencing each other.

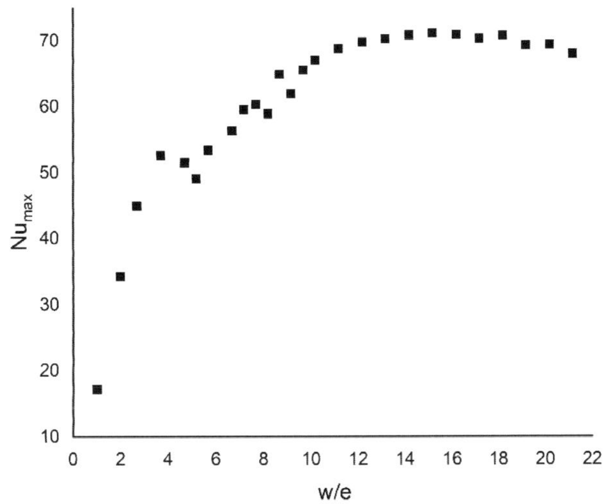

Figure 6. The maximum Nusselt number vs. a distance between the ribs.

Conclusion

The turbulent flow dynamics and heat transfer in a flat channel in the presence of periodic transverse ribs were numerically simulated by the RANS and LES methods at a Reynolds number of 12600. It is shown that with a change in the distance between the ribs, different types of roughness are implemented, which in turn affects heat transfer. It is found that at w/e = 10, the maximum Nusselt number almost stops changing, since adjacent ribs have a weak effect on each other. When passing from the d-type roughness to the k-type roughness, heat transfer on the lower wall of the channel increases significantly.

Acknowledgments
The study was financially supported by the RSF (Project No. 21-19-00162).

Reference

[1] Eaton J K and Johnston J P 1981 *AIAA J* **19** 1093
[2] Miau J J, Lee K C, Chen M H and Chou J H 1991 *AIAA J* **29** 1140
[3] Sano M, Suzuki I, Sakuraba K 2009 *J. Fluid Sci. Technol.* **4** 188
[4] Terekhov V V and Terekhov V I 2017 *J. Applied Mechanics and Tech. Physics* **58** 254
[5] Park H, Jeon W-P, Choi H, Yoo J Y 2007 *Phys. Fluids* **19** 105103
[6] Dyachenko A Yu, Zhdanov V L, Smulskii Ya I, Terekhov V I 2019 *Thermophysics and Aeromechanics* **26** 509
[7] Leonardi S, Orlandi P, Antonia R A 2007 *Phys. Fluids* **19** 125101
[8] Weihing P, Younis B A, Weigand B 2014 *International Journal of Heat and Mass Transfer* **76** 509

Experimental research of metal hydride heat storage reactor processes

D V Blinov [1,2], V I Borzenko [1] and A V Bezdudny[1]

[1] Joint Institute for High Temperatures, Russia, 125412 Moscow, Izhorskaya 13-2
[2] National Research University "MPEI",
Russia, 111250 Moscow, Krasnokazarmennaya, 14

h2lab_bezdudny@mail.ru

Abstract. The results of the experimental research of thermal, mass exchange and dynamical characteristics of processes inside the low temperature metal hydride (MH) thermal energy storage system are presented. Single stage pressure driven MH heat storage system of closed cycle concept was studied and tested. Intermetallic compound (IMC) $LaFe_{0.1}Mn_{0.3}Ni_{4.8}$ in the quantity of 5 kg was used as main hydrogen storage/heat emitter element in the reactor. Nominal maximum hydrogen capacity of the reactor is 850 st.l. with though resulting effective volume of cycled hydrogen ended up to be around 240-250 st.l. The reactor type and intermetallic alloy, which were used in the series of experiments, proved to be somewhat suitable for the task, but more advanced heat exchange design along with selection of different type of IMC promise to increase the cycled effective volume along with the system dynamics, resulting in greater thermal energy power output.

1. Introduction

Decreasing supplies of fossil fuels, growing concern about global warming, and the accelerating rise in energy demand have made the power-producing areas to shift their focus from conventional sources of energy to renewable ones. The Paris Agreement, signed by 195 countries in a struggle against global warming, demands that each of the participating nations plans and implements actions to mitigate climate change. When a perfect scenario would make provision for the complete transition from fossil fuels to renewable energy, only about 23% of the potential global renewable energy needs to be harnessed [1], leaving vast moment for improvement and innovation.

Along with other renewable sources, solar energy is the most exuberant and should be given the highest possible priority [2]. Mainly, Solar Photovoltaic (SPV) and Concentrated Solar Thermal Power (CSTP) systems are used to harness solar energy [3].

However, more like SPV power, CSTP is affected by seasonal and location characteristics of solar insolation, which also varies significantly during the course of a single day at a particular site, therefore limiting the period of effective electricity generation to about 7 to 8 h per day. One very promising option for resolving this issue in the case of CSTP is the integration of a thermal energy storage system (TESS). Thermal energy storage systems become relevant when the heat supply and its demand do not overlap. Heat can be stored in the form of sensible, latent and chemically stored heat. Mainstream thermal energy storage systems mostly utilize Phase Change Materials (PCMs) [4-6] and sensible heat storage materials such as cast steels, concrete, fire bricks, etc. From different variants of thermal energy, thermo-chemical energy acts as the promising one due to the high energy density of chemical reactions comparing to sensible or latent heat. There is a variety of accessible thermochemical energy storage materials such as salt hydrates, carbonates, metal hydroxides, metal hydrides etc. [7]. The key advantages of reversible thermo-chemical thermal energy storage systems over the two aforementioned traditional ones can be summarized as follows:

- Higher energy density due to use of chemical reaction heat
- Stable chemical compounds preserved at an ambient temperature without diminishing provide longe-term storage
- Can be adopted for heat transport over long distance
- Regeneration of the stored heat is possible with the lack of stored heat exergy deteriorating
- Can be adopted for various useful heat transformation operations like heat pumps, refrigeration, thermal amplification, heat transforming, reversible low-temperature energy storage, etc.

Metal hydride thermal energy systems are considered as reversible chemical heat storage systems, able to store thermal energy in the estimate range of 2 MJ/kg [8]. In metal hydride thermal energy storage system, useful heat is released when the hydrogen is introduced into the system at certain pressure and temperature. The rate of thermal energy release and the temperature can be managed by altering the hydrogen supply pressure and gas flow parameters. When the external heat is available, hydrogen could be regenerated from metal hydride and stored back in to a container. Main scientific and technical questions of development and constructing efficient MH devices though are tied with handling the thermal and mass transfer peculiarities inside the MH volume. Inefficient ways of managing the heat transfer inside the IMC volume combined with high thermal effect of reaction result in crisis events characterized by rapid decreasing of reaction dynamics [9]. Thus, intensification of thermal transfer between IMC and coolant is crucial to achieve maximum thermal energy power output possible.

2. Materials and metal hydride reactors

Varying an IMC composition allows one to compose an alloy of needed sorption/desorption pressure in a temperature range of 0 to 100°C to meet needed requirements for reversible thermal energy storage. AB_5-type hydrides can achieve necessary parameters to be utilized in a model thermal energy storage system. $LaFe_{0.1}Mn_{0.3}Ni_{4.8}$ composition was chosen on the basis of readily availability in our facility along with its acceptable pressure-content-temperature characteristics, which allow us to utilize it in a thermal energy storage system in a comparatively low temperature range (0-100°C).

PCT-isotherms of sorption and desorption processes were obtained via Siverts method (figure 1) [10].

Experimental research of thermal processes in the MH hydrogen accumulation and purification systems were conducted employing experimental rig 12-04 of JIHT RAS in the laboratory of hydrogen energy technologies [11-16].

Experimental rig (figure 2) consists of gas ramp, electromagnetic valves (EV), pressure sensors (PS), gas flow regulator (FR) and metal hydride reactor RSP-3 №3 (figure 3). Measuring equipment and gas flow regulators are operated via PC block based on LabView system.

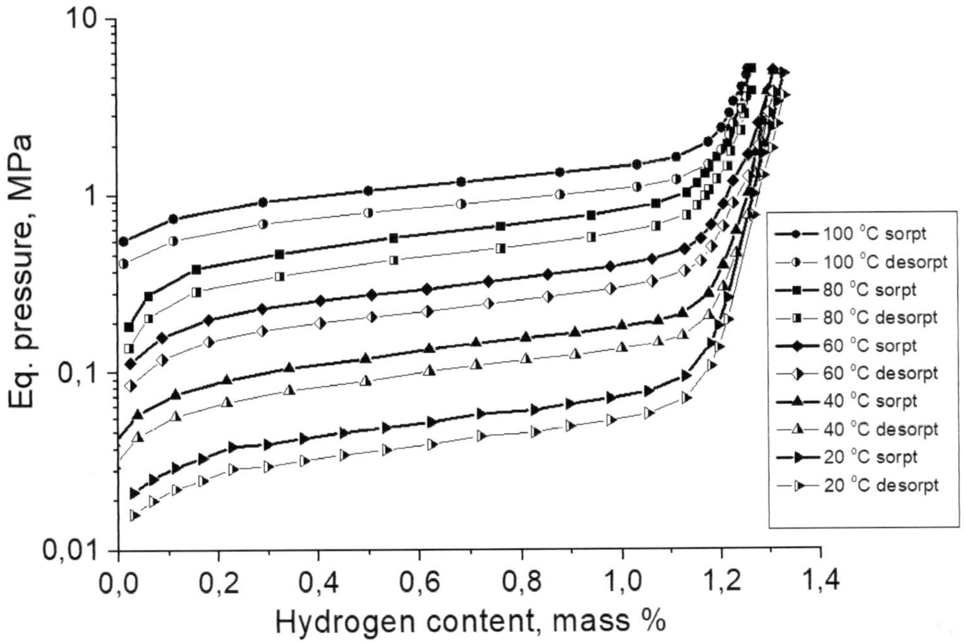

Figure 1. PCT curves of the AB_5 alloys used at a temperature of 25.0 °C.

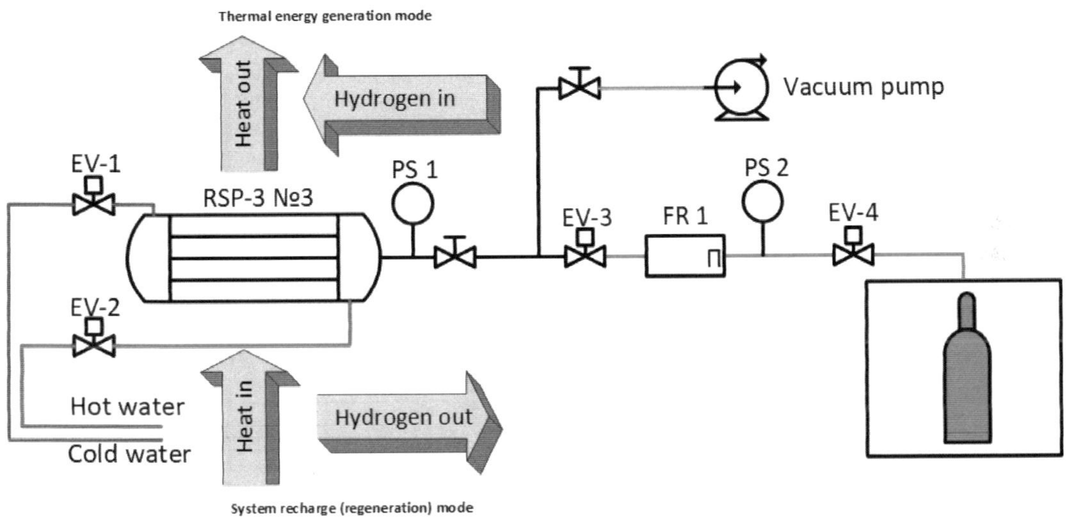

Figure 2. Diagram of the part of experimental rig 12-04 of the JIHT RAS utilized in the research along with basic schema of the concept.

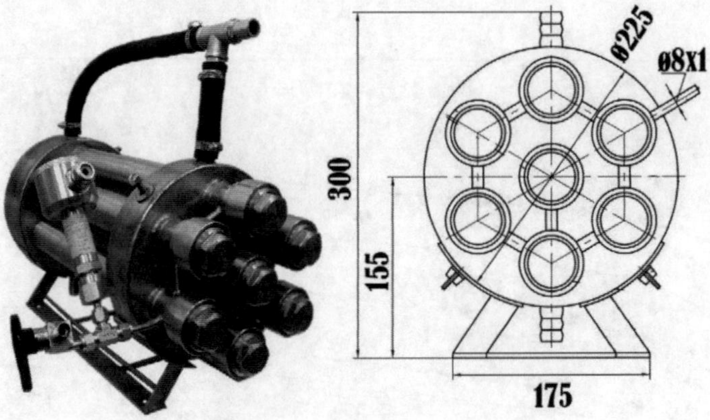

Figure 3. Reactor Type 3 (RSP-3) (pipe board).

3. Results and discussions

Analysis of performance of MH single stage pressure driven thermal energy storage system should be focused on its main characteristics – working capacity of hydrogen. Due to inherent properties of MH, in according to its PCT-diagram, one cannot fully discharge the whole capacity into a closed system. Eventually, the pressure levels inside the MH device and a gas tank would reach equilibrium, defined by the temperature one can put the MH device into, after which no more gas could be put into desorption process. This so called "first-stage pressure" should define most of the working capacity amount. But in order to obtain accurate data about working capacity the sorption-desorption cycle should be repeated at least once.

First stage of experiments consisted in desorption of fully charged RSP-3 reactor (850 st.l of hydrogen in MH with an approximately 1.4 st.l of free volume above the MH bed) into a standard 40. l gas cylinder. Gas tank and gas pipeline were preliminary vacuumed up to $1.2 \cdot 10^{-5}$ bar. Coolant fluid (water) flow was set to 0.110 st.l/sec. with a temperature of, at first, 83.0°C with an additional rising up to 90.0°C. Such temperatures allowed us to reach up to 7.6 atm in the gas line, which is an equivalent of 304 st.l. of hydrogen inside the gas cylinder.

Second stage of experiments consisted of discharging the hydrogen from the gas tank into preliminary cooled RSP-3 reactor. System temperature was cooled to the temperature of 13.0°, after reaching which the gas valves were opened with a virtually no limits on the gas flow. Coolant fluid flow was set to 0.110 st.l/sec. Total amount of hydrogen absorbed reached ~238 st.l., with maximum temperature of exhaust water of 17.1°C. The obtained experimental data is represented on figure 4.

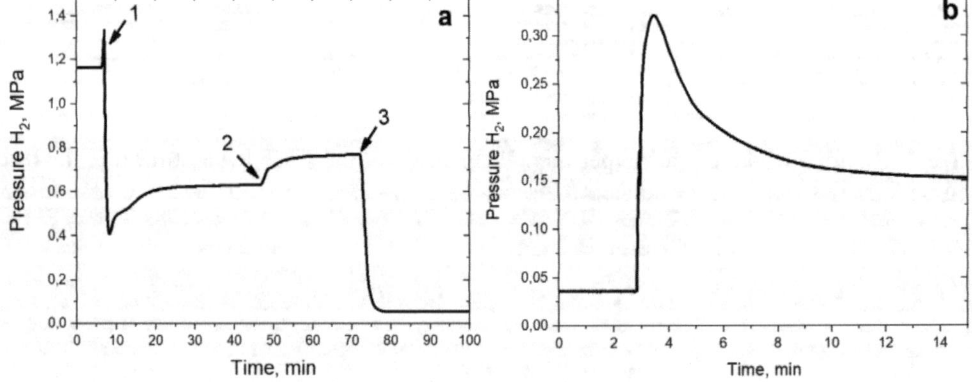

Figure 4. a – first stage. b – second stage.

On figure 4a position "**1**" points out the moment when EV-3 and EV-4 are opened to let built-up pressure inside the MH device drop down into the gas storage volume. Position "**2**" points out the moment of a water temperature switched to 90.0°C, and position "**3**" points out the end of the desorption process with closing of all vents, leaving the gas tank on achieved pressure.

As could be seen on figure 4b, the pressure equilibrium during the sorption was established on ~0.15 MPa, resulting in 60 st.l of dead weight hydrogen inside the gas cylinder. Thus, the working capacity of this particular setup could be evaluated as ~240 st.l H_2, but one more cycle was conducted to reaffirm the results.

Third stage (first part of the new cycle) consisted in desorption of partially charged reactor RSP-3 into the gas cylinder with 60 liters of hydrogen in it. Coolant flow was of 0.117 l/min, with a temperature of, again, 80.0 and 90.0°C. Overall system pressure reached almost 8 atm. Thus 260 st.l. of hydrogen were discharged into the gas tank.

Fourth stage (second part of the new cycle) was conducted the same way the second one was, but with water cooled down to 12.0°C. Amount of hydrogen absorbed (~238 st.l) and max coolant temperature achieved (~17.0°C) made clear that the cyclic stable mode of operation was successfully achieved.

The thermal energy storage system was able to reach 2 kW peak power output during the initial part of sorption process. Changes of instantaneous power over time are represented below in the figure 5.

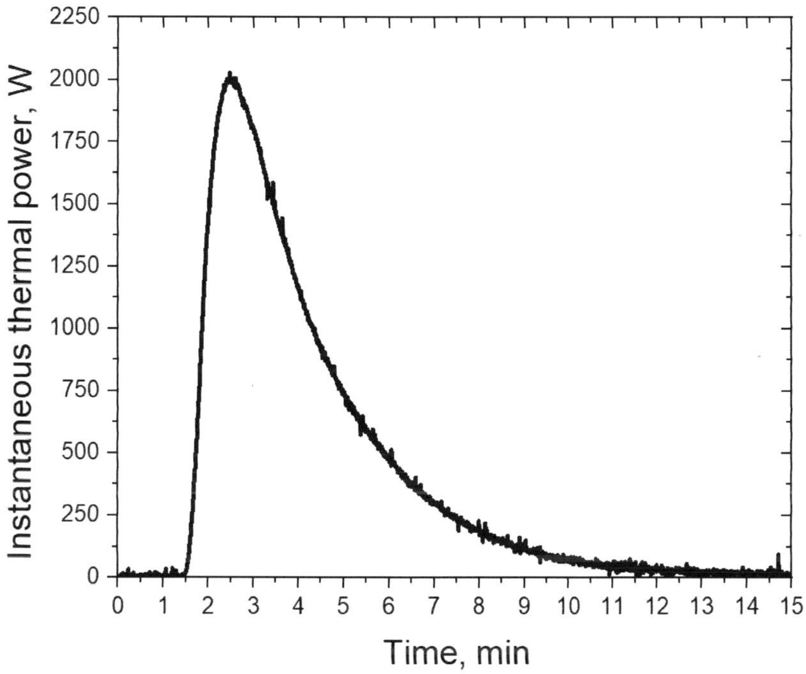

Figure 5. Thermal power of the experimental device.

4. Conclusion

Though only 28% of total reactor capacity was employed into cyclic work, the concept of single stage pressure driven system worked within expected parameters. The dynamics of reaction could be vastly improved by implementing more efficient design of MH reactor, where much emphasis is to be put on thermal exchange intensification. The IMC, adopted for this study, is hardly suitable for a practical application, so another type of alloy should be put into consideration or even be developed anew with a goal to maximize cyclic working capacity within estimated operational conditions.

References

[1] Peng X, Root T W, Maravelias C T 2017 Storing solar energy with chemistry: the role of thermochemical storage in concentrating solar power. *Green Chem*;19:2427–38
[2] Abbott D Keeping the energy debate clean: how do we supply the world's energy needs? Proc IEEE 2010;98(1):42e66
[3] Desideri U, Campana P E 2014 Analysis and comparison between a concentrating solar and a photovoltaic power plant. *Appl Energy*;113:422–33
[4] Farid M M, Khudhair A M, Razack SAK, Al-Hallaj S 2004 A review on phase change energy storage materials and applications. *Energ Convers Manag*; 45:1597e615
[5] Jinjia W, Yasuo K, Satoshi H, Hiromi T 2005 Study on a PCM heat storage system for rapid heat supply. *Int J Appl Thermal Eng*; 25:2903e20
[6] Shanmugasundaram V, Brown J R, Yerkes K L 1997 Thermal Management of high heat-Flux Sources using phase change materials. *Proceedings of the 32nd Thermo physics Conference*; AIAA-97e2451:23e25
[7] Kuravi S, Trahan J, Goswami D Y, Rahman M M, Stefanakos E K Thermal energy storage technologies and systems for concentrating solar power plants. *Prog Energy Combust Sci* 2013;39:285e319
[8] Reiser A, Begdavonic B, Schliche K 2000 The application of Mg based metal hydrides as heat energy storage systems. *Int J Hydrogen Energy*;25:425e30
[9] Dunikov D, Borzenko V and Malyshenko S 2012 Influence of impurities on hydrogen absorption in a metal hydride reactor *International Journal of Hydrogen Energy* **37** 13843-8
[10] Borzenko V I, Romanov I A, Dunikov D O and Kazakov A N 2019 Hydrogen sorption properties of metal hydride beds: Effect of internal stresses caused by reactor geometry *International Journal of Hydrogen Energy* **44** 6086-92
[11] Khayrullina A G, Blinov D and Borzenko V 2019 Novel kW scale hydrogen energy storage system utilizing fuel cell exhaust air for hydrogen desorption process from metal hydride reactor *Energy* **183** 1244-52
[12] Khayrullina A, Blinov D and Borzenko V 2019 Air heated metal hydride energy storage system design and experiments for microgrid applications *International Journal of Hydrogen Energy* **44** 19168-76
[13] Borzenko V, Blinov D and Glagoleva A 2018 Stand-Alone, Logistics - Free kW-Scale Hydrogen Energy Storage System. *In: 2018 International Multi-Conference on Industrial Engineering and Modern Technologies, FarEastCon*
[14] Dunikov D, Borzenko V, Blinov D, Kazakov A, Lin C-Y, Wu S-Y and Chu C-Y 2016 Biohydrogen purification using metal hydride technologies *International Journal of Hydrogen Energy* **41** 21787-94
[15] Blinov D V, Borzenko V I, Dunikov D O and Kazakov A N 2018 Experimental investigations of thermal processes in the flow-throw hydrogen purification reactor *Journal of Physics: Conference Series* **1128** 012120
[16] Blinov D V, Borzenko V I, Dunikov D O, Romanov I A 2014 Experimental investigations and a simple balance model of a metal hydride reactor. *Int J Hydrogen Energy*; 39: 19361-19368

Metal hydride hydrogen storage and purification technologies

D V Blinov[1,2], V I Borzenko[1], A V Bezdudny[1] and A N Kazakov[1]

[1] Joint Institute for High Temperatures,
Russia, 111116 Moscow, Krasnokazarmennaya, 17A
2 National Research University "MPEI",
Russia, 111250 Moscow, Krasnokazarmennaya, 14

h2lab@mail.ru

Abstract. The results of the development of metal hydride (MH) reactors for the storage and purification of hydrogen of various types are presented. Two methods of metal hydride purification of hydrogen are presented. The use of the MH method of flow-through purification of hydrogen has high hydrogen recovery rates at high volume contents of hydrogen in the mixture (≥10% vol.), while the method of periodic evacuation of accumulated impurities is most effective at low hydrogen contents in the mixture (<10% vol.).

1. Introduction

Growing industrialization and increasing of world energy carrier consumption dynamics inevitably enlarges the influence on the global ecosystem and pushes the humankind to research for new approaches to increase energy generation and consumption effectiveness. Interest in hydrogen as an alternative type of fuel had existed for years. Above all, it is connected with high prices for traditional types of energy resources, political characteristics of forming the energy market along with many different ecological aspects of utilizing traditional fuel [1].

Transition to truly ecologically pure hydrogen production is possible only with the use of renewable energy sources (RES). Utilization of locally available RES (including the biohydrogen ones) combined with storing energy in hydrogen with its subsequent use in fuel cell could be the way to a development of complex energy generation and utilization systems of relatively small power scale (below 20 kWt).

Hydrides of intermetallic compounds (IMCs), liable to selective hydrogen absorption, are to be perspective materials for the development and construction of hydrogen storage and high purity fuel supply systems for FC. Stationary autonomous energy systems are mainly focused on compactness, operational safety and productivity rather than on weight characteristics. That is why the MH storage and purification systems coupled with RES could be perspective for the development of such systems. MH device could be a thermal energy storage system at the same time, due to the existence of high thermal reaction of hydrogen sorption-desorption processes (25-70 kJ/mol H_2), thus consumer heat supply could be rationally organized [2,3].

Main scientific and technical problems of development and constructing highly efficient MH hydrogen storage devices are connected with a managing of thermal and mass transfer inside the MH filling volume. Finely dispersed powder of IMC in the MH reactor has a size of 1-10 μm and possess relatively low thermal conductivity (0.1-1 Wt/m·K), which depends on free hydrogen pressure and absorbed hydrogen concentration) [4]. Ineffective way of organization of heat input-output inside the

IMC filling with a high thermal effect of reaction leads to emerging of crisis events characterized by rapid decreasing of reaction dynamics [5]. Gaseous admixtures are also make great impact on the hydrogen sorption kinetics. Even the presence of inert gases in the main volume of hydrogen leads to blocking of access for hydrogen to the IMC surface, thus forcing one to exploit the techniques of removing accumulated admixtures in the MH devices [6].

Another existing type of scientific-technical problems bound to the development of effective technologies of MH storage and purification devices integration with FC-based energy systems, hydrogen sources (electrolyzer, bioreactor) and the source of initial energy (wind, solar systems). Model integrated energy supply systems are to become crucial for the solving of these problems with fine-tuning of all the possible work modes, especially the emergency ones for all components of the energy system [7,8].

2. Materials and metal hydride reactors

Operational characteristics of developing MH devices, as part of integrated energy systems, should be consistent with the requirements of FC and electrolyzer (operation pressure, hydrogen purity, hydrogen flow rate). Tweaking an IMC composition allows one to make an alloy of needed sorption/desorption pressure in a temperature working range of 0 to 100°C to meet the requirements for hydrogen storing and cleansing. AB_5-type hydrides can achieve needed for the development of hydrogen storing systems parameters. As for AB_5-type alloys – it allows us to provide hydrogen desorption processes with a use of FC/electrolyzer waste heat.

Melting of IMC samples for our research work was undertaken in electric arc furnace in the argon atmosphere. PCT-isotherms of sorption and desorption processes were obtained via Siverts method (figure 1) [9].

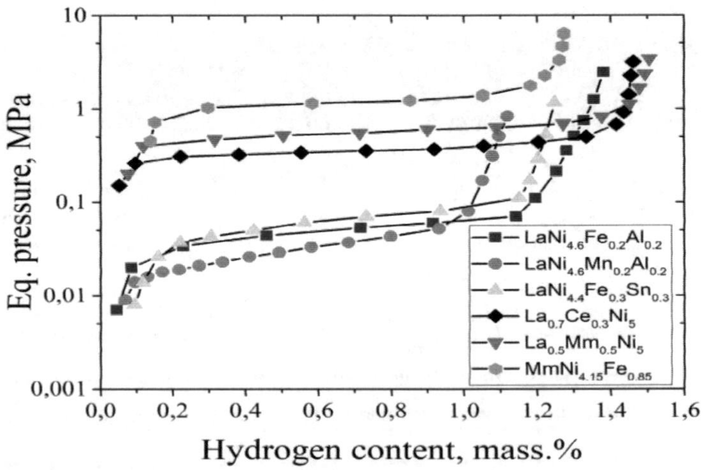

Figure 1. PCT curves of the AB_5 alloys used at a temperature of 25 °C (50 cycles).

Experimental research of thermal processes in the MH hydrogen accumulation and purification systems were conducted employing experimental rig 12-04 of JIHT RAS in the laboratory of hydrogen energy technologies. Complex experimental rig is designed to:

- Experimental research thermal and mass exchange processes in the porous bed of IMC during the sorption of pure and polluted hydrogen.
- Utilize gathered data to verify mathematical models of thermal and mass transfer inside the MH hydrogen accumulation and purification devices.
- Experimental research different modes of operation of developed MH devices several types and several IMC loads.
- Research sorbing and thermophysical properties of different IMCs.

- Research problems of integration of hydrogen production sources, purification and accumulation systems and FCs.
- Refinement of automation and operation of energy systems with MH hydrogen accumulation and purification subsystems.

Experimental rig (figure 2) consists of gas ramp, solid-polymer electrolyte electrolyser of 10 st.l H_2/min gas flow, MH block of fine purification, made of 3 reactors RSP-3, MH hydrogen accumulation system RS-1 to provide 5 kWt FC with fuel, flow-through purification reactors RSP-8, gas analyzer (GA), gas flow regulators (FR) and gas flow meters (FM), electromagnetic valves (EV), pressure sensors (PS) and vents. Block of fine purification, RSP-8 reactors and RS-1 systems are in the metallic box equipped with air exhaust vent with air exchange multiplicity from 12 to 15. The placement and the box are equipped with sensors of dangerous accumulation, paired with emergency exhaust vent to turn it on if the hydrogen concentration in the room reaches 1%. Measuring equipment and gas flow regulators are operated via PC block based on LabView system.

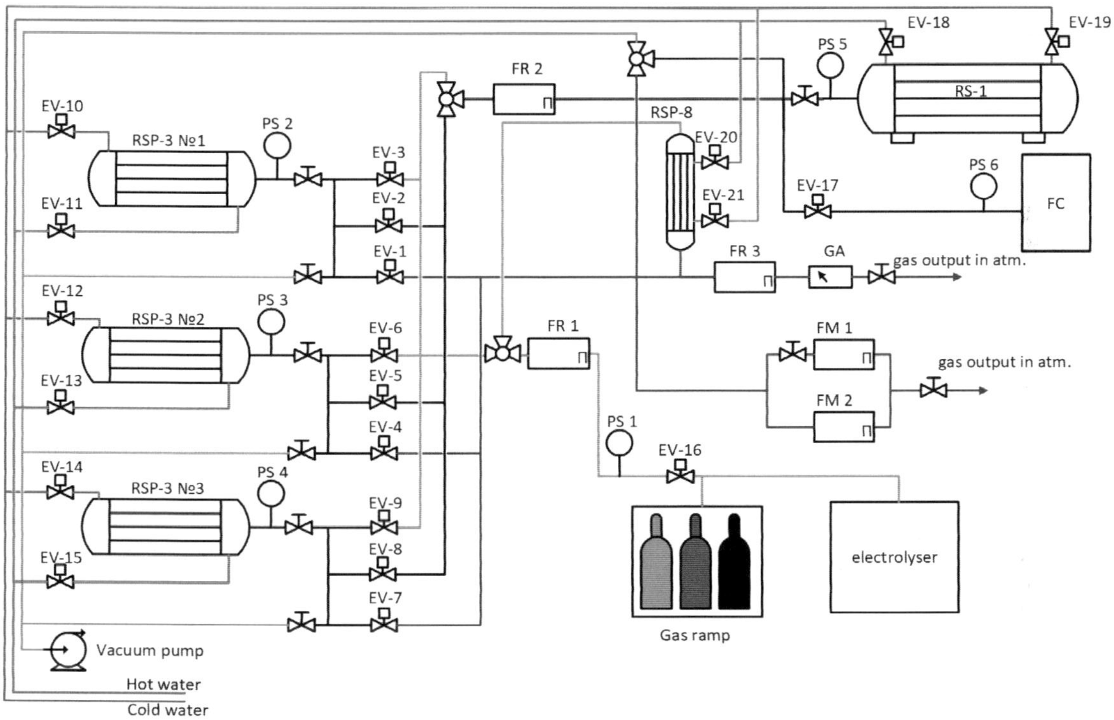

Figure 2. Schematic diagram of the experimental rig 12-04 of the JIHT RAS.

Several different types of reactors, varied by their thermal exchanger design, were created for the fine purification block subsystem (figure 3). Mathematical models of thermal and mass exchange processes during pure and polluted hydrogen sorption in the MH bed were used for the development of inner and outer thermal exchange patterns [10,11].

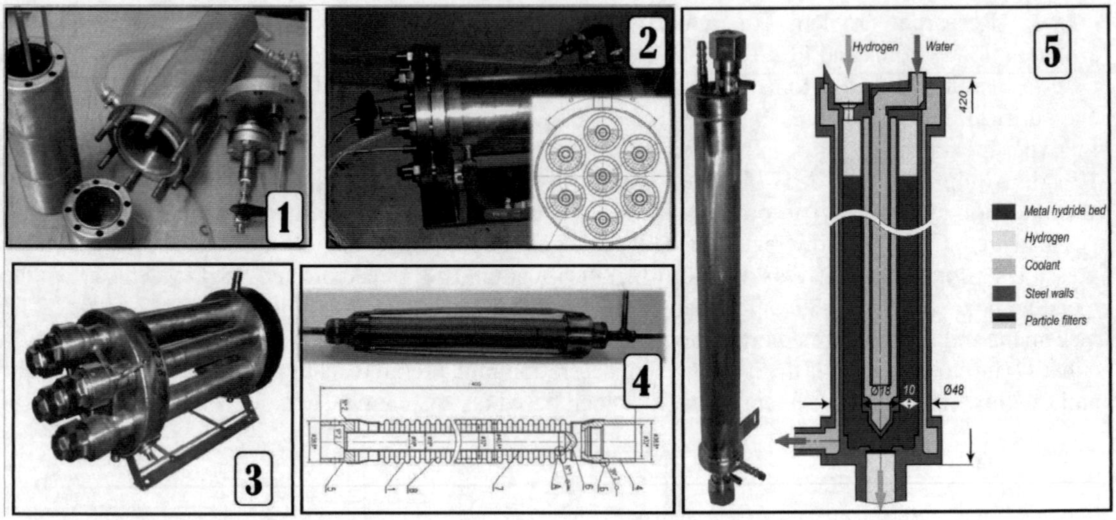

Figure 3. Hydrogen storage and purification reactors: 1 – Type 1 (cartridge) [12]; 2-Type 2 (pipe board); 3-Type 3 (RSP-3 reactor) (pipe board) [13]; 4-Type 4 (bellows); 5-Type 5 (RSP-8 flow-through purification reactor) [14].

3. Results and discussions

Analysis of performance of MH hydrogen accumulation systems is usually focused on its practical side – integral dynamics of hydrogen sorption, change in number of absorbed hydrogen with respect to time. Efficiency of different types of MH reactors was evaluated by conducting a series of experiments of sorption process for the pure hydrogen with the flow restriction maxed at 120 st.l/min. Reactor types represented on figure 3 were used. The experimental data obtained (figure 4) revealed, that the most effective, in terms of inner thermal exchange optimization, method is an application of thin (less than 10mm) MH layer and introduction of channels for the heat carrier inside the filament.

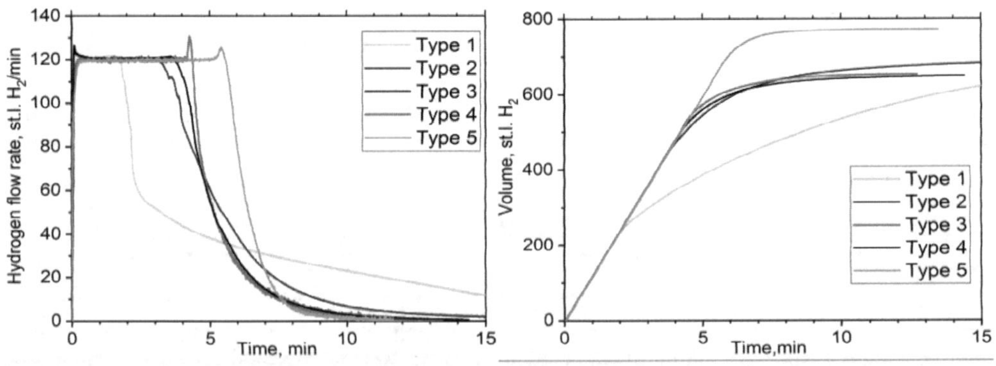

Figure 4. Hydrogen sorption for five types of reactors.

To solve the target of extracting the hydrogen from the gas mixture 2 purification methods were used: recurrent evacuation method utilizing MH reactors type 3 (RSP-3) and flow- through metod utilizing MH reactor type 5 (RSP-8). Research of hydrogen purification experimental processes were conducted by the use of inert gases as admixtures (N_2 and CO_2).

Hydrogen purification technique via RSP-3 reactors consisted in recurrent evacuation of accumulated in the reactor free volume mixture. The evacuation could conducted in different operating modes, depending on the specific goal of optimization: constant evacuation time interval, constant target pressure after which the evacuation starts or constant volume of evacuated gas. For the further development of the purification subsystem the "constant target pressure (atmospheric)" mode was chosen. (figure 5 and Table 1).

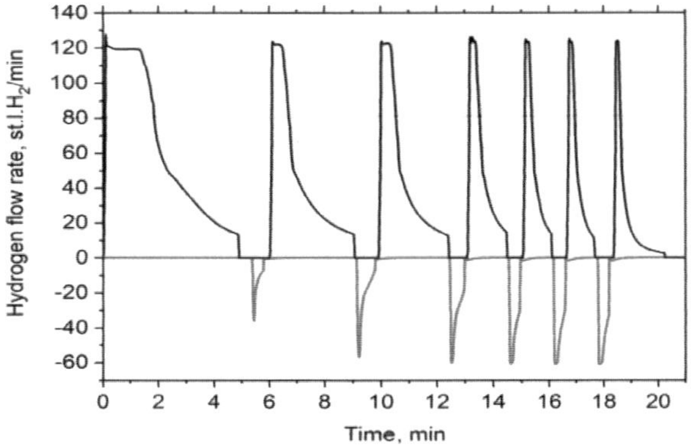

Figure 5. Hydrogen purification using the RSP-3: hydrogen with an impurity (3% vol. N_2), evacuation to atmospheric pressure. Black – gas injection, red - evacuation an impurity.

Table 1. Hydrogen losses during purification by the RSP-3 reactor.

Impurity, % vol.	**Losses, %**
0	0
3	17
10	22
25	33

Flow-through metod utilizing MH reactor type purification technique is basically a flushing the mixture of hydrogen and other gases through the porous filament of IMC. Experiments were conducted by means of RSP-8 reactor (figure 3). Hydrogen is recurrently pushed from the top of RSP-8 to its bottom part through the bed, while the most part of existing hydrogen is being absorbed in the IMC and the rest of it is exhausted from the bottom of the reactor along with non-absorbable admixtures. figure 6 represents integral coefficient of hydrogen extraction during the continuous-flow purification.

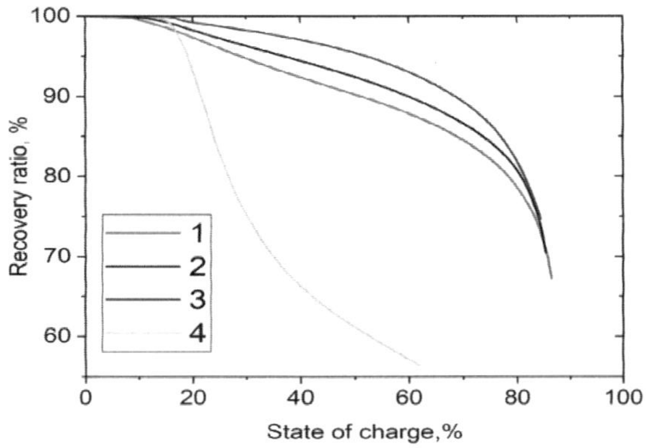

Figure 6. The change in the Recovery ratio of hydrogen depending on the degree of charging of the RSP-8 reactor and the restriction of the flow rate of the H_2/CO_2 mixture (50/50% vol.): 1 -20 st.l / min; 2 -3 st.l / min; 3-10 st.l / min: 4 -3 st.l / min without reactor cooling.

Utilizing the continuous-flow purification with RSP-8 reactors demonstrates high rates of hydrogen extraction (the efficiency reaches 85% along with relatively high current reactor capacity ≤ 80% (figure 6)).

4. Conclusions

Hydrogen stores in the MH accumulation and purification systems in a bound solid-state in a hydrides of metals and alloys with a high volumetric density and relatively low pressure (below tens of atm.) which proves to be convenient and safe to use it in the autonomous systems of energy provision.

IMCs used in the MH systems possess peculiarity to separately absorb hydrogen, allowing them to be used in the development of hydrogen extraction systems and combined MH accumulation and purification systems. MH continuous-flow purification technique has high hydrogen extraction rate while the portions of hydrogen in mixture is relatively high (≥10% vol.) and it allows to subsequent utilization of purified hydrogen in a FC fuel supply system. On the other hand, the periodic evacuation technique is most effective while the portions of hydrogen in mixture does not exceed 10% vol.

External cooling of MH reactors proved to be ineffective. Intensification of processes inside the MH filament should be carried out via adjusting the powder layer thickness (less than 10mm) and utilizing thermal exchangers of "tube bundle" and "tube in tube" types.

Acknowledgment

The work has been funded by Russian Science Foundation (project № 17-19-01738).

References

[1] The Future of Hydrogen, Seizing Today's Opportunities/ *International Energy Agency: Paris, France. 2019*

[2] Manickam K, Mistry P, Walker G, Grant D, Buckley C E, Humphries T D, Paskevicius M, Jensen T, Albert R, Peinecke K and Felderhoff M 2019 *International Journal of Hydrogen Energy* **44** 7738-45

[3] Nyamsi S N, Lototskyy M and Tolj I 2018 *International Journal of Hydrogen Energy* **43** 22568-83

[4] Sun D-W and Deng S-J 1990 *International Journal of Hydrogen Energy* **15** 331-6

[5] Dunikov D, Borzenko V and Malyshenko S 2012 *International Journal of Hydrogen Energy* **37** 13843-8

[6] Sandrock G D and Goodell P D 1984 *Journal of the Less Common Metals* **104** 159-73

[7] Khayrullina A G, Blinov D, Borzenko V 2019 *Energy* **183** 1244-52

[8] Han G, Kwon Y, Kim J B, Lee S, Bae J, Cho E, Lee B J, Cho S and Park J 2020 *Applied Energy* **259** 114175

[9] Borzenko V I, Romanov I A, Dunikov D O, Kazakov A N 2019 *International Journal of Hydrogen Energy* **44** 6086-92

[10] Artemov V I, Minko K B, Yan'kov G G 2016 *International Journal of Hydrogen Energy* **41** 9762-8

[11] Nashchekin M D, Minko K B, Artemov V I 2019 *Case Studies in Thermal Engineering* **14** 100485

[12] Dunikov D O, Borzenko V I 2020 *Journal of Physics: Conference Series* **1683** 052018

[13] Blinov D V, Borzenko V I, Dunikov D O, Romanov I A 2014 *International Journal of Hydrogen Energy* **39 (33)** 19361-8

[14] Borzenko V I, Blinov D V, Dunikov D O, Leontiev A I 2018 *Journal of Physics: Conference Series* **1128** 012126

Power spectra of pressure pulsations in the processes of evaporation/boiling of a liquid at low pressures

A E Brester[1,2], V I Zhukov[1,2], D A Shvetsov[2] and A N Pavlenko[2]

[1] Novosibirsk State Technical University,
Russia, 630073 Novosibirsk, K. Marx Prosp., 20
[2] Kutateladze Institute of Thermophysics, SB RAS,
Russia, 630090 Novosibirsk, Lavrentiev Prosp., 1

brester94@gmail.com

Abstract. This paper presents experimental data on pressure fluctuations during evaporation/boiling of a thin liquid film under conditions of reduced pressure. The experimental data were obtained as a result of studying heat transfer on a smooth horizontal surface in a wide range of changes in the height of the liquid layer. Using the fast Fourier transform, the power spectra of pressure pulsations versus frequency were obtained. It was found that the power spectra of pressure pulsations differ depending on the mode of evaporation/boiling in the system.

1. Introduction

Currently, much attention is paid to fluctuations (noise) of various quantities measured in the system to determine the characteristics or state of the system. Of particular interest are fluctuations with a power spectrum inversely proportional to the frequency $S \sim 1/f$, where S and f are power spectrum [em] and the frequency [Hz], respectively. Most often in the literature, $1/f$ noise is understood as fluctuation processes with a power spectrum proportional to $1/f^{\alpha}$, where the exponent α varies within certain limits. An inversely proportional power spectrum is found in various systems, for example, in physical, chemical, mechanical, and biological systems [1-8].

In physics, $1/f^{\alpha}$ spectra of current pulsations in semiconductor materials, metal films, and radiophysical components are known [1-3]. In chemical technologies, $1/f^{\alpha}$ power spectra of pulsations can be used to interpret the data obtained on gas sensors [4], also, using power spectra, one can trace the kinetics of material synthesis [5]. In geophysics, $1/f^{\alpha}$ power spectra are used to describe seismic activity, drought, and floods [6, 7]. In biology, $1/f^{\alpha}$ spectra are observed in fluctuations in the potential energy of plastocyanin, an electron transport protein containing copper [8].

In the last few decades, the object of intensive research is the processes associated with the influence of fluctuations on the phase transition processes [9]. Since heat and mass transfer in two-phase systems is characterized not only by the average values of the process parameters but also by chaotic fluctuation deviations from the average values of these parameters. In particular, a strong increase in fluctuations occurs in critical and transient regimes of heat and mass transfer. The dynamics and evolution of random pulsations can be characterized by the dependence of the fluctuation power spectrum on frequency [10, 11].

The authors of [10, 11] showed that using the power spectra of voltage pulsations on a wire heater, it is possible to determine the moment of transition from one boiling mode to another, determining the change in the exponent α in the power-law dependence of the power spectrum of voltage pulsations on frequency ($S(f) = 1/f^{\alpha}$).

The purpose of this work is to determine the transition to the crisis mode of evaporation/boiling by the nature of pressure pulsations by analyzing the power spectra of pressure pulsations obtained by the fast Fourier transform method.

2. Experiments

The experiments were carried out on an experimental heat-exchange vacuum installation, the working chamber of which is shown in figure 1.

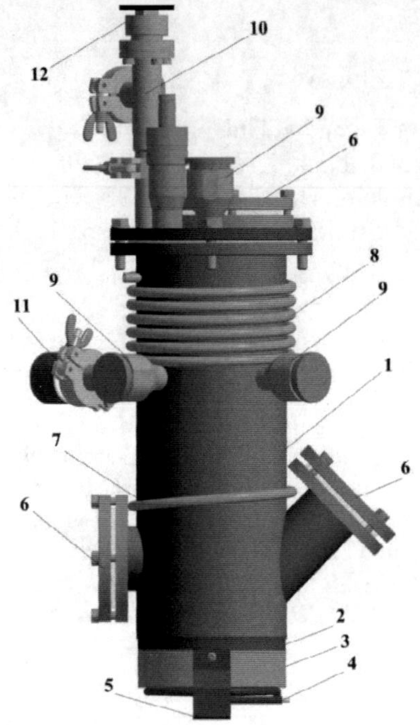

Figure 1. Working chamber of the experimental setup:
1 – case; 2 – bottom; 3 – brass plate; 4 – electric heater; 5 – heater mounting bracket; 6 – viewing windows; 7 – heating coil; 8 – cooling coil; 9 – vacuum inlet; 10 – branch pipe for mounting the pumping system; 11 – pressure sensors Setra 730; 12 – valve.

The case (1) of the working chamber is a cylindrical vessel made of steel 12X18H10T with an inner diameter of 120 mm, a height of 300 mm, and a wall thickness of 1 mm. A cooling coil (8) is located on the outer surface of the upper part of the working chamber. The chamber was cooled with water flowing through the coil. To reduce heat losses due to overflows along the walls of the chamber from the bottom to the cooling coil, as well as for a more uniform temperature distribution along the bottom of the chamber, a coil for heating (7) the side vertical wall of the chamber is located above the viewing window (6). The bottom (2) of the chamber is also made of steel 12X18H10T. There are five holes in the bottom with a pitch of 2 mm for installing thermocouples. Between the bottom and the electric heater (4), there is a brass plate (3) for a more even distribution of temperatures from the electric heater to the bottom. The electric heater is fixed with a bracket (5). For visual observation from above and from the side, there were observation windows on the working chamber (6). The pressure in the working chamber was measured with a Setra-730 ionization-deformation pressure sensor (11) and maintained by constant regulation using a valve (12). The measurement error of this sensor is ± 0.5% of the current reading. To enter the operating mode, the sensor needs to warm up, the warm-up time is at least 15 minutes, and the sensor response time is less than 20 ms. The data from the sensor were read out using the National Instruments module into the LabVIEW application software package with a frequency of 1 kHz. A more detailed description of the setup is given in [12, 13].

In the experiments, n-dodecane was used as a working fluid. Before the start of the experiments, the calculated amount of the working fluid was poured onto the bottom of the working chamber, which was necessary to create a layer of the required height. After that, the liquid was degassed with a decrease in pressure in the volume of the working chamber. During degassing, a characteristic bubbling was observed caused by the removal of dissolved air bubbles. In the course of the experiments, several stationary heat transfer modes were realized, at which the temperatures were recorded along with the thickness of the heated bottom, the pressure above the liquid layer in the volume of the working chamber, and at the same time, the process was filmed with a high-speed video camera. In the experiments, boiling curves were obtained at a constant pressure. The experiments were carried layer heights, $h = (1.7 - 4.0)$ mm out at a pressure above the layer P (P/P_{cr}): 133 ($7.4 \cdot 10^{-5}$), where P and P_{cr} are pressure over liquid layer and the critical pressure [Pa], respectively.

3. Results and discussion

It was shown in [14] that nucleate boiling was absent in liquid layers on a smooth heating surface at a pressure of $P \leq 10^3$ Pa. Heat transfer was carried out due to intense evaporation from the upper layer of the liquid during the formation in it, under the influence of the under the action of the vapor recoil force, structures in the form of "funnels" and "craters".

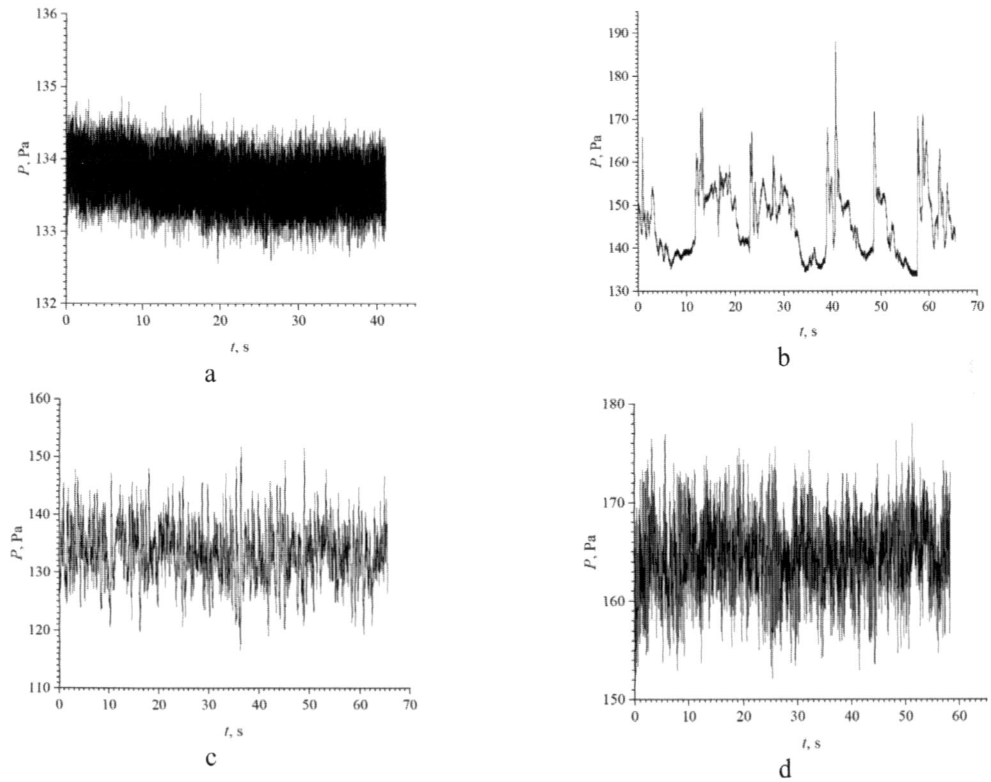

Figure 2. The change in pressure versus time, obtained during the evaporation/boiling of n-dodecane at an operating pressure in the chamber of 133 Pa and a layer height of 2.5 mm:
a) convection mode, heat flux $2.5 \cdot 10^3$ W/m^2; b) mode of formation of structures in the form of "funnels", $5.8 \cdot 10^3$ W/m^2; c) mode of formation of structures in the form of "craters", $2.31 \cdot 10^4$ W/m^2; d) pre-crisis mode, $7.45 \cdot 10^4$ W/m^2.

Figure 2 shows the experimental data of pressure fluctuations obtained at various heat fluxes for the evaporation of a 2.5 mm thick n-dodecane film at a pressure of P = 133 Pa. At low heat fluxes, the convective regime of heat transfer prevailed in the system, at this time convective rolls were observed in the liquid. The pressure pulsations are evenly distributed in time and have the appearance of white noise (figure 2(a)). With an increase in heat flux, uniformly distributed pressure pulsations begin to be interrupted by regular intensive pressure surges (figure 2(b)), these pulsations correspond to the regime of formation of structures in the form of "funnels" and "craters", where maximum pressure pulsations correspond to the formation of single craters. With a further increase in the heat flux, the pulsations become more chaotic (figure 2(c)), and these pulsations correspond to the regime of the formation of structures in the form of "only craters". In pre-crisis modes of evaporation/boiling, crater formation takes on a higher frequency character, this phenomenon can be seen in the pressure pulsation graph (figure 2(d)). In addition, in the pre-crisis state, the formation of structures in the form of "craters" can be accompanied by a short-term drying of the heating surface.

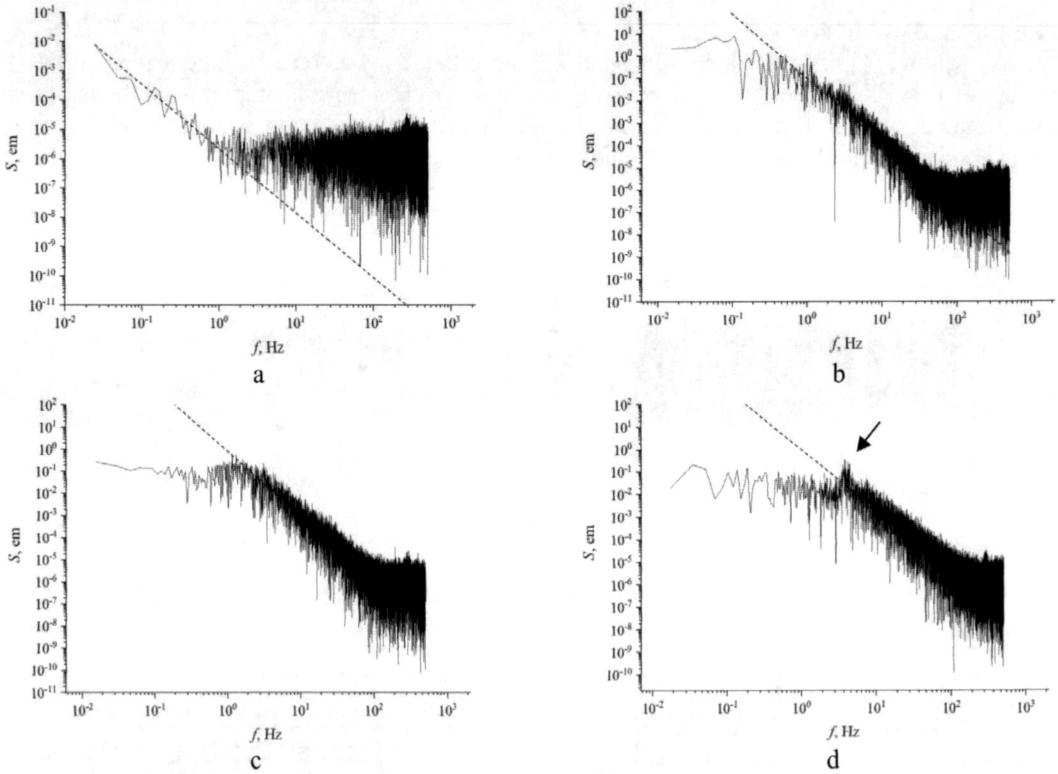

Figure 3. Power spectra of pressure pulsation versus frequency obtained during the evaporation/boiling of n-dodecane at an operating pressure of 133 Pa in the chamber and a layer height of 2.5 mm:
a) convection mode, heat flux $2.5 \cdot 10^3$ W/m^2; b) mode of formation of structures in the form of "funnels", $5.8 \cdot 10^3$ W/m^2; c) mode of formation of structures in the form of "craters", $2.31 \cdot 10^4$ W/m^2; d) pre-crisis mode, $7.45 \cdot 10^4$ W/m^2.

Using the fast Fourier transform, the power spectra of pressure fluctuations versus frequency were obtained (figure 3) during the evaporation/boiling of n-dodecane at an operating pressure in the chamber of 133 Pa and a layer height of 2.5 mm. In the modes of evaporation/boiling in the entire range of variation of heat fluxes, the obtained power spectra of pressure pulsations had the form $1/f^{\alpha}$. For low heat fluxes, spectra with $1/f^{\alpha}$ were observed in the low-frequency district and had one inflection point (figure 3(a)) onward, the inflection point in the low-frequency zone corresponds to "inflection point №1". At these heat fluxes, convective rolls were observed, and low-frequency fluctuations could appear in the system, which cannot be determined visually. In the modes of evaporation with the formation of structures in the form of "funnels" and "craters" in the liquid layer, the obtained spectra had the form $1/f^{\alpha}$ in the higher-frequency district and had two inflection points (figure 3(c), (d)) onward, the inflection point in the high-frequency zone corresponds to "inflection point №2". In the transition to the pre-crisis regime of evaporation/boiling, the power spectra of pressure fluctuations versus frequency are similar to the spectra obtained in the regime of formation of structures in the form of "only craters". However, the obtained spectra had a distinctive feature, in the form of a characteristic peak (arrow in figure 3(d)) in comparison with the regime of formation of structures in the form of "only craters".

During the transition to the crisis mode of evaporation/boiling, the power spectra of pressure fluctuations versus frequency at liquid heights of 1.7 mm and 4.0 mm also had a characteristic peak (arrow in figure 4(a), (b)).

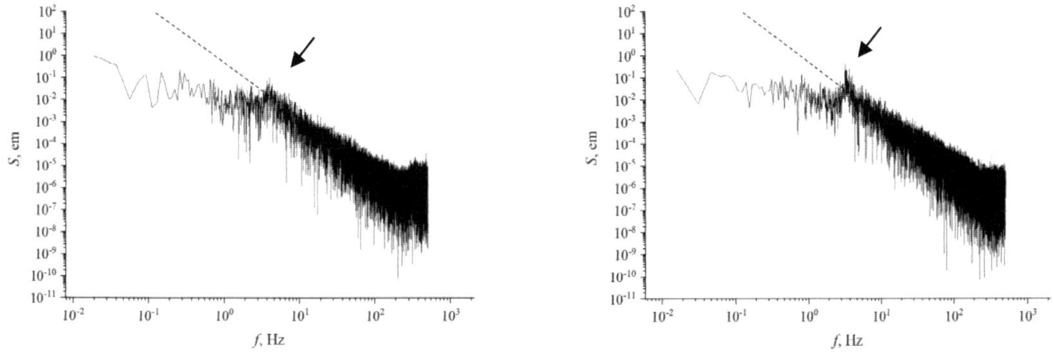

a b

Figure 4. Power spectra of pressure pulsations versus frequency for the pre-crisis regimes of evaporation/boiling of n-dodecane at an operating pressure in the chamber of 133 Pa: a) liquid layer height 1.7 mm, heat flux $5.16 \cdot 10^4$ W/m^2; b) 4.0 mm, $7.17 \cdot 10^4$ W/m^2.

The frequency value at which peaks and breaks observed in the power spectra of pressure pulsations are presented in table 1.

Table 1. Parameters of inflection points and peaks in the power spectra of pressure pulsations.

h, mm	Regimes	Inflection point №1, Hz	Inflection point №2, Hz	Characteristic peak, Hz
2.5	convection (fig. 3(a))	1.27	-	-
	"funnels" and "craters" (fig. 3(b))	1.43	77	-
	"only craters" (fig. 3(c))	1.48	127	-
	pre-crisis (fig. 3(d))	3.85	226	3.85
1.7	pre-crisis (fig. 4(a))	3.85	220	3.85
4.0	pre-crisis (fig. 4(b))	3.22	296	3.22

4. Conclusion

Analysis of experimental data shows that the processes of evaporation/boiling are accompanied by pressure pulsations, the nature of which changes depending on the magnitude of the heat flux. One of the possible tools for extracting useful information from pressure pulsations during evaporation/boiling of thin layers of n-dodecane under conditions of reduced pressure in a wide range of changes in layer heights is the analysis of the power spectra of pressure pulsations obtained using Fourier analysis.

It is shown that the processing of the pressure fluctuation spectrum makes it possible to identify the relationship between the behavior of the spectral power function of pressure pulsations in different heat transfer regimes and the degree of approximation to the critical heat flux during the development of crisis phenomena.

Acknowledgment
The study was funded by a grant from the Russian Science Foundation (project No. 19-19-00180).

References

[1] Maneglia Y and Bauza D 1999 *Microelectron. Eng.* **48** 189
[2] Samnakay R, Balandin A A and Srinivasan P 2017 *Solid. State. Electron.* **135** 37
[3] Kazakov K A 2008 *Phys. B Condens. Matter* **403** 2255
[4] Nicolae I, Miu D and Viespe C 2019 *Sensors Actuators, A Phys.* **295** 302
[5] Parkhutik V, Budnikov E Y and Timashev S F 2000 *Mater. Sci. Eng. B Solid-State Mater. Adv. Technol.* **69** 53
[6] Telesca L, Lasaponara R and Lanorte A 2006 *Phys. A Stat. Mech. its Appl.* **361** 699
[7] Vallianatos F, Makris J, Saltas V, Telesca L and Lapenna V 2006 *Commun. Nonlinear Sci. Numer. Simul.* **11** 745
[8] Carlini P, Bizzarri A R and Cannistraro S 2002 *Phys. D Nonlinear Phenom.* **165** 242
[9] Domb C 1996 *The critical point*. Taylor & Francis, London 376
[10] Skokov V N, Koverda V P, Reshetnikov A V, Skripov V P, Mazheiko N A and Vinogradov A V 2003 *Int. J. Heat Mass Transf.* **46** 1879
[11] Skokov V N, Koverda V P, Vinogradov A V, Reshetnikov A V 2010 *High Temp.* **48** 706
[12] Zhukov V I, Pavlenko A N, Nagaitseva Y V and Weiss D 2015 *High Temperature* **53** 690
[13] Zhukov V I and Pavlenko A N 2018 *Int. J. of Heat and Mass Transfer* **117** 978
[14] Zhukov V I and Pavlenko A N 2018 *Technical Physics Letters* **44** (6) 508

Experimental investigations of the processes in gas generator of hydrogen-air energy storage

N I Chukhin and A I Schastlivtsev

Joint Institute for High Temperatures of Russian Academy of Sciences (JIHT), Russia, 125412 Moscow, Izhorskaya, 13, Bd. 2

h2lab@mail.ru

Abstract. This paper describes the results of experimental investigation of the sample of the hydrogen-air gas generator unit with the expected average power of 65 kW. In total 5 test runs were made. Two tests showed that the mass flow and outlet gas temperature was in an agreement with the designed parameters. Additional attention should be paid to the cooling system design for the combustion chamber. In future such a gas generator in couple with the suitable gas turbine unit could be a part of the renewable energy accumulation system e.g. of hydrogen-air energy storage.

1. Introduction

The power generation and accumulation systems demand high reliable, efficient and ecological energy conversion technologies as gas turbines are likely to play a significant part of the future power generation scenarios. The hydrogen may effectively accumulate the energy generated from the wind or solar power stations as well as biomass gasification. For this there is a necessity in constructing of hydrogen combustion chamber which allows the stable, reliable and safe combustion occurring with the low NO_x emission level. Also according to [1] there is a large difference between the hydrogen physical properties and the carbon fuels. So the current essential task is the development of dry-low-NO_x combustion (DLN). The experimental unit in here included 0.3 mm micromix combustion chamber and it showed its low NO_x percentage over operation time. The test burner was scaled for the gas turbine of 335 kW electrical and pneumatic power. Gaseous hydrogen was injected through injectors perpendicularly to air flow by means of guiding panels for the air. The comparison of three combustion models showed the good agreement of these models to one another. But the Eddy Dissipation Concept which combined chemical kinetics and turbulent effects got the closest results to the experimental NO_x emissions data.

In [2] the combustion dynamics was described of the lean-premixed hydrogen-air flames. The high purity hydrogen was used as a fuel. For the comparison the high purity methane fuel was used in this case. The combustion test rig consisted of 60 injector multinozzles. The combustion chamber was made of a fused quartz pipe which was cooled by the air. The varying combustor construction is a key feature of the experimental rig (1200-1800 mm). Cooling water system also was set up for the piston centered in the combustion chamber. The experiments showed that the hydrogen flames oscillated in this nozzle system in isolation and with only weak interactions. Finally obtained results let the authors describe the behavior of lean hydrogen-air flames and show the self-induced instabilities at a wide range of conditions. All these results are also a part of the modern experimental investigations for development of hydrogen gas turbine systems.

The hydrogen combustion experiments are also vital for the nuclear power industry [3] as there hydrogen can be released in a reactor core melt accident. This described study was carried out using new Hydrogen Middle scale Test facility which can assess the efficiency of the hydrogen mitigation. The experimental setup consisted of a stainless steel tank with two domes connected at bottom and top positions. The paper concentrated on the combustion of the hydrogen-air mixture at three elevation level and also three concentration level (to 10% vol.). Four photodiode sensors in the axial direction could track the movement of the flame front. For the concentrations less than 10% the hydrogen was partially burned and this process was local. At 10% vol. H_2 concentration the hydrogen burned almost completely. All directions of flame propagation were a significant feature for this concentration. In [4] a thorough summary of the experiments and computational modes for hydrogen distribution and hydrogen combustion for containment hydraulics has been elaborated. What for the computer codes Lumped Parameter Codes and Computational Fluid Dynamics codes were described.

The results of experiments on the combustion regimes of lean and stoichiometric hydrogen-air mixture (up to 30 % H_2) are analyzed in [5] in a closed rectangular chamber. There were a smooth channel, a channel with metal grid filling a quarter of space and a channel completely filled with a metal grid. This was made for modeling a gap between an assembly of a fuel cell and a housing. Also five types of flame propagation regimes were systemized. But the main goal was to measure the hydrogen-air combustion behavior, flame acceleration and deflagration-to-detonation transition in a closed geometry.

Numerical studies were also devoted to the flame propagation of the lean 15% hydrogen-air mixture in the confined vessel of 4 cm diameter [6]. The results showed that mixing of water micro-droplets to the hydrogen-air could lead to the significant acceleration of the flame and the combustion. This effect was more significant for larger water droplets. The author also mentioned that the evaporation of relatively large water droplets can cool the combustion products which in turn can reduce NO_x emissions. Another study [7] was devoted to the initiated ignition of the lean hydrogen-air mixtures at atmospheric pressure by high-speed colour cinematography alongside with the numerical simulation of the process.

In [8] the combustion scheme for hydrogen-oxygen mixture stoichiometric combustion in the counter-current flow of water steam was implemented along with the combustion chamber of a high-temperature steam superheat. The superheated steam temperature at the combustion chamber outlet reached 1350 K. Also the optimum modes for hydrogen-oxygen mixture outflow were determined for its steady combustion in the counter-current water steam flow core. The geometry of combustion chamber flow section allowed to regulate the relative flow of the steam superheat in a wide range. The resulted steam temperature at the combustion chamber outlet provides prospects for development of steam turbines and combined power plants.

The goal of present work is obtaining experimental parameters of gas-dynamic processes of hydrogen-air gas generator in a wide range of fuel, oxidizer and ballasting air consumption. Such a generator was designed for using in the cycle with the gas turbine unit of 50 kW capacity which in its turn could be a part of the hydrogen-air accumulation system for the renewable energy sources e.g. [9].

2. Experimental details

The experimental hydrogen-air gas generator unit (figure 1) for process research is made of stainless steel and consisted of a mixing unit, a combustion chamber, a simulator nozzle, a flange, and an air casing with an O-ring. The whole length of unit is 389 mm without the auxiliary equipment, the combustion chamber length is 259 mm with the larger inner diameter 81 mm. The experimental testing unit for hydrogen combustion with a capacity of up to 200 kW [10] of the Joint Institute for High Temperatures of the Russian Academy of Sciences was upgraded for this investigation.

Figure 1. Experimental hydrogen-air gas generator unit.

As for the combustion chamber in these experiments the process of combustion (in figure 2) can be classified neither diffusion nor kinetic. A premixed part of hydrogen and oxygen is ignited in the central channel with the ignition spark first. After that the main part of hydrogen coming from the 16 circumferential orifices is being ignited due to the turbulent mixing. The necessary part of the experimental unit is the air casing for creating low-temperature wall layer and reducing the heat density of the edge. As it can be seen in figure 2 the air casing has a ribbed surface with a high heat transfer rate, the ability to resist mechanical stress and endurance to the high pressure. The air casing was designed for the mass flow 280 g/s in order to cool effectively the gas generator walls. Air goes through the gap between the casing and the wall and only afterwards enters the combustion chamber. The outlet gas temperature 690 °C was defined from the high-temperature strength condition for the gas turbine blades.

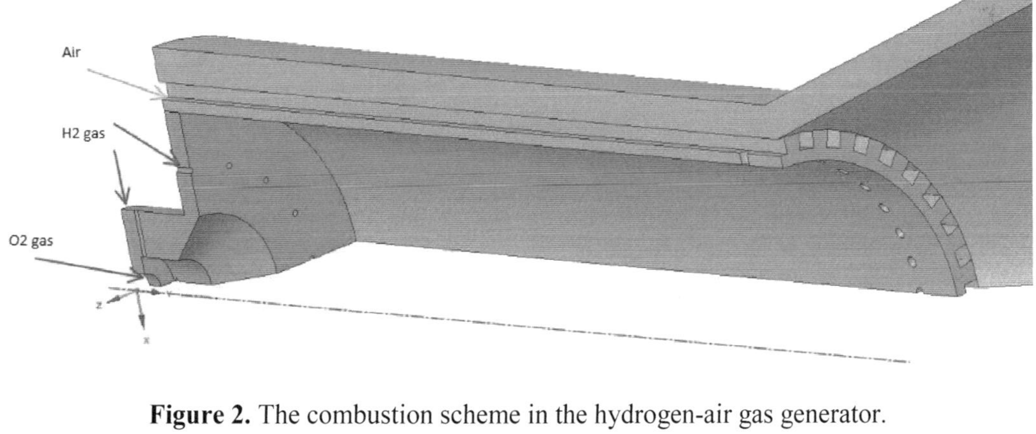

Figure 2. The combustion scheme in the hydrogen-air gas generator.

The used automation system made it possible to ensure the safety of the hydrogen-air gas generator operation and its emergency shutdown. The experimental unit was connected to the main pipelines for supplying hydrogen, oxygen and air from the corresponding gas tanks. The entire experimental unit was tested by high pressure beforehand. During the tests of the experimental unit two starts for tuning and two control starts were carried out with an expected power of 60 and 72 kW.

- In the first test the optimal starting mode of the hydrogen-air gas generator was determined and the cyclogram was adjusted since the combustion broke for 10 s due to a sharp supply of oxygen.
- Taking this into account, in the second test the supply of increased flow rates of hydrogen and oxygen was made earlier. But due to the failure of the air pressure reducing valve and a decrease in the consumption of cooling air the gas temperature at the outlet increased to 1394 °C which of cause exceeded for a significant value the designed outlet temperature.
- The third control start was performed with an increased hydrogen flow rate and the outlet gas temperature reached 634 °C.
- In the fourth also control start the hydrogen consumption was increased more and the outlet gas temperature reached 723 °C. During this test, outlet gas was also taken into a special tank for analysis for the content of nitrogen oxides (NO_x). Analysis of this gas composition showed the content of nitrogen oxides in the concentration of 14.7 ppm which suited the expectable values.

In table 1 we summarize the main average experimental parameters for the tests.

Table 1. The average parameters of the hydrogen-air gas generator control starts.

	3rd start	4th start
Hydrogen pressure, atm	18.3	10.87
Oxygen pressure, atm	29.1	28.63
Air pressure, atm	26.8	29.22
Hydrogen mass flow, g/s	0.5	0.6
Oxygen mass flow, g/s	4	4.8
Outlet gas temperature, °C	634	723

For the fourth start which was successful the mass flows and the measured temperatures are shown in the following graphs in figure 3 and 4. During all tests a significant decrease in the temperature of the air pressure reducing valve was noted, which in its turn at the increase of cooling air consumption could lead to the freezing and problems in operation with the high-pressure air tanks.

After all the control runs had been carried out, an additional study was made to study the emergency operation mode of the hydrogen-air gas generator unit and determine the most heat-stressed areas of the combustion chamber. For this the cooling air consumption was reduced by 85%. As a result the outlet gas temperature exceeded 1500 °C and the internal cooled air casing was partially melted. After examining air casing it was obvious that the most heat-stressed zone is located at a distance of 7-9 cm from the fire bottom of the mixing unit. Consequently in order to increase the reliability of the combustion chamber in this area it is necessary to intensify cooling or to reduce heat load by increasing the diameter of the combustion chamber in this area.

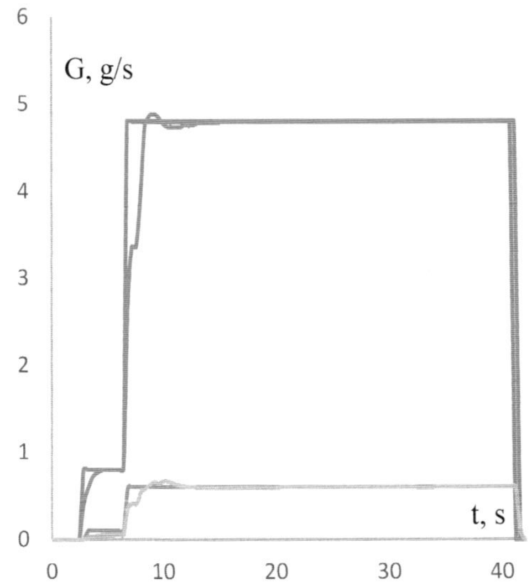

Figure 3. Hydrogen and oxygen mass flows over time. ___ O_2 set point, ___ O_2 mass flow, ___ H_2 set point, ___ H_2 mass flow.

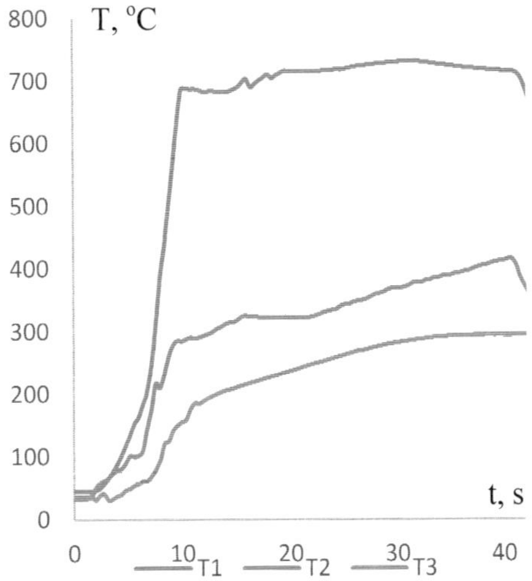

Figure 4. Temperatures changes over time. T1 – outlet gas temperature, T2 – mixing unit wall temperature, T3 – cooling air temperature.

3. Conclusion

First the modern works devoted to the hydrogen combustion were described briefly. It was mentioned that quite a lot of attention is paid nowadays to the lean hydrogen mixtures combustion and flame propagation analyses and hydrogen combustion in mesoscale. In the field of nuclear power engineering also several experimental and numerical investigations are described. For the main topic the results of series of experiments with the novel construction of mixing unit for the hydrogen-air gas generator of kilowatt level are presented in the second chapter. It was concluded that the construction of the cooling air casing should be improved to some extent but for the designed outlet gas temperature of 690 °C it worked without fails. Choosing the reducing pressure valve, one also should take into account the gas consumption flow rate for the combustion in order to avoid freezing. Analysis for the NO_x concentration at the outlet gas showed also permissible value. So as most of the runs here were successful such gas generator unit could be a key element for development of the hydrogen-air energy storage system for the different renewable energy sources.

References

[1] Haj Ayed A, Kusterer K, Funke H H-W, Keinz J, Striegan C and Bohn D 2015 Experimental and numerical investigations of the dry-low-NOx hydrogen micromix combustion chamber of an industrial gas turbine *Propulsion and Power Research* **4(3)** pp 123–131
[2] Lee T and Kim K T 2020 Combustion dynamics of lean fully-premixed hydrogen-air flames in a mesoscale multinozzle array *Combustion and Flame* **218** pp 234–246
[3] Hu P and Zhai S 2021 Experimental study of lean hydrogen-air mixture combustion in a 12m3 tank *Progress in Nuclear Energy* **133** p. 103633
[4] Punetha M, Yadav M K, Khandekar S, Sharma P K, Ganju S 2020 Intrinsic transport and combustion issues of steam-air-hydrogen mixtures in nuclear containments *Int. J. of Hydrogen Energy* **45(4)** pp 3340–3371
[5] Kuznetsov M and Grune J 2019 Experiments on combustion regimes for hydrogen/air mixtures in a thin layer geometry *Int. J. of Hydrogen Energy* **44(17)** pp 8727–8742

[6] Yakovenko I S and Kiverin A D 2021 Intensification mechanisms of the lean hydrogen-air combustion via addition of suspended microdroplets of water *Int. J. of Hydrogen Energy* **46** pp 1259–1272

[7] Rubtsov N M, Kotelkin V D, Seplyarskii B S, Chernysh V I, and Tsvetkov G I 2011 Investigation into the combustion of lean hydrogen–air mixtures at atmospheric pressure by means of high-speed cinematography *Mendeleev Communications* **21** pp 215–217

[8] Guryanov A I, Piralishvili S A, Guryanova M M, Evdokimov O A and Veretennikov S V 2020 Counter-current hydrogen–oxygen vortex combustion chamber. Thermal physics of processing *J. of the Energy Institute* **93(2)** pp 634–641

[9] Chukhin N I Schastlivtsev A I and Shmatov D P 2020 Feasibility of hydrogen-air energy storage gas turbine system for the solar power plant in Yakutsk region. *J. Phys.: Conf. Ser.* **1652** p 012039

[10] Schastlivtsev A, Dunikov D and Borzenko V 2019 Experimental study of the processes in hydrogen-oxygen gas generator *Int. J. of Hydrogen Energy* **44(18)** pp 9450- 9455

Non-trivial capillary hydrodynamics and strong evaporative cooling in multilayer two-dimensional materials: application in thermal management systems of electronic and energy devices

A S Dmitriev
Thermo Nano Laboratory, Low Temperature Department,
National Research University «Moscow Power Engineering Institute»
Russia, 111250 Moscow, Krasnokazarmennaya, 14

asdmitriev@mail.ru

Abstract: Studies of evaporation from graphene nanostructures with nontrivial capillary hydrodynamics are being carried out. Data were obtained for the rates of evaporation and heat fluxes during cooling of experimental heated cells. Models were built and compared with experimental data. It is shown that the existing models overestimate the evaporation rates and heat fluxes and heat transfer coefficients.

1. Introduction

The problems of thermal stabilization and thermal management of various energy (fuel cells, solar photo panels, solar thermal power engineering, etc.) and electronic (processors, nanotransistors, semiconductor micro- and nanolasers, etc.) cannot be solved (in particular, "thermal wall "of Moore's law) without using more radical cooling methods than are currently used [1-3]. Today it is believed that the most effective cooling method is the liquid-vapor phase transition, when the latent heat of vaporization is used. However, the limiting factor of this process is the rate of fluid supply to the interface [3]. More recently, it was found that the use of layered 2D materials (graphene, molybdenum disulfide, etc.) leads to nontrivial capillary hydrodynamics in nanoscale pores and channels [4, 5], which allows a sharp increase in the rate of supply of a cooling evaporating liquid to the interface, practically removing the limiting factor of the fluid flow rate.

The aim of this study is to compare the calculated and experimental cooling rates of heated silicon substrates using a layer of graphene material, which is a partially oriented graphene nanoflakes 1.5-3.5 nm thick with lateral dimensions 10-30 μm, obtained by supercavitational cleavage of high-purity graphite in distilled water (Graphene Institute Ltd, Russia).

In this paper, modern data on the capillary thermohydrodynamics of nanopores and nanochannels are presented, the mechanisms of strong evaporative cooling using two-dimensional materials are considered, and examples of the use of such approaches to promising thermal management are given. In natural phenomena, strong evaporative cooling occurs very often, therefore, many mechanisms of capillary thermohydrodynamics and strong evaporation can be "spied" from nature, which only in recent years has become an obvious fact.

The nanoporous structure is very difficult to implement in conventional materials, however, the possibility of its creation is directly related to two-dimensional materials (in several atomic layers), which at present can be implemented, for example, using graphene nanoflakes (thickness 2-4 nm, lateral size - from 20 to 100 μm) in the manufacture of bulk structures. It is important that in such structures the transverse size of nanochannels can range from 2 to 8-10 nm, which naturally leads to an abnormally high velocity of water movement, as shown by direct experiments with graphene nanomembranes [3,4] and our experiments on evaporation from such structures under the action of solar radiation [2,4].

In addition to traditional models for describing capillary thermohydrodynamics with an interphase boundary, new physical effects appear in some problems - taking into account the slip of the liquid phase near the boundary, peculiarities of surface wetting with nano-roughness, electroosmotic, thermoosmotic flows and thermophoresis [4].

2. Materials and method

The experimental cell consisted of a silicon heated substrate, a container inside which an evaporating liquid was supplied (in this work, data are given only for water and aqueous graphene nanofluids), a graphene structure block placed inside steel meshes of copper or stainless steel, and also consisting of a nanoporous structure steam condensation unit. The condenser could force condensed liquid into the cooling unit (or similar to heat pipes) (figure 1).

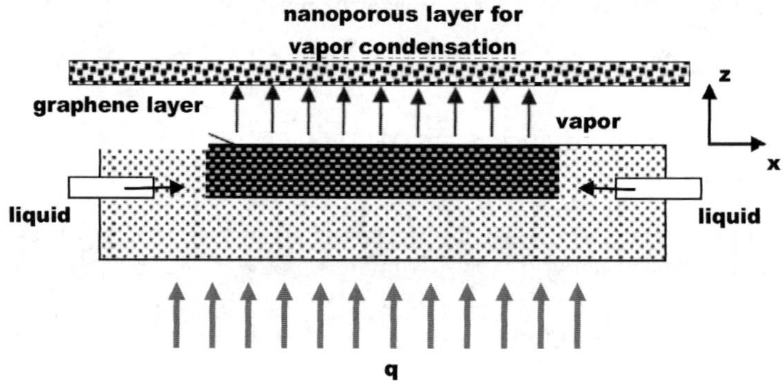

Figure 1. Schematic of an experimental cell for studying evaporation from nanoporous graphene structures.

Graphene nanoflakes with an average size of 10-30 μm (thickness - 1.5-3.5 nm) were received from Graphene Institute Ltd (Russia) (figure 2). Chemically clean graphite microparticles ~300 μm were used. As a mesh we used cooper M1(Cu>99.6%) or stainless steel mesh with cell side size 0.3 mm and wire diameter 0.1 mm supplied by TDMC, Russia.

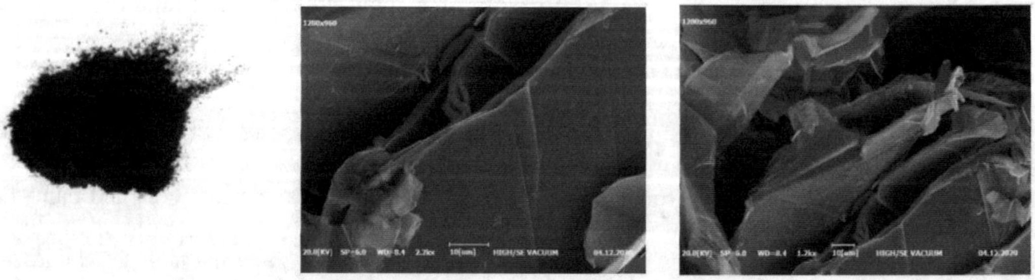

Figure 2. Oriented graphene nanoflakes for evaporation.

A nanoporous oriented layer of graphene nanoflakes before being placed inside the experimental cell is shown in figure 3. The geometrical dimensions of the evaporating block were chosen differently, however, a high graphene layer of about 2 mm and a diameter of 10-12 mm was found to be optimal. For more efficient operation, the graphene layer was saturated with water or other liquids using drop dispensers (figure 3). The temperature of the silicon substrate was varied from 25 to 65 ^{0}C. It is clear that the evaporation rates and cooling efficiency increased with an increase in the substrate temperature. The substrate temperature was measured with thermocouples and heat flow sensors. The temperature was measured with an accuracy of about 0.5 ^{0}C. The evaporation rate was measured by an accurate electronic balance inside the nanoporous capacitor, as well as by the liquid flow rate through the heating unit.

Figure 3. Graphene structure (wick) with mesh for placement inside the evaporation unit.

3. Evaporation rate data

In addition to pure liquids, we were the first to study the evaporation of graphene nanofluids inside nanoporous structures. Preliminary data show that such fluids heat up and evaporate more efficiently, as shown by the empirical data in figure 4.

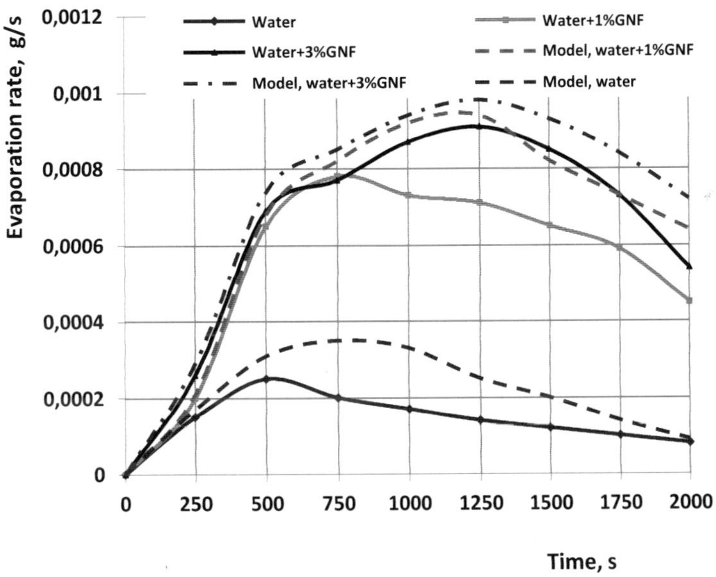

Figure 4. Evaporation rates of various liquids using graphene nanochannels and nanopores.

The physical mechanisms of the increased evaporation rate are not yet completely clear and require additional study. At the same time, the use of graphene nanofluids together with graphene nanoporous structures makes it possible to create an active system of evaporative cooling of a continuous cycle, which is extremely important for a large number of applications. The obtained experimental data on the evaporation of water and aqueous graphene nanosuspensions with mass concentrations of 1% and 3% are shown in figure 4.

4. Comparison with mathematical models

A feature of evaporation from oriented graphene nanostructures is nontrivial capillary hydrodynamics - a strong influence on the movement of the meniscus of the interface of the boundary conditions on the liquid - graphene plate surface (almost complete water sliding without tangential resistance - (Navier boundary condition and significant slip length) [4]. In this work, we used models close to those developed in the [5].

4.1 Capillary-driven flow

The imbibition of liquid into a nanoporous structure occurs when the solid–vapor specific surface energy (σ_{sv}) is greater than the solid–liquid specific surface energy (σ_{sl}). This usually occurs when the contact angle is less than 90^0. The driving force for imbibition is characterized by the Young–Laplace equation

$$\Delta p = p_l - p_v = 2\sigma_{lv}\Gamma$$

where Δp is the local capillary pressure, p_l is the local liquid pressure at the liquid–vapor interface, p_v is the pressure of the vapor phase, σ_{lv} is the liquid–vapor surface tension, and Γ is the local mean curvature of the liquid surface.

In contrast to the work [5], where the Brinkman equation (or an analogue, the Darcy equation) was used for the flow of a liquid inside a porous structure, in this study we describe the flow of a liquid inside a graphene structure based on non-slip flow. In this case, the slip length significantly exceeded the characteristic scale of nanopores L_s>>δ (L_s~50 nm, δ ~3-5 nm).

4.2 Heat transfer

Heat in thermal management devices is generated due to Joule heating and is transferred to the liquid vapor interface through the combined solid-liquid matrix by the heat conduction mechanism. Further, heat is dissipated by convection, heat conduction and thermal radiation into the environment. Typically thermal management systems use a heat transfer coefficient α=q(T_s-T_∞), q is the steady state heat flux dissipated by the graphene nanostructures (wick), T_s is the temperature at the solid interface, and T_∞ is the temperature of the far field ambient.

The effective thermal conductivity of the graphene nanostructures is dependent on the working fluid, the solid material and nanostructures and operating conditions. In the simplest case, the following model can be adopted (the study showed that the complication of the model insignificantly affects the final result)

$$\lambda_{eff} = \lambda_s \frac{2 + \lambda_l\lambda_s^{-1} - 2\phi(1 - \lambda_l\lambda_s^{-1})}{2 + \lambda_l\lambda_s^{-1} + \phi(1 - \lambda_l\lambda_s^{-1})}$$

where λ_s, λ_l - the thermal conductivity of the solid and liquid, ϕ - the porosity. When using nanofluids, the quantity is replaced by the effective thermal conductivity of the nanofluid ($\lambda_l \rightarrow \lambda_{l,eff}$).

4.3 Vapor flow

The mass flow of vapor during evaporation can be determined by the following relationship

$$\vec{j}_v = \rho_m \varphi \vec{u}_v - \rho_m D_v \nabla\varphi$$

where j_v is the vapor mass flux vector, ρ_m is the mixture (air + vapor) density, φ is the vapor mass fraction, and D_v is the diffusivity of vapor in the gas (or mixture of gases). The evaporative mass flux is related to the evaporative heat flux by $q_v = j_v\alpha_{lv}$ equivalent thermal resistance is $R_v = (T_s - T_\infty)(j_v\alpha_{lv}S)^{-1}$.

Using the above model, the rates of evaporation of various liquids (water and water suspensions of graphene) were calculated. These calculated data are presented in figure 4, which also shows experimental data in the study of evaporation from graphene oriented nanostructures.

Unfortunately, within the framework of the used model, there is no good agreement between theoretical and experimental data, which speaks, first of all, of the shortcomings of the existing models of evaporation from nanostructures. Note that the results of the calculations give data exceeding the evaporation rate found experimentally.

It should be especially emphasized that for the case of liquid sliding inside graphene nanochannels and evaporation from the surface of the meniscus inside the channel, an interesting situation can arise - the liquid inside the channels can move much faster (slip boundary condition) than the liquid-vapor interface. As far as we know, such cases have not been previously considered within the framework of mathematical models.

In our opinion, it is precisely in these processes that the nontrivial capillary hydrodynamics of liquids inside the nanochannels manifests itself. Therefore, in addition to the importance of further constructing models of vapor transport in the external environment, as noted in, it is necessary to more consistently consider capillary nanohydrodynamics inside channels, especially when it concerns two-dimensional materials such as graphene and its structure.

4.4 Cooling

The fraction of heat dissipated by evaporation was determined by subtracting the perceptible cooling and parasitic heat loss from the heating source. The heat flux used for cooling was calculated from the ratio of the energy balance

$$q_c = [j_v + j_L] C_p (T_{out} - T_{in}) S_{Cool}^{-1}$$

where q_c - heat flow for cooling, j_v - vapor mass flux, j_L - evaporation loss, C_p - specific heat capacity of the working fluid, S_{Cool} - cooling surface area. The table shows the calculation results and experimental data on the dependence of heat fluxes during cooling on the type of working fluids and the heat transfer coefficient.

Table

Working fluids	α, W/cm^{2}K	Maximum heat flux, W/cm^2	Publication
Water	9.6	140	[6,7]
Water	14.8	180	[6,7]
Water	38	390	[6,7]
R-245fa	39.7	550	[6,7]
Pentane	18.4 (th – 20.2)	230 (theory – 244)	[this work]
Water	24.7 (th – 20.2)	297 (th –316)	[this work]
Water+GNF1%	31.2(th – 34.8)	340 (th – 369)	[this work]
Water+GNF3%			

5. Results and discussions

A preliminary comparison with the rate of evaporative cooling in other systems, as well as using heat pipes, showed that the efficiency of evaporative cooling using oriented graphene nanostructures is several times higher. The use of active cooling systems in thermal management systems (thermal interface materials and efficient capillary hydrodynamics with strong evaporation from nanopores) is currently the most attractive approach to thermal stabilization of energy-intensive electronic and energy systems.

Acknowledgments
The work is supported by the Russian Science Foundation (Grant No. 17-19-01757).

References

[1] Alexander S. Dmitriev 2019 Hybrid Graphene Nanocomposites: Thermal Interface Materials and Functional Energy Materials, Graphene Production and Application, Sadia Ameen, M. Shaheer Akhtar and Hyung-Shik Shin, *IntechOpen*
[2] Singh H and Myong R 2018 *Adv. Mat. Sci. and Eng.* 9565240
[3] Zhu Y, Zhan K and Hou X 2018 *ACS Nano* **12** 908–911
[4] Lei W, Rigozzi M and McKenzie D 2016 *Rep. Prog. Phys.* 79 025901
[5] Wilke K, Barabadi B, Lu Z, Zhang T and Wang E 2017 *Appl. Phys. Lett.* 111 171603
[6] Vaartstra G, Zhang L, Lu Z et al 2020 *J. Appl.Phys.* **128** 130901
[7] Hanks D, Lu Z, Sircar J et al 2020 *ACS Appl. Mater. Interfaces* **12**, 7232−7238

Numerical and experimental study of abnormal enhancement of separated turbulent flow and heat transfer in inclined oval-trench dimples on the plate and on the narrow channel wall

S A Isaev[1,5], S V Guvernyuk[2], N I Mikheev[3], I A Popov[4] and D V Nikushchenko[5]

[1] Saint-Petersburg State University of Civil Aviation,
Russia, 196210 Saint-Petersburg, Pilots Str., 38
[2] Institute of Mechanics Lomonosov MSU,
Russia, 119192 Moscow, Michurinsky pr., 1
[3] Kazan Scientific Center RAS,
Russia, 420111 Tatarstan, Kazan, Lobachevsky Str., 2/31
[4] Tupolev Kazan National Research Technical University – Kazan Aviation Institute,
Russia, 420111 Tatarstan, Kazan, K.Marks Str., 10
[5] Saint-Petersburg State Marine Technical University,
Russia, 190121 Saint-Petersburg, Lotsmanskaya, 3

isaev3612@yandex.ru

Abstract. The numerically found abnormal enhancement of separated flow and heat transfer in inclined oval-trench dimples (OTDs) on the plate and on the wall of the narrow channel was experimentally confirmed at the stands of the Research Institute of Mechanics (Lomonosov Moscow State University), the Kazan Scientific Center RAS, and the Kazan State Research Technical University – the Kazan Aviation Institute. The measured static pressure drops in a single OTD at Re = 6.7×10^4 and 16.7×10^4, the velocity profiles of accelerating laminar (Re = = 1000) and turbulent (Re = 4300) flows in the narrow channel with two rows of 26 OTDs, the estimate of Nusselt numbers on the channel wall with single-row OTDs are in good agreement with the numerical predictions within the RANS approach.

1. Introduction

The researchers paid attention to surface vortex generators as tools for heat transfer enhancement, allowing them to provide that heat transfer increases more quickly than hydraulic losses grow [1,2]. Spherical dimples are the simplest in the technology of applying dimples [3,4]. However, the separated flow around the dimples is accompanied by weak return and secondary currents, and a significant decrease in heat transfer occurs inside the dimples [5-9]. New-type surface vortex generators are suggested – inclined oval trench dimples (OTDs) with a significant lengthening of the cylindrical part (of the order of 3-6 of width size). As shown in [10], when a single OTD of depth 0.39 is arranged on the narrow channel wall (of height 1 and width 7.5) at an inclination angle of 45°, swirling turbulent flows are created. The maximum secondary flow velocity is of the order of 0.8 of the bulk velocity of the flow.

The phenomena of abnormal enhancement of separated flow and heat transfer on the OTD entrance portion, as well as of the core flow acceleration in the channel [11] are discovered for laminar and turbulent flows of air in narrow channels with packages of inclined single-row OTDs. The periodic areas of stabilized flow in narrow channels and in micro channels of height 1 are considered. The Reynolds numbers are assigned as equal to 10^3 and 10^4. As a rule, the OTD inclination angle is selected as equal to 45°. For the package of sparsely arranged OTDs, where a single dimple has a length of 7.05, a width of 1.05, and a depth of 0.3, and it is located in the center of the periodic computational domain (of length 6) of the channel of width 7 on the heated inner surface, the absolute minimum relative friction value increases four times and relative heat transfer (in relation to the parameters in the plane-parallel channel) increases 5 times. The maximum absolute value of the secondary (cross) flow velocity appears to be of the same order as the maximum flow velocity in the plane-parallel channel. For inclined single-row OTDs of width 1, length 4.5, and of depth over 0.25, the phenomenon of the laminar flow acceleration with a 1.5-fold growth of the maximum velocity in the flow core is found in the periodic computational domain of the narrow micro channel of width 6 and length 4 with a single dimple inclined at 45° . The influence of the packing of dimples in one row is studied in the periodic computational domain (of lengthening 8) of the narrow (9 to 1) plane-parallel channel. Its heated wall is provided with the inclined OTD of length 7.05, width 1.05, and depth of 0.25 at the rounding radius of edges equal to 0.21. The distance H between the centers of the dimples is varied from 2 to 8. Like the laminar flow acceleration in a narrow channel with inclined OTD, when compaction of a packet of inclined OTD at an angle of 65°, the turbulent flow acceleration effect is obtained, showing that the maximum velocity in the flow core increases by a factor of 1.39 at the step H equal to 2 between the dimples. The packing of single-row dimples significantly augmented abnormal enhancement of separated turbulent flow and heat transfer on the entrance portion of the OTD inclined at an angle of 65° in the periodic computational domain of the narrow channel on its heated wall. At H=2, it is characterized by a 4-fold increase in the maximum absolute value of the relative friction projection onto the direction of the middle portion of the OTD and an almost 6.5-fold increase in relative heat transfer (in relation to the parameters in the plane-parallel channel). The maximum absolute value of the secondary (cross) flow is about 10% higher than the value of the maximum flow velocity in the plane-parallel channel. The maximum absolute value of the recirculating flow velocity almost three times exceeds the same velocity of the backflow in the spherical dimple, reaching a value of 0.89 of the bulk velocity in the channel. The inclination of OTDs within the range of 0-60° in the package of sparsely arranged in-line dimples on the stabilized section of the narrow channel has a drastic impact on abnormal enhancement of separated and swirling flows in the dimple. This results in the 3-4-fold decrease in relative negative friction at the angles from 40° to 60°, in the increase in the highest velocity of the backflow and the secondary flow from 0.8 to 1.18 with respect to the flow bulk velocity in the channel.

Work [12] consider the numerical simulation of laminar and turbulent heat transfer enhancement in the space between the ribs of the air condenser – the finite-size narrow channel with a package of inclined single-row OTDs at Re=1000 and 6000 (the symmetry conditions on the lateral sides). 31 dimples are applied on the channel wall section of length 80. Each dimple has a length of 4.5, a width of 1, an edge rounding radius of 0.2, and an inclination angle of 45°. The OTD depth is varied from 0 to 0.35. The channels with the sparsely (21) and densely (31) arranged OTDs placed on the heated wall are compared. The influence of the dimple depth on the thermal and thermal–hydraulic performance is assessed. In the laminar regime, the variant with the 31st OTD of depth 0.25 (the growth of hydraulic losses does not exceed 46% at an almost 2-fold increase in heat transfer in comparison to the plane-parallel channel) is preferred. A decision was suggested for packages of air condensers composed of dimpled ribs and intermediate plates with variable gaps between them and the plates. In the turbulent regime, the effect of abnormal enhancement of separated flow and heat transfer, earlier observed in the area of stabilized heat transfer in the inclined OTDs, which increases with the growth of their depth, has been confirmed. The relative negative friction reaches a value of -3.5 and the maximum relative values of the local Nusselt numbers appear to be of the order of 14. The highest total relative heat transfer reaches a value of 2.4 at a 70% growth of relative hydraulic losses. The earlier discovered effect of turbulent flow acceleration in the narrow channel with the inclined OTDs placed on the wall has also been confirmed. In the dimpled channel, the local longitudinal velocity increases more than one and a half. The secondary flow velocity in the separation zones achieves a characteristic flow velocity at the channel inlet.

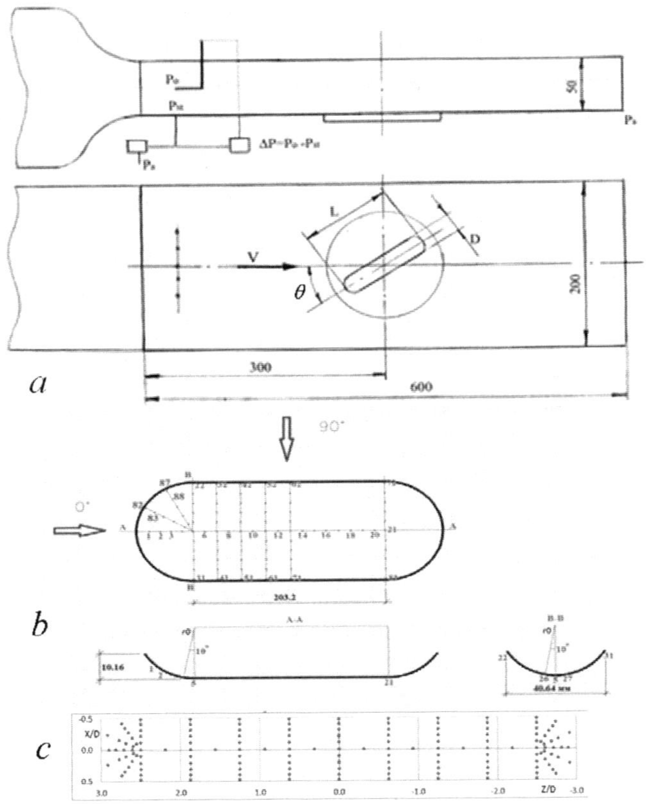

Figure 1. Projections of the working section of the wind tunnel with the inclined OTD model (*a*), the layout of drain holes meant for static pressure measurement and the dimple contours (*b*), the layout of points for static pressure measurement (*c*).

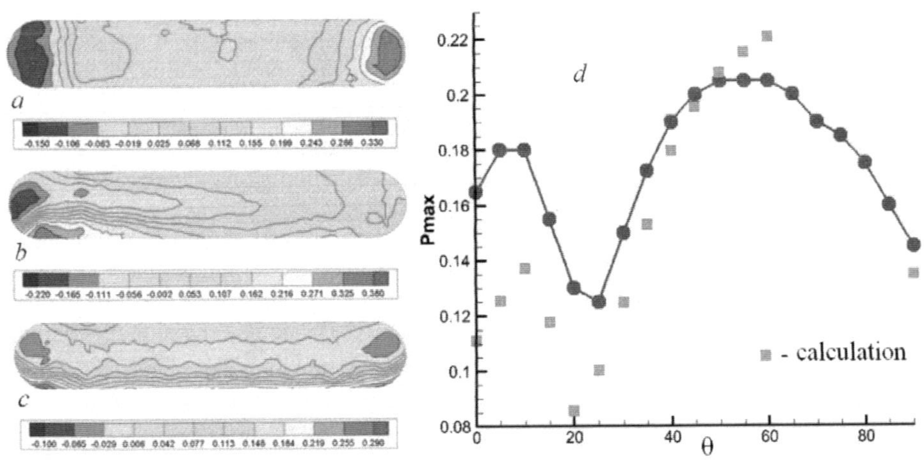

Figure 2. Fields of the static pressure coefficient at the inclination angles 0°(*a*), 45°(*b*), 90°(*c*) and the predicted and measured maximum static pressure distributions (*d*).

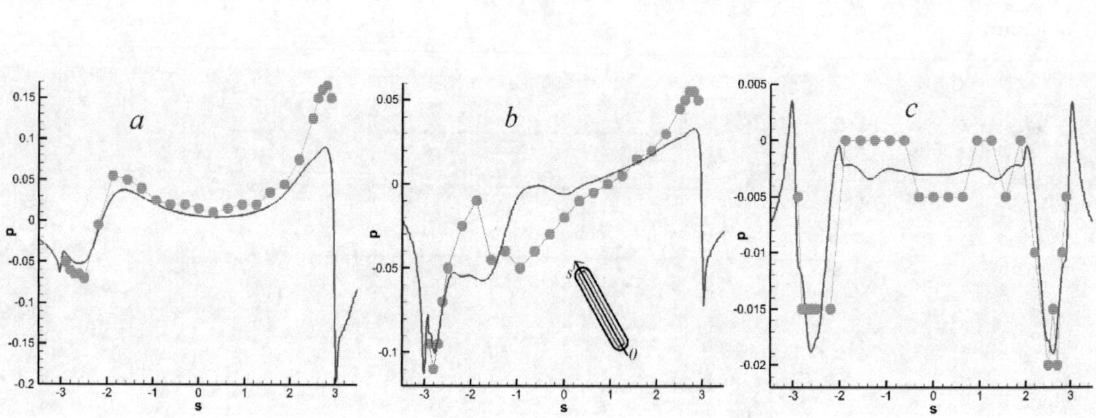

Figure 3. Predicted (lines) and measured (points) static pressure distributions on the middle portion of the dimple at the inclination angles 0°(*a*), 45°(*b*), 90°(*c*).

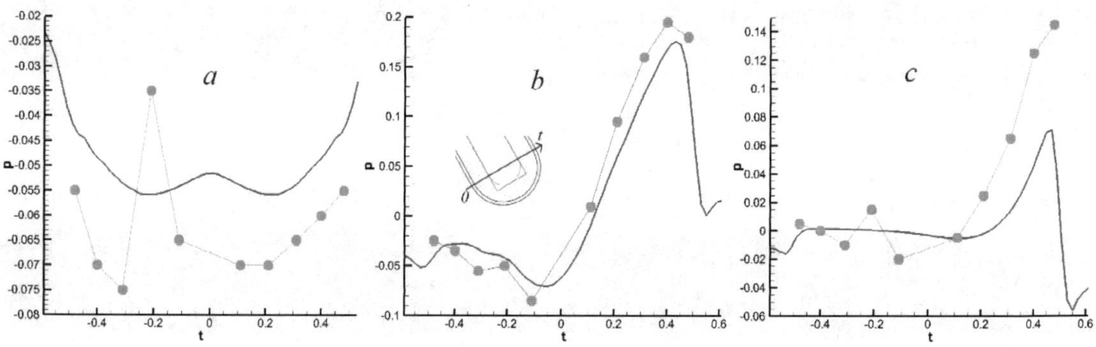

Figure 4. Predicted (lines) and measured (points) static pressure distributions on the transverse entrance portion of the dimple at the inclination angles 0°(*a*), 45°(*b*), 90°(*c*).

The cause of the hydrodynamic and thermophysical effect of abnormal enhancement of separated flow and heat transfer has been established and consists in creating a large pressure drop between the flow stagnation zone on the windward slope of the inclined OTD (of the of 0.5-0.6) and the closely spaced rarefaction region ((-0.3 –(-0.6)) at the place where a tornado-like vortex is generated on the OTD entrance spherical portion.

2. Results and Discussion

The main objective of the experiments at the Research Institute of Mechanics (Moscow State University) was to measure static pressure distributions on the characteristic portions of an inclined OTD. A series of experiments was carried out in the wind tunnel A4, in the working section of which the plate with the OTD of length 6, width 1, and depth 0.25 (Figure 1) was positioned (Figure 1). The external flow Reynolds number was equal to 67000 and the boundary layer thickness in the vicinity of the OTD was 0.16. The OTD inclination angle was varied from 0 to 90°. The experimental distributions of Cp as shown in figure 2 clearly illustrate the influence of the inclination angle on the internal flow structure in the OTD; in particular, they indicate the range 25° <θ <85°, within which the main rounding of the OTD has a double configuration consisting of the localized zones of the extreme values of the opposite-sign pressure (in the terminology [11], this corresponds to the OTD 'operating' regimes).

The dependences of the absolute maximum values and absolute minimum values of the pressure coefficient over the OTD entire surface, when the inclination angle φ is varied, show that the maximum rarefaction at Cp =-0.22 is seen within the range 40° < θ < 45° and the maximum positive value of

Cp = 0.41 is observed within the range 55° < θ < 60°. The numerical predictions of the turbulent flow of air (at Re=10000) around a plate with a single OTD at the inclination angle from 0 to 90° are obtained with the use of the VP2/3 code. The simplified model of the OTD with a radius of 0.2 (of the dimple width) is considered. It is found that the numerical predictions quite qualitatively and quantitatively correlate with the experimental data obtained at the Research Institute of Mechanics (Moscow State University). figure 3 shows that the maximum pressure values in the external flow stagnation zone on the windward slope on the OTD entrance portion and on the OTD exit portion in the longitudinal middle section of the channel agree well (figure 3). Figure 4 illustrates that the minimum negative pressure values are also clearly seen in the rarefaction region where a tornado-like vortex (transforming into the swirling flow) starts forming. Thus, the numerical and experimental confirmation of the controlling mechanism of abnormal enhancement of separated flow and heat transfer in the inclined OTD is obtained, the inclinations angles are found, at which this mechanism is successful. It is important to emphasize that the external Reynolds number range is significantly expanded, within which this interesting phenomenon is observed. The dimpled channel experiment carried out on the setup of the KazSC RAS is aimed at confirming the acceleration phenomenon in the flow core with a noticeable excess of the maximum velocity in comparison to the maximum velocity in the plane-parallel smooth-wall channel.

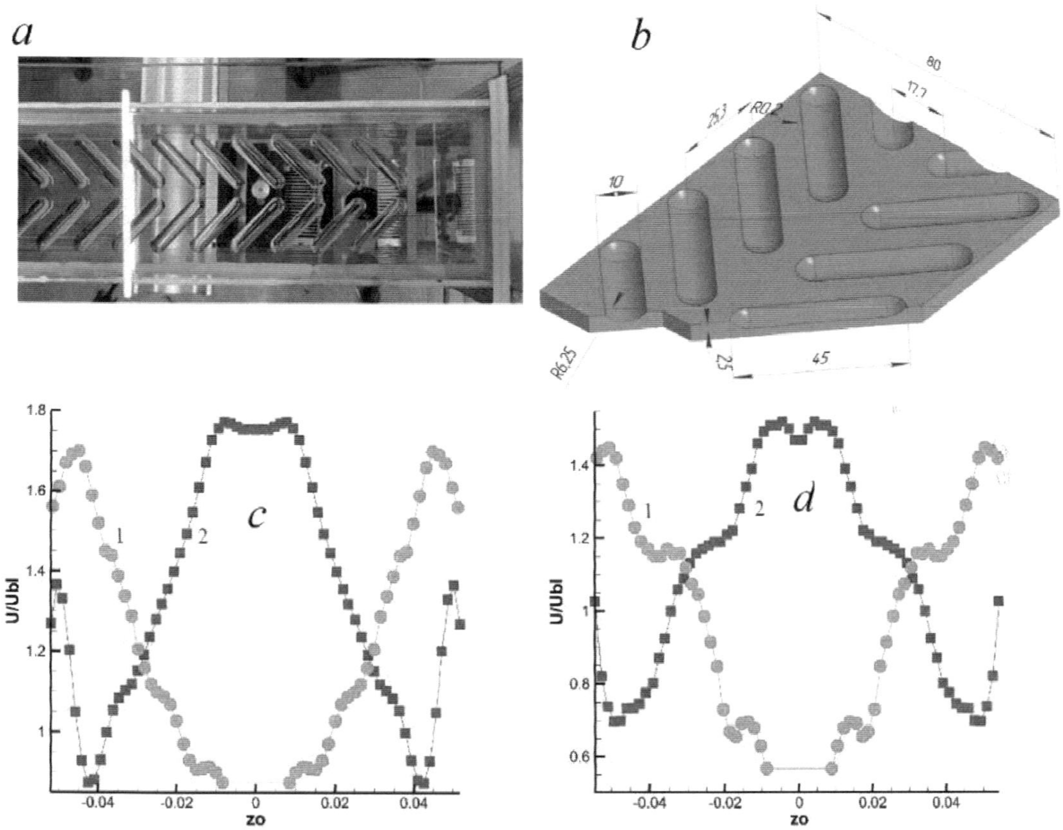

Figure 5. Setup for measuring the velocity profiles in the channel with two-row OTDs (*a*), the dimpled wall fragment at the inclination angle of 45° (*b*), and the U/Ubl profiles on the middle portion of the 22nd dimple in the laminar (*c*) and turbulent (*d*) regimes. 1 – θ = 135°; 2 – 45°.

In the case of the optically transparent channel (of height 10 mm, width 10 mm, length 800 mm) with 26 two-row oval trench dimples placed on the wall at two Reynolds numbers 1000 and 4300, the vector fields of the flow velocity were measured. The experimental data on the flow structure and turbulence were generalized and the smooth-wall channel characteristics were compared (Figure 5). The dimple width of 10 mm is chosen as the characteristic size. The length of dimples is equal to 4.5. The depth of dimples is 0.25, and the step between the dimple centers is 2.53 (the dense packing of dimples). Two locations of OTDs are considered: at the angle of 45° to the channel symmetry plane, the dimples are turned to the side walls (in this case, the entrance portions of dimples are near the symmetry plane) and at the angle of 135° to the channel symmetry plane, the dimples are turned from the side walls (in this case, the entrance portions of the dimples are near the walls). The velocity profile of the longitudinal velocity component in the channel symmetry plane strongly depends on the orientation of dimples relative to the flow: when orienting the wedge towards the flow, the velocity near the channel axis is maximum and exceeds approx. 1.5 times the average velocity in the channel cross-section; when orienting the wedge along the flow the axial velocity is from 50 to 75% of the bulk velocity.

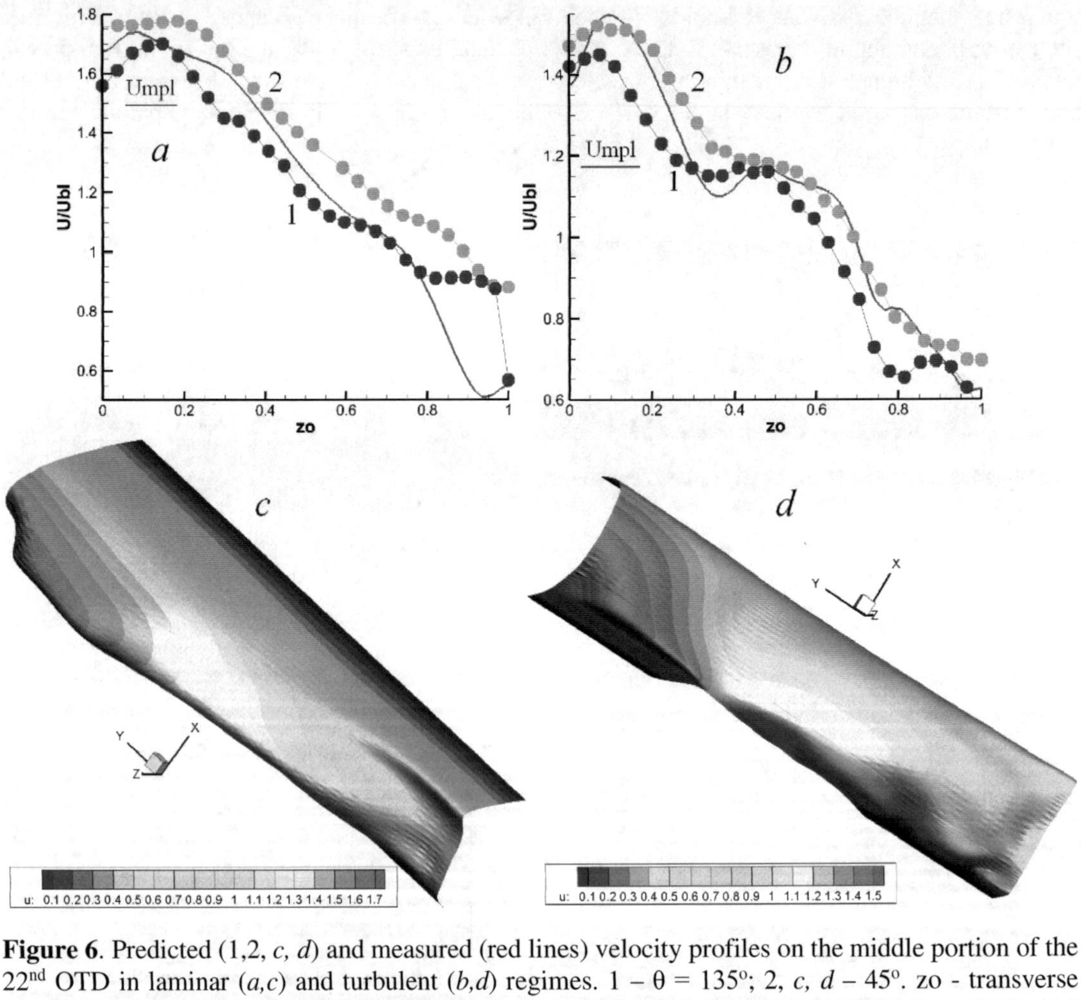

Figure 6. Predicted (1,2, *c, d*) and measured (red lines) velocity profiles on the middle portion of the 22nd OTD in laminar (*a,c*) and turbulent (*b,d*) regimes. 1 – θ = 135°; 2, *c, d* – 45°. zo - transverse coordinate along the channel half-width (calculation with lateral symmetry conditions).

Thus, the profile of the longitudinal velocity component in the channel cross-section in the vicinity of the 22nd dimple has a maximum value in the vicinity of the symmetry plane for the OTD oriented at the angle of 45°. The velocity profile changes radically at the OTD inclination angle of 135°. In this case, the velocity profile with two peripheral maximum longitudinal velocity values located near the side walls appears in the channel. The maximum velocity exceeds approx. 1.3 times the bulk velocity in the channel (Figure 5). To compare the experimental longitudinal velocity data with the numerical simulation results, the laminar (Re=1000) and turbulent (Re=4000) flows in the narrow channel with 26 single-row dimples, close in geometry to the experimental analogs, have been computed. In the computations, the sharp edges are rounded along the radius equal to 0.2. On the side boundaries of the computer analog, the symmetry conditions are assigned. The predicted profiles for laminar and turbulent flows in the channel agree fairly well with the experimental ones at the considered inclinations angles of dimples (Figure 6). The distance from the wall is chosen such as to capture a maximum velocity in the considered cross-section. Thus, shear flow with a maximum velocity above the dimple entrance develops on the longitudinal strip of the channel above the package of single-row dimples. The maximum shear flow velocity exceeds the maximum velocity in the plane-parallel channel and thus confirms the numerically discovered phenomenon of the flow acceleration in the dimpled channel.

3. Conclusion

The phenomenon of abnormal enhancement of separated flow and heat transfer in inclined OTDs is confirmed experimentally. The computational models are verified. It was established that the interference of the incident flow with the profiled wall and the emerging powerful tornado-like vortex structure hidden in the depression causes a large pressure drop and high velocity gradients in the OTD.

Acknowledgments
This research was sponsored by the Russian Science Foundation (the grant number 19-19-00259).

References

[1] Dzyubenko B V, Kuzma-Kichta Y A, Leontiev A I, Fedik I I and Kholpanov LP 2016 *Intensification of heat and mass transfer on macro-, micro-, and nanoscales* (New York: Begell House)
[2] Leontiev A et al. 2017 *Vortex technologies for energy* (Moscow: MEI)
[3] Ligrani P, Oliveira M and Blaskovich T 2003 *AIAA J.* **41** (3) 337
[4] Rashidi S, Hormozi F, Sunden B and Mahian O 2019 *Applied Energy* **250** 1491
[5] Hwang S D, Kwon H G and Cho H H 2008 *Int. J. Heat and Fluid Flow* **29** (4) 916
[6] Hwang S D, H G Kwon and Cho H H 2010 *Energy* **35** 5357
[7] Kwon H G, Hwang S D and Cho H H 2011 *Int. J. Heat and Mass Transfer* **54** 1071
[8] Xiao N, Zhang Q, Ligrani P and Mongia R. 2009 *Int. J. Heat and Mass Transfer* **52** (7-8)
[9] Tay C M, Chew Y T, Khoo B C and Zhao J B 2014 *Exp. Thermal and Fluid Science* **52** 278
[10] Isaev S A, Schelchkov A V, Leontiev A I, Gortyshov Yu F, Baranov P A and Popov I A 2017 *Int. J. Heat and Mass Transfer* **109** 40
[11] Isaev S, Gritckevich M, Leontiev A and Popov I 2019 *Acta Astronautica* **163** (Part.A) 202
[12] Isaev S A, Leontiev A I, Milman O O, Nikushchenko D V and Sudakov A G 2020 *AIP Conf. Proc.* 2211 020003

Heat transfer and skin-friction in a nonequilibrium adverse pressure gradient

N A Kiselev, Yu A Vinogradov and A G Zditovets

Institute of Mechanics, Lomonosov Moscow State University,
Russia, 119192 Moscow, Michurinsky Prospect, 1

kiselev.nick.a@gmail.com

Abstract. The paper presents the results of an experimental study of influence of a weak and moderate nonequilibrium adverse pressure gradient (APG) on the parameters of the dynamic and thermal boundary layers. The Reynolds number based on the momentum thickness at the beginning of the APG region was Re^{**}=5500. The section of the channel with APG was a slotted channel with an opening angle of the upper wall of 0-14°. The values of the relative (referred to the parameters in a zero pressure gradient flow at the same Re^{**}) friction and heat transfer coefficients, as well as the Reynolds analogy factor depending on the longitudinal pressure gradient, are obtained. The values of the relative friction coefficient decreased to c_f/c_{f0}=0.7 and those of the heat transfer to St/St_0=0.9. A maximum value of the Reynolds analogy factor $(St/St_0)/(c_f/c_{f0})$=1.16 was reached for the pressure gradient parameter β=2.9. The ratio of the heat transfer and drag coefficients of the dimpled to smooth surfaces remained approximately constant regardless of the presence or magnitude of a adverse pressure gradient.

1. Introduction

Turbulent Boundary Layers (TBL) with a adverse pressure gradient (APG) can be found almost everywhere. The scientific interest is caused by the presence of additional (in addition to the shear stress on the wall) parameters that control the development of the boundary layer. This causes significant difficulties in modeling, both numerical and experimental.

The experimental study of isothermal, incompressible, equilibrium two-dimensional turbulent boundary layers on flat surfaces was first undertaken by Clauser [1]. In this paper, a dimensionless pressure gradient parameter β is introduced. It was defined as the ratio of the force caused by the longitudinal pressure gradient and the force caused by the shear stress on the wall: $\beta=(\delta/\tau_w)\cdot(\mathrm{d}p/\mathrm{d}x)$, where δ is a characteristic boundary layer thickness, τ_w is the shear stress on the wall, and $\mathrm{d}p/\mathrm{d}x$ is the longitudinal pressure gradient. Also Clauser showed that the displacement thickness (δ^*) can be taken as the characteristic boundary layer thickness. In subsequent works (including this one), the pressure gradient parameter is determined as follows $\beta=(\delta^*/\tau_w)\cdot(\mathrm{d}p/\mathrm{d}x)$. In addition, in Clauser's works, the definition of the equilibrium of the boundary layer was given - this is the constancy of the pressure gradient parameter β.

However, in the absence of proper geometry profiling, which presupposes the condition of constant pressure gradient parameter - under the conditions of real currents (diffusers, aerodynamic profiles, etc.), the currents will probably never be in equilibrium. In this regard, modern research in the field of APL is increasingly devoted to non-equilibrium boundary layers.[2–4].

The difficulty in studying such boundary layers lies in the need to take into account the prehistory of the development of the boundary layer, which affects the area of the trace of the boundary layer, together with the local equivalence of its inner area.

In the present work, experimental studies of the influence of a nonequilibrium APG on the heat and momentum transfer on a smooth and dimpled surface are carried out.

2. Description of the experimental setup

Experimental studies were carried out on a subsonic wind tunnel with a working part in the form of a slotted rectangular channel [5–8] (width - B=300 mm, height - H=30 mm, $(L+L_{APG})$=1190 mm see figure 1). The flow velocity at the channel entry was measured by a Pitot—Prandtl tube and was remained constant (V_{in}=50 m/s).

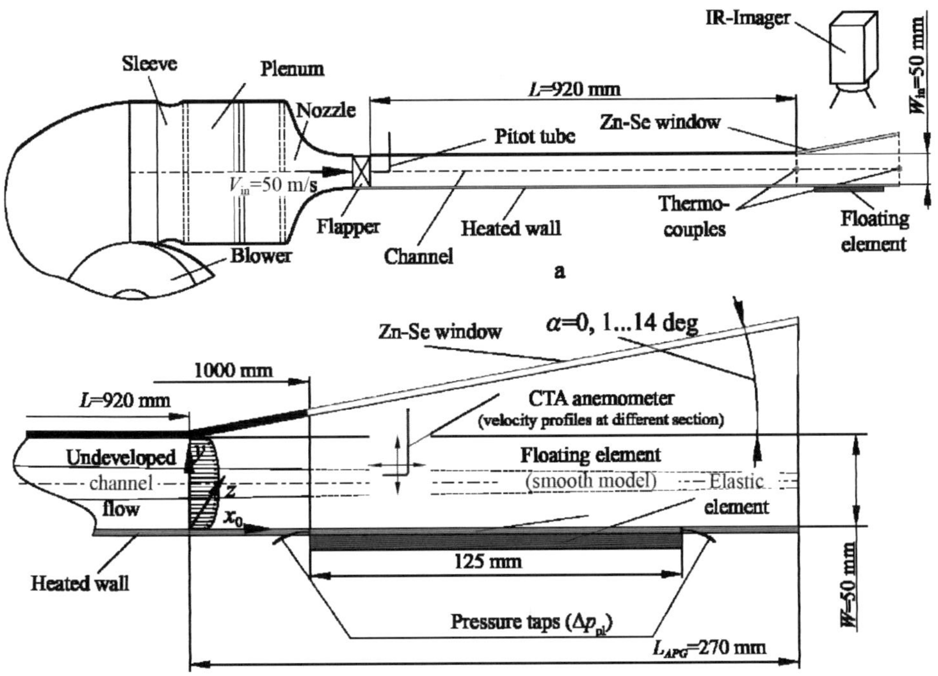

Figure 1. Schematics of the setup (a) and working channel (b).

The first section of the working channel had a length of 920 mm and was made in the form of a slot. The second section, 270 mm long, was used to create the APG. To create a thermal boundary layer on the lower wall of the channel, it was made heated. To study the friction and drag coefficients of smooth and dimpled models, they were suspended on floating elements in a section with an APG on the bottom wall. The longitudinal pressure gradient was created by changing the geometry of the channel: the slope of the rectilinear upper wall in the range of angles 0-14° with a step of 1°.

The value of the longitudinal pressure gradient dp/dx, Pa/m was determined using static pressure taps on the bottom wall. Velocity profiles were measured using a hot-wire anemometer DISA 55M01 CTA equipped with a Dantec Dynamics small-size single-wire 55P81 probe with thermal compensation (the wire length was 1.25 mm and its thickness was 5 μm).

By weighing the models on floating elements [6,7,9], the friction coefficient of the smooth and the drag coefficient of the dimpled models were determined as follows. In the experiment, the total force arising on the floating elements was recorded, including the resultant of friction/drag forces and the pressure drop acting on the ends of the elements. By registering the pressure drop over the length of the floating element, the contribution of this component to the total force on the floating element is determined. Floating elements occupied the APG section from x_0=80 mm to x_0=205 mm. The width of the elements was 100 mm. The dimpled surface was set parallel to the smooth model. The staggered array of the spherical dimples was made by milling a dimple on an initially smooth model. The longitudinal and transverse spacing of the dimples was 8 and 9 mm, respectively. The diameter of the sphere of the instrument was 16 mm, and the depth of the dimples was 1 mm. The print diameter of the spherical dimple was 7.75 mm.

To determine the heat transfer coefficient (Stanton number, St=$\alpha/(\rho \cdot c_p \cdot V)$, α – heat transfer coefficient, W/(m^{2}K), ρ – density of air, kg/m^3, c_p - heat capacity at constant pressure, J/(kg·K), V – core velocity above the models, m/s), the transient heat transfer method was used. At the initial moment of experiment the IR camera recorded the temperature field of the model surface. Inhomogeneity of the initial temperature of the dimpled surface did not exceed 5 K, smooth one - 2 K. The process of model cooling was recorded during 60s. The one-dimensional equation of transient cooling of a wall of finite thickness was used. The value thus obtained was averaged over the entire model surfaces.

The uncertainty of measurements of the local friction coefficients by using a hot-wire anemometer was 2.5%. The uncertainty of measurements of the friction and drag coefficients using floating elements was 3.1-9.6% (large values correspond to large β). The uncertainty of measurements of the heat transfer coefficients did not exceed 5.6%, relative heat transfer coefficients - 7.9%. The uncertainty of measurements of the pressure gradient parameter β did not exceed 7.6%.

3. Results of experimental studies

Static pressure profile (expression $c_{p0}=(p_i-p_{out})/(\rho \cdot V_{in}^2/2)$, p_{out} corresponds to the static pressure in the last section considered, x_0=265 mm) and the pressure gradient dp/dx are shown in figure 2. The profile

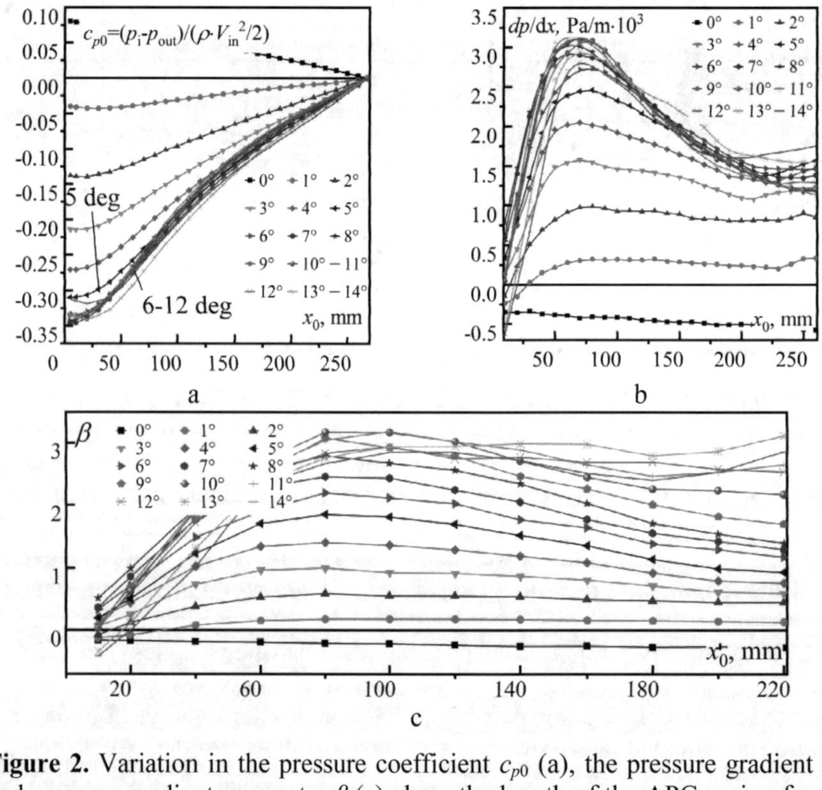

Figure 2. Variation in the pressure coefficient c_{p0} (a), the pressure gradient (b) and pressure gradient parameter β (c) along the length of the APG region for the channel expansion angles 0…14°.

of the pressure gradient parameter β is shown in figure 2, c. For small opening angles the value of β remains almost constant in the area of installation of smooth and dimpled models. For large opening angles, a maximum is observed in the range x_0=80-120 mm in accordance with the dp/dx distribution along the channel. The parameter of the pressure gradient β in the area of installation of the models increases up to the opening angles of 12° and then remains almost constant.The influence of the prehistory [4] of the development of the boundary layer in a nonequilibrium pressure gradient can be traced by analyzing the velocity profiles with the same parameters of the pressure gradient (figure 3 for β=1 and β=2), but different longitudinal coordinates. As the duration of the action of the longitudinal pressure gradient increases, the velocity profile in the wake region becomes less and less filled and approaches the equilibrium one. The inner part of the boundary layer remains practically unchanged (within the experimental operating parameters).

The imposition of a longitudinal pressure gradient leads to a decrease in the friction coefficient (from c_f/c_{f0} =1.03 at β=-0.25 to c_f/c_{f0}=0.75 at β=2.9 for average over the x_0=80-200 mm values (figure 4)).

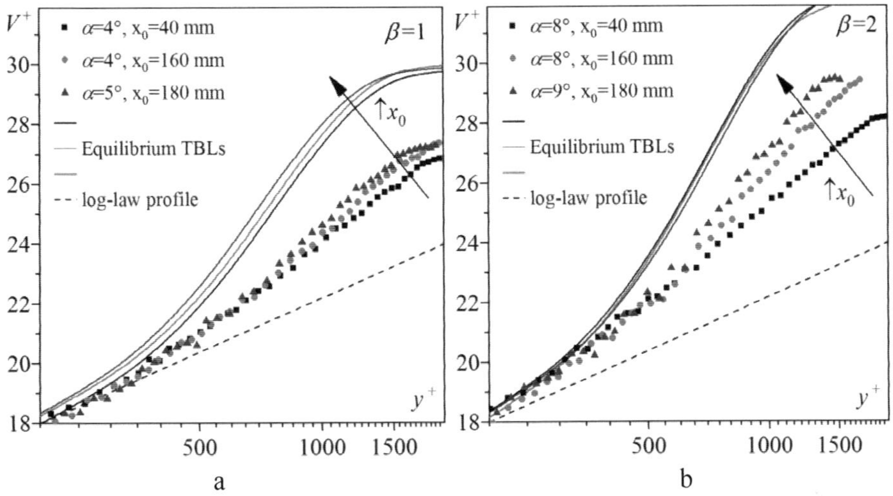

Figure 3. Influence of flow prehistory on the velocity profile in APG TBL: a, β=1; b, β=2. Equilibrium boundary layer profiles are plotted using the corresponding V^* according to [10].

The dependence of averaged over the x_0=80-200 mm region relative heat transfer coefficient St/St$_0$ on the pressure gradient parameter β is presented in figure 4. Data of heat transfer coefficient, calculated in paper [11] for the Reynolds number Re**=5000 and the turbulent Prandtl number Pr$_t$=1.0 are also presented in figure 4.

As can be seen from figure 4, the change in St/St$_0$ is not as significant as c_f/c_{f0}. In the encompassed range of operating parameters, the thermal boundary layer is less affected by the positive pressure gradient. Nevertheless, the data obtained allow one to assert that in the APG flows the Reynolds analogy factor (St/St$_0$)/(c_f/c_{f0}) is higher than in zero pressure gradient flow.

In general, the results obtained on the heat transfer and friction coefficients fit into the generally accepted trend of violation of the Reynolds analogy towards heat transfer in flows with a adverse pressure gradient (figure 5). Maximum value (St/St$_0$)/(c_f/c_{f0})=1.16 is achieved at β=2.9. Also in figure 5, the uncertainties of the experimentally obtained values of (St/St$_0$)/(c_f/c_{f0}) are plotted.

According to the results of the experimental study, the drag coefficient of the dimpled surface, referred to the friction coefficient of a smooth surface in the same APG condition does not depend on the presence or magnitude of a APG (within the measurement accuracy): the value c_x/c_f≈1.05-1.12 for all values of β, which corresponds to the value obtained during channel flow (figure 5).

The relative heat transfer coefficient related to the heat transfer coefficient of a parallel-standing smooth model also remains practically constant within the parameters covered in the experiment: St_d/St=1.08-1.15. Consequently, a further violation of the Reynolds analogy towards heat transfer on dimpled surfaces in the presence of a APG is observed (in comparison with a smooth surface) (figure 5).

4. Summary and conclusion

An experimental investigation of the influence of weak and moderate non-equilibrium adverse pressure gradient on the boundary layer parameters and the heat transfer and friction/drag coefficients of a smooth and dimpled surface is carried out. On the basis of this investigation the following conclusions concerning the effect of the non-equilibrium streamwise adverse pressure gradient on the heat transfer and friction/drag coefficients and the Reynolds analogy factor can be made.

Both local and average relative friction coefficients c_f/c_{f0} of smooth surface essentially depend on the dimensionless pressure gradient parameter: they diminish with increase in β in the entire parameter range covered in the experiment.

On the smooth surface the dimensionless relative heat transfer coefficient St/St_0 diminishes in the presence of APG but this reduction is not so considerable as that of c_f/c_{f0}.

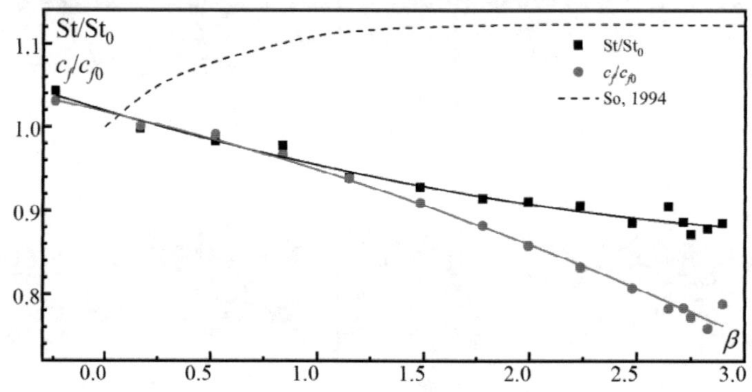

Figure 4. Dependence of the relative heat transfer and friction coefficients on the pressure gradient parameter β.

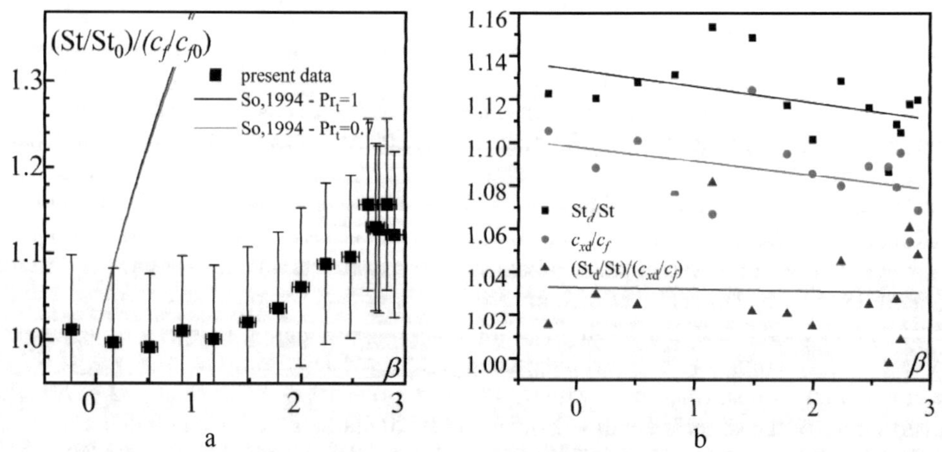

Figure 5. Dependence of the Reynolds analogy factor of a smooth surface (a) and heat-hydraulic efficiency of dimpled surface (b) on the pressure gradient parameter β.

The Reynolds analogy factor increases with increase in β in APG flows. The maximum value $(St/St_0)/(c_f/c_{f0})$=1.16 is reached at the maximum value of the pressure gradient parameter β=2.9 . Moreover, further violation of the Reynolds analogy towards heat transfer on dimpled surfaces in the presence of APG is observed.

Acknowledgments
The study was carried out with the financial support of the Russian Science Foundation under grant 19-79-10213.

References

[1] Clauser F H 1954 Turbulent Boundary Layers in Adverse Pressure Gradients *J. Aeronaut. Sci.* **21** 91–108
[2] Leont'ev A I, Lushchik V G and Reshmin A I 2016 Heat transfer in conical expanding channels *High Temp.* **54** 270–6
[3] Lushchik V G and Reshmin A I 2018 Heat Transfer Enhancement in a Plane Separation-free Diffuser *High Temp.* **56** 569–75
[4] Bobke A, Vinuesa R, Örlü R and Schlatter P 2017 History effects and near equilibrium in adverse-pressure-gradient turbulent boundary layers *J. Fluid Mech.* **820** 667–92
[5] Kiselev N A, Burtsev S A and Strongin M M 2015 A Procedure for Determining the Heat Transfer Coefficients of Surfaces with Regular Relief *Meas. Tech.* **58**
[6] Burtsev S A, Kiselev N A and Leontiev A I 2014 Peculiarities of studying thermohydraulic characteristics of relief surfaces *High Temp.* **52** 869–72
[7] Leontiev A I, Kiselev N A, Burtsev S A, Strongin M M and Vinogradov Y A 2016 Experimental investigation of heat transfer and drag on surfaces with spherical dimples *Exp. Therm. Fluid Sci.* **79** 74–84
[8] Kiselev N A, Leontiev A I, Vinogradov Y A, Zditovets A G and Strongin M M 2019 Effect of large-scale vortex induced by a cylinder on the drag and heat transfer coefficients of smooth and dimpled surfaces *Int. J. Therm. Sci.* **136** 396–409
[9] Titov A A, Leontiev A I, Vinogradov U A, Zditovets A G and Strongin M M 2010 Experimental Investigation of Skin Friction Drag and Heat Transfer on the Surfaces With Concavities in Compressible Fluid Flow *2010 14th International Heat Transfer Conference, Volume 2* (ASME) pp 597–601
[10] Mellor G L and Gibson D M 1966 Equilibrium turbulent boundary layers *J. Fluid Mech.* **24** 225–53
[11] So R M C 1994 Pressure gradient effects on Reynolds analogy for constant property equilibrium turbulent boundary layers *Int. J. Heat Mass Transf.* **37** 27–41

Modeling of heat transfer in an element with anisotropic porosity

N N Kozhukhov, E A Kozhukhova, D A Konovalov, D A Prutskikh and V I Perunova

Voronezh State Technical University,
Russia, 394006 Voronezh, 20 letiya Oktyabrya st., 84

k0zhukhov@yandex.ru

Abstract. The paper presents the results of modeling the heat transfer process in a channel filled with a porous medium. Moreover, the porosity changes according to the specified law along one of the coordinates. It is shown that an element with anisotropic porosity has better heat transfer characteristics in comparison with an element with a homogeneous porous structure.

1. Introduction

The use of porous structures makes it possible to intensify the heat exchange process in the channels of power plants for various purposes, for example, in turbine blades, in devices for cooling systems for radio electronics, satellite communication systems and many other technical devices that require compactness and high thermal efficiency [1, 2]. A significant contribution to the heat transfer process is made by the heat transfer surface area, which is increased in comparison with the empty channel. Unfortunately, this leads to an increase in the hydraulic resistance in the channel, and, accordingly, in the pressure drop, which does not affect the energy consumption in the best way during the transportation of the coolant. This aspect is one of the most important in deciding on the use of porous elements for thermal stabilization of heat-stressed elements.

It is possible to reduce the hydraulic resistance in a porous medium in several ways: the use of interchannel transpiration of the coolant, the use of a porous medium with a local or uniform change in the permeability tensor [3, 4]. In this work, the anisotropy of a porous medium is considered in the form of a smoothly varying porosity. Moreover, its value at the base of the channel exposed to thermal action is large compared to the area in the central part of the channel. This solution will allow to increase the speed of the coolant near the heated surface, therefore, to intensify the heat transfer process. But, on the one hand, an increase in the height of the channel region with higher porosity leads to a decrease in the heat exchange surface area, and on the other hand, to a decrease in the pressure drop. In this regard, we will consider several functions responsible for the law of porosity variation near the heat-stressed boundary.

2. Statement of the problem

Consider a channel 40×5 mm in size, filled with a porous medium, through which a coolant flows (figure 1). The flow regime is laminar. The lower part of the channel is exposed to heat flow. The upper part of the channel is thermally insulated. The internal space of the channel is a structure, the porosity of which changes along the ordinate axis according to the specified piecewise continuous function with smoothing using a continuous second derivative [5]. Let us consider 2 variants of such a function, determined by the systems of equations (1), (2). Function graphs are shown in figure 2.

$$\varepsilon(y)\big|_{1}^{3.2} = 0.227y + 1.027, \varepsilon(y)\big|_{3.2}^{6} = 0.3; \tag{1}$$

$$\varepsilon(y)\big|_{1}^{2} = -0.5y + 1,3, \varepsilon(y)\big|_{2}^{6} = 0.3. \tag{2}$$

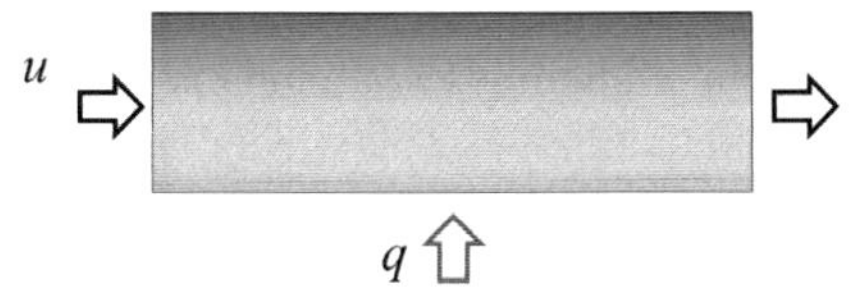

Figure 1. Formulation of the problem.

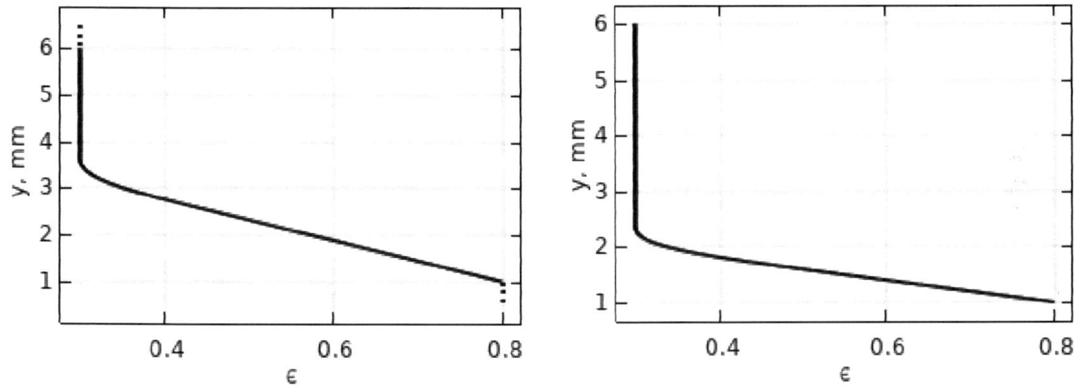

Figure 2. Piecewise continuous functions of porosity change.

For a mathematical description of the hydraulic subproblem, we use a system consisting of the Brinkman equation and continuity

$$\nabla \cdot [-p\mathbf{I} + \mathbf{K}] - \left(\mu\kappa^{-1} + \beta\varepsilon\, \rho|\mathbf{u}| + Q_m\varepsilon^{-2}\right)\mathbf{u} + \mathbf{F} = 0, \tag{3}$$

$$\rho\nabla \cdot \mathbf{u} = Q_m, \tag{4}$$

where p – pressure, Pa; μ – dynamic viscosity coefficient, kg/(m·s); κ – porous medium permeability; β – inertia coefficient, 1/m; ε – porosity; ρ – density; $\mathbf{u}$ – velocity vector, m/s; Q_m – mass source; $\mathbf{F}$ – additional force vector.
Permeability was determined according to the Kozeny-Karman ratio by the formula

$$\kappa = \frac{1}{180}\varepsilon^3 d_p^2 (1-\varepsilon)^{-2}. \tag{5}$$

Here d_p – diameter of spherical particles that make up a porous structure, m.

The thermal subproblem was solved, like the hydraulic one, in a stationary two-dimensional formulation. For the solution, the temperature difference between the porous matrix and the coolant was taken into account, i.e. a two-temperature heat transfer model was used. In addition, the problem

took into account the non-isothermality of the coolant flow, i.e., its density was a function of pressure and temperature. Below are the equations used in general terms for:
- heat transfer in a porous matrix

$$d_z \rho c_p \mathbf{u} \cdot \nabla T + \nabla \cdot \mathbf{q} = d_z Q + q_0 + d_z Q_{\text{ted}};$$
$$\mathbf{q} = -d_z k \nabla T; \tag{6}$$

- heat transfer agent

$$d_z \rho c_p \mathbf{u} \cdot \nabla T_2 + \nabla \cdot \mathbf{q} = d_z Q + q_0 + d_z Q_{\text{vd}};$$
$$\mathbf{q} = -d_z k \nabla T_2; \tag{7}$$

- accounting for the two-temperature model

$$Q_s = \frac{q_{sf}}{\theta_p}\left(T_f - T_s\right);$$
$$Q_f = \frac{q_{sf}}{1-\theta_p}\left(T_s - T_f\right); \tag{8}$$
$$q_{sf} = a_{sf} h_{sf}.$$

To determine the thermal characteristics, the value of the element depth is taken as mm. The material of the porous matrix is aluminum. The heat carrier is water. Boundary conditions at the channel entrance: temperature 20 °C, speed - was in the range from 0.03 to 0.06 m / s. The excess pressure at the outlet was taken to be zero. To determine the regularity of heat transfer between the porous matrix and the coolant, the following expressions were used to determine the specific surface area

$$a_{sf} = \frac{6(1-\varepsilon)}{d_p} \tag{9}$$

and the interstitial heat transfer coefficient between the porous matrix and the heat carrier

$$q_{sf} = \frac{\lambda_{\text{f}}}{d_{\text{p}}}\left[2 + 1.1 Pr^{1/3}\left(\rho_{\text{f}} \frac{v_0 d_{\text{p}}}{\mu_{\text{f}}}\right)^{0.6}\right], \tag{10}$$

$$Pr = \frac{\mu_{\text{f}} c_{\text{pf}}}{\lambda_{\text{f}}}. \tag{11}$$

The components for equations (8) - (11) were determined as follows. So, the values of porosity, thermal conductivity, dynamic viscosity were calculated as average values over the entire area of the element. Pressure - as an average value at the boundary of the coolant inlet.

3. Solution

3.1. Modeling the operating modes of the element

The solution of equations (3) - (8) and the obtaining of thermohydraulic characteristics for an element with anisotropic porosity was carried out using one of the engineering analysis packages based on a numerical method for solving differential equations. At the beginning of the calculation, for each of the porosity options, the operating modes were determined in which the Reynolds number did not exceed 2300, and the coolant temperature was 100 C. This was controlled in order to consider the solution of the problem within the framework of formulating a laminar flow regime and preventing a phase transition inside the porous element, as well as in order to adequately compare the proposed options for anisotropic porosity.

The hydraulic performance was estimated based on the relationship between the drag coefficient and the Reynolds number. For this, in each of the 5 modes, both of these parameters are determined by the following formulae

$$\xi = \frac{2d\Delta p}{l\rho_{\mathrm{f}} v^2}, \tag{12}$$

$$\mathrm{Re} = \frac{vd\rho_{\mathrm{f}}}{\mu_{\mathrm{f}}\varepsilon}. \tag{13}$$

Here d – hydraulic channel diameter; Δp – pressure drop; l – porous channel length.

The results of the study are presented as a relation $\xi(\mathrm{Re})$ in figure 3. Relationships $\varepsilon = f_1(y)$ and $\varepsilon = f_2(y)$ correspond to the laws of porosity change (1) and (2), respectively.

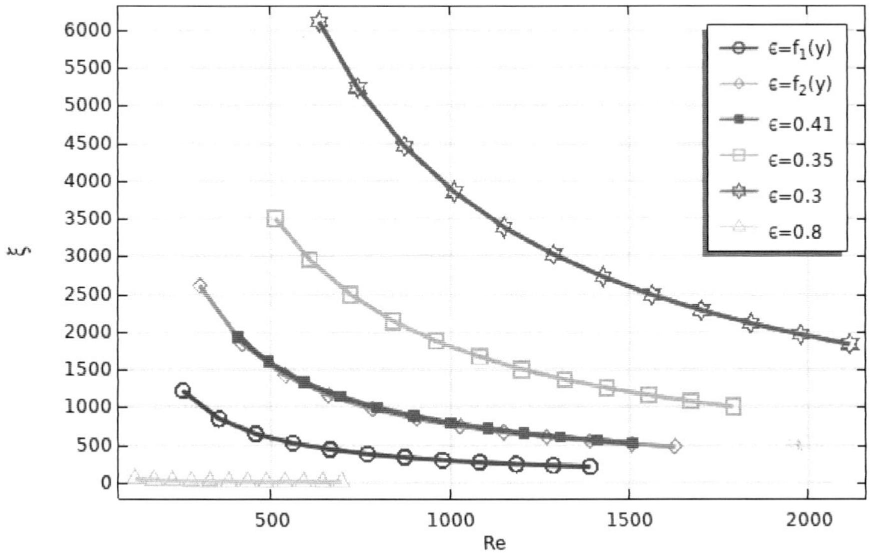

Figure 3. Relationship between the drag coefficient and the Reynolds number for different variants of the porous element.

Porosity values $\varepsilon = 0.41$ and $\varepsilon = 0.35$ determined as averages over the computational domain with a change in porosity according to (1) and (2), respectively. A $\varepsilon = 0.41$ and $\varepsilon = 0.35$ taken as the extreme values of the range of porosity variation when determining relations (1) and (2).

To analyze the thermal characteristics of an anisotropic element for each operating mode, the Nusselt number was determined by the formula

$$\mathrm{Nu} = \frac{\alpha_m d}{\lambda_{\mathrm{f}}}. \tag{14}$$

Here $\alpha_m = \dfrac{q_{\mathrm{s}}}{t_{\mathrm{w}} - t_{m\mathrm{f}}}$. t_{w} – average temperature of the channel base, °C; $t_{m\mathrm{f}}$ – average temperature of the coolant in the channel, °C.

The relationship between the Nusselt number and the Reynolds number for two variants of anisotropic porosity and several values of uniform porosity is shown in figure 4.

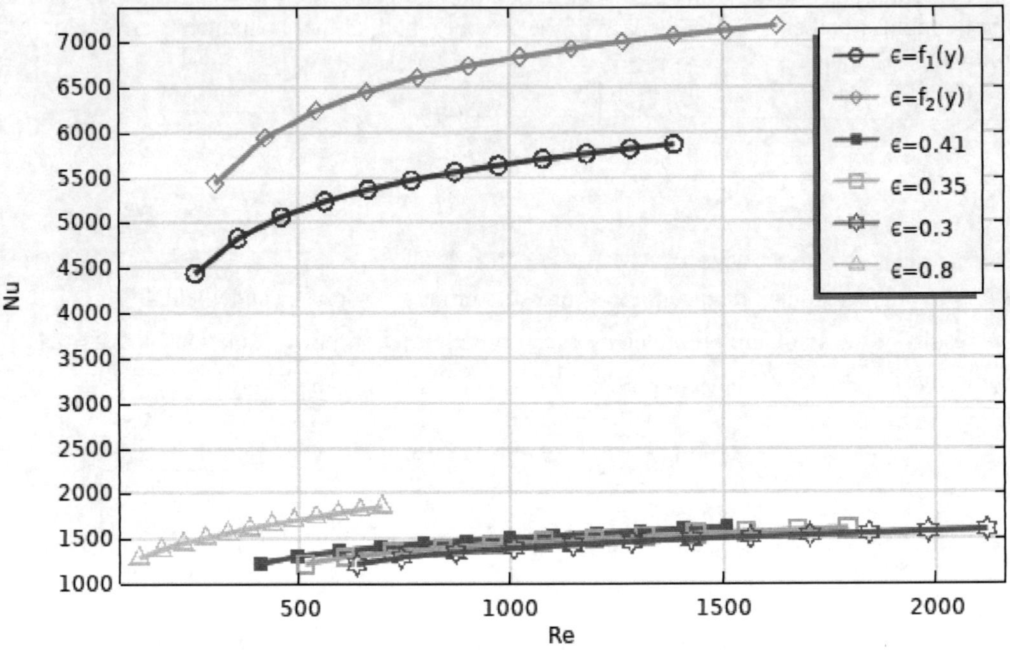

Figure 4. Relationship graph of $\mathrm{Nu}(\mathrm{Re})$ for different types of porous element.

3.2. Findings

Analysis of the relationships shown in figures 3 and 4 shows that the drag coefficient for the first variant of anisotropy is lower than for the second. While the intensity of heat transfer in the channel with the first option of anisotropy is lower compared to option 2. Comparing the difference between the drag coefficient and the heat transfer rate for one of the flow regimes corresponding, for example, to the value $\mathrm{Re} = 800$, you can see that ξ differs by 2.5 times. In this case, the difference in the intensity of heat transfer is 1.2. The use of isotropic porosity in the channel leads to an increase in ξ for $\varepsilon = 0.35$ and $\varepsilon = 0.3$ compared to anisotropic. When $\varepsilon = 0.41$ value ξ comparable to the porosity anisotropy corresponding to law (2). The channel has the lowest resistance coefficient with $\varepsilon = 0.8$, but such a configuration is inferior in heat exchange to a channel in which the porosity corresponds to relationships (1) or (2). Heat transfer for a channel with any variant of isotropic porosity is less intense than for a channel with anisotropic porosity.

4. Conclusion

The results of the study of two models of anisotropic porosity for a flat channel show that with a sharper shift of the porosity gradient to the heat-stressed boundary, the heat transfer intensity increases, and the ratio of the values of the resistance coefficients in this case exceeds the ratio of the heat transfer intensity. In this case, for small values of the Reynolds number (Re <500), the difference in the intensity of heat transfer is smaller in comparison with the regimes corresponding to Re> 500. The use of porosity anisotropy leads to a more significant intensification of heat transfer into the channel as compared to an isotropic porous medium. Thus, the use of anisotropic porous media makes it possible to increase the efficiency of heat exchange elements.

References

[1] Konovalov D A, Drozdov I G, Shmatov D P, Dakhin S V and Kozhukhov N N 2013 *Development and Modeling of Microchannel Cooling Systems* (Voronezh: Voronezh State Technical University Press) p 222

[2] Konovalov D A, Kozhukhov N N and Drozdov I G 2016 Modeling of heat and mass transfer processes in microchannel heat exchangers of space technology control systems *Bull. of Reshetnev Siberian State University of Science and Technology* **17** № **1** 137-46

[3] Pelevin F V 2018 Heat transfer in metallic mesh materials during interchannel transpiration and two-dimensional intergrid motion of the coolant *Thermal physics of high temperatures* **56** № **2** 219-28

[4] Ryazhskikh V I, Konovalov D A, Kozhukhov N N, Ryazhskikh A V, Nikolenko A V and Troshin A Yu 2020 Numerical simulation of the hydrodynamics of the flow of a heat carrier in a flat channel filled with an anisotropic porous medium *J. Phys.: Conf. Series* **1683** 022038

[5] Kozhukhov N N, Kozhukhova E A, Konovalov D A, Prutskikh D A and Perunova V I 2020 Modeling the hydrodynamics of the coolant flow in a porous element with one-dimensional gradient porosity *Physical and technical problems of energy industry, ecology and energy saving* 82-9

Modelling the hydrodynamics of a two-phase flow in the non-Newtonian fluid

E A Kozhukhova, N N Kozhukhov, V I Ryazhskikh, A M Naumov and A V Netreba

Voronezh State Technical University,
Russia, 394006 Voronezh, 20-letiya Oktyabrya st., 84

k0zhukhov@yandex.ru

Abstract. The paper presents the results of modelling the flow of droplets in the flow of a non-Newtonian fluid. The task is solved in a nonstationary formulation. We determined the time of stabilization of the speed profile of the droplets and the length of the initial hydrodynamic sector. We compared the specified characteristics for the main flow of the non-Newtonian fluid and for the set of droplets in the flow. We show the coincidence of the lengths of the initial hydrodynamic sections determined by the stabilization of speeds for the droplets and the non-Newtonian medium.

1. Introduction

Among many works that are devoted to studying the behaviour of non-Newtonian fluids under different conditions of operation of heat engineering installations, [1-3] can be noted as fundamental ones. This is due to the essential dependence of the rheological properties of the non-Newtonian fluid on the speed of its shift. The change in the temperature of the medium further complicates the prediction of its behaviour, and, consequently, the operation modes of heat engineering devices and systems using non-Newtonian media.

The lengths of the initial hydrodynamic segment for such media and the usual ones subjected to Newton's law are different. The heat exchange process in the medium in the initial segment in scientific works and engineering techniques is considered to a lesser extent. The reason for this lies in the fact that the length of such a segment is sufficiently small, and the main process, to a greater extent affecting the work of the equipment, happens in the hydrodynamically established region, i.e. after it. And it is so for Newtonian media. With non-Newtonian media, the length of the initial segment may be much larger, and in this case, it is impossible not to take into account mechanisms different from the laws in the region of steady flow, i.e. after the initial segment. In addition, the specified circumstance can affect manufactured products or even the work of an industrial enterprise.

2. Statement of the problem

To analyze the hydrodynamic pattern, let us consider a pipe cross-section with a length of 30 m in which the non-Newtonian medium flows with the initial speed specified at the input $v_f = 0.2$ m/s (Figure 1). In many ways, non-Newtonian media flow is determined by their rheological characteristics. For an aqueous solution of surfactants, such characteristics were defined earlier: fluid consistency coefficient $m = 9.3319$ Pa·s, flow behaviour index $n = 0.073$, lower shear rate limit $\dot{\gamma} = 0.01$ 1/s, reference shear rate $\dot{\gamma}_{ref} = 1$ 1/s [4]. To determine the speed profile of the non-Newtonian medium, we will use the analysis of the speed profile formed by the particles of the second phase, which represents the Newtonian medium in the form of watercolour paint with the following characteristics: the particle diameter $d_p = 0.3$ mm, the density $\rho_p = 1031$ kg/m^3, the dynamic viscosity of the particles $\mu_p = 4.5$ mPa·s, the particle surface tension $\sigma_p = 30$ mN/m.

At the initial time, 100 drops are located at the boundary of the input of the non-Newtonian medium into the channel (are shows conventionally in Figure 1). In a planned further experimental study, it will be possible to compare the resulting profile in this model with a steady speed profile of the drops of paint for various modes, pipe sections and time segments.

The problem was considered in a nonstationary statement for a laminar mode of fluid flow, taking into account the gravitational force. Paint particles were taken as drip liquid, and their volume fraction was less than 1%. The effect of particles on the main flow was taken as insignificant. At the initial time, the speed of the second phase (drops) was taken $v_p = 0$ m/s.

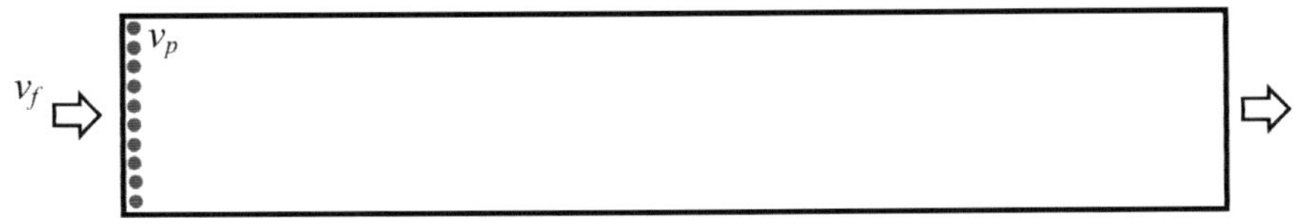

Figure 1. Problem statement.

Let us make a mathematical model of the movement of an aqueous solution of surfactants, i.e. non-Newtonian fluid, in a pipe. To do this, we use the system of equations consisting of the movement equation and the equation of continuity. In general, for the stationary problem, they look as follows

$$\rho\ \mathbf{u}\cdot\nabla\ \mathbf{u} = \nabla\cdot[-p\mathbf{I} + \mathbf{K}] + \rho\mathbf{g}, \tag{1}$$

$$\rho\nabla\cdot\mathbf{u} = 0. \tag{2}$$

Here, ρ – the density; $\mathbf{u}$ – the speed vector; $\mathbf{I}$ – the unit vector.

$$\mathbf{K} = m\left(\frac{\dot{\gamma}}{\dot{\gamma}_{ref}}\right)^{n-1}(\nabla\mathbf{u} + (\nabla\mathbf{u})^{\mathrm{T}}), \tag{3}$$

$$\dot{\gamma} = max(\sqrt{2\mathbf{S}:\mathbf{S}}, \dot{\gamma}_{min}), \tag{4}$$

$$\mathbf{S} = 0{,}5\left[\nabla\mathbf{u} + (\nabla\mathbf{u})^{\mathrm{T}}\right]. \tag{5}$$

Here, m – the fluid consistency coefficient; $\dot{\gamma}_{ref}$ – the reference shear rate; $\dot{\gamma}_{min}$ – the lower shear rate limit. In equation (4) «:» - the contraction operator defined as $\mathbf{a}:\mathbf{b} = \sum_n\sum_m a_{nm}b_{nm}$.

We describe the movement of the second phase - paint drops with the following system of differential equations of the first order for the nonstationary problem

$$m_p\frac{d\mathbf{v}}{dt} = F_t, \tag{6}$$

$$\frac{dd_p}{dt} = \frac{2R}{\pi\rho_p d_p^2}. \tag{7}$$

Here, $\mathbf{v} = \frac{d\mathbf{q}}{dt}$; $\mathbf{q}$ – the vector of the coordinates of the particles; t – the time; d_p – the particle diameter; R – the accretion rate; ρ_p – the density of the particles.

Let us consider the action of the two forces on the particles: resistance and gravity. Let us describe the law of resistance to the movement of particles with the following equation

$$\mathbf{F}_D = \frac{1}{\tau_p} m_p (\mathbf{u} - \mathbf{v}), \tag{8}$$

where $\mathbf{u}$ and $\mathbf{v}$ - the speed of the first and second phase, respectively.
The characteristic time of the particle response, determining the resistance as a function of the particle's properties and the particle's number of Reynolds, is determined by the following dependence

$$\tau_p = \frac{4\rho_p d_p^2}{3\mu C_D Re_r}, \tag{9}$$

The coefficient of resistance is determined by the Hadamard-Rybczynski formula

$$C_D = \frac{8}{Re_r} \cdot \frac{2+3\kappa}{1+\kappa}. \tag{10}$$

Here, the particle's number of Reynolds $Re_r = \frac{\rho |\mathbf{u} - \mathbf{v}| d_p}{\mu}$, μ – the dynamic viscosity of the main non-Newtonian medium. $\kappa = \frac{\mu_p}{\mu}$, μ_p – the dynamic viscosity of the second phase, i.e. of the particle.

We determine the gravity force from the following equation

$$\mathbf{F}_g = m_p \mathbf{g} \frac{\rho_p - \rho}{\rho_p}, \tag{11}$$

where ρ_p – the density of the particles.

3. Solution

3.1. Modelling the movement of particles in the flow of an aqueous solution of surfactants

Let us solve the task in two stages. At the first stage, using the system of equations for the movement of the first phase (non-Newtonian medium) (1)-(2), we define its fields of pressures and speeds for the steady flow. At the second stage, we solve the problem of movement of the second phase (Newtonian medium) in the nonstationary formulation using (6)-(11). To solve this, we use one of the numerical packages for engineering analysis of hydrodynamic systems. To determine the stabilization of the speed profile, we draw it for sections from 10 mm to 3000 mm. Figure 2 shows the dynamics of changes in the speed profile of the main flow. The speed changing along the three central lines is represented in the logarithmic coordinates for better visual assessment of the stabilization of speeds. Moreover, the axial line corresponds to the line in the center of the flow, and two more lines are located at distances corresponding to distances of 21 and 79 drops from the central one. From both drawings, we can see that the speed is stabilized at a distance of about 1000 mm from the input. This value corresponds to the length of the initial hydrodynamic segment.

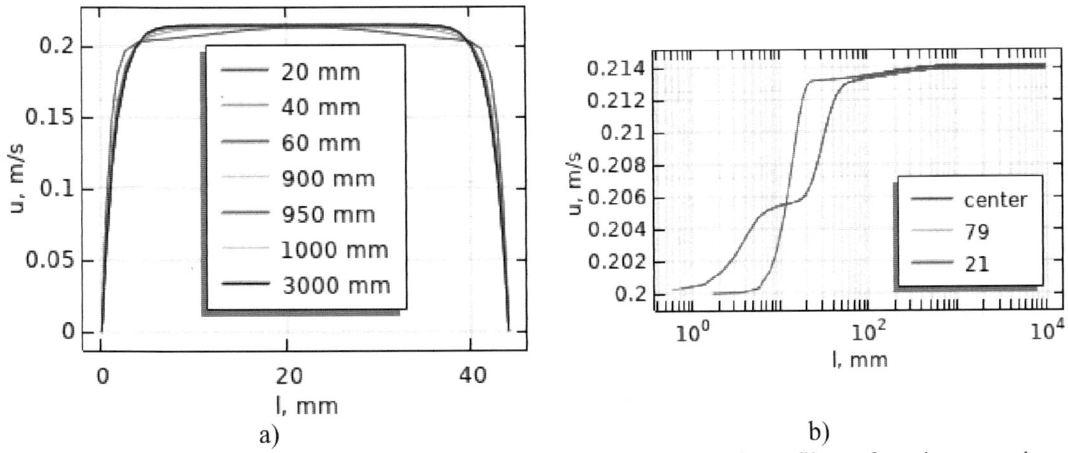

Figure 2. Changes in the speeds of the main flow: a) – speed profiles of various sections; b) – speed along the flow lines.

In order to visually identify this profile in the planned experimental study, we compare the dependencies obtained with similar dependencies for the drops. To do this, we estimate the speed of several drops in the vicinity of the central flow lines and determine the length of the initial hydrodynamic segment. This can be determined by the drop's speed stabilization segment in Figure 3. The first of them shows that the speed stabilization time is about 5 s. Using this time value, we can see that the distance that the drops passed during this time is about 1000 mm. This value practically coincides with the value for the main flow.

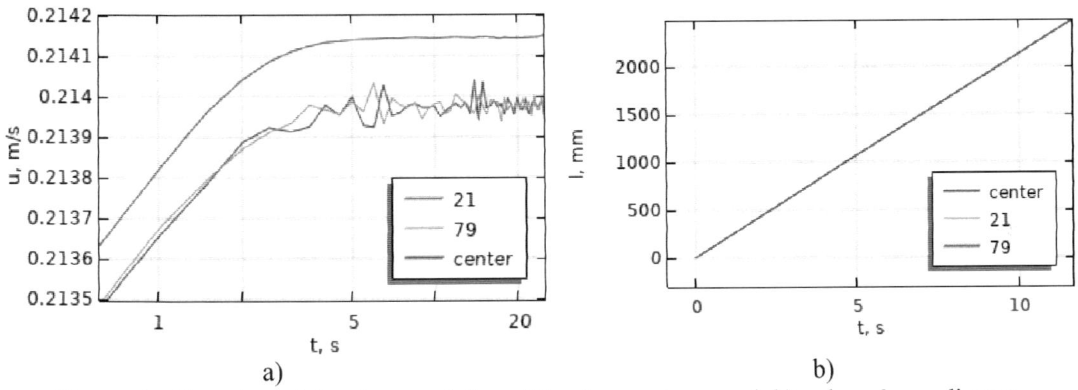

Figure 3. Changing of the characteristics of the drops: a) – speed; b) – drops' coordinates.

For a number of drops from the central set, i.e. from drop 21 to drop 79, we define the speed of each of them when the cross-section is located at a distance equal to the length of the initial hydrodynamic segment for the non-Newtonian medium (1000 mm). To do this, we use the results of the solution of the non-stationary problem of the flow of drops in the non-Newtonian medium. We find the speed for each of the drops using the speed determination itself

$$v_{pn} = \frac{1}{t_n}, \tag{12}$$

where t_n – the time, which it takes the n-th drop to reach 1 m distance.

Calculating the speed value for each of the drops by formula (12), we can construct a speed profile for the section at a distance of 1000 mm from the input of the pipe.

3.2. Findings

The speed profiles for each of the phases of the flow are parabolic. This does not contradict the classical ideas of the hydrodynamic pattern of the flow of the liquid medium [5]. The comparative analysis shows that their difference is insignificant. The nature of the stabilization of the speed of the first and second phases for the central line and extreme central flow lines have differences. This is explained by the fact that the speed on the peripheral lines of the central flow slows down due to an increase in the thickness of the boundary layer. However, the speeds for all the lines of the central flow are stabilized at a distance of about 1000 mm. The speed value along the specified lines increases and takes the value of 0.214 m/s, i.e. more than the value that is set at the input of the pipe. After passing 1000 mm, the speed of the central flow in the pipe does not change. Thus, it can be assumed that the length of the initial hydrodynamic area for this non-Newtonian fluid is 1000 mm.

4. Conclusion

The results of the study of the two-phase flow of non-Newtonian fluid show that to identify the hydrodynamic pattern of the flow in a round tube, we can use the second phase in the form of drops of paint. The calculated length of the initial hydrodynamic segment, estimated by the parameters of the main flow (non-Newtonian fluid) and droplets (Newtonian fluid) is the same. And such characteristics of the non-stationary flow process as the time and the distance passed by the drops in the future physical experiment will allow us to evaluate the adequacy of the presented mathematical model.

References

[1] Chhabra R P and Richardson J F 1999 *Non-Newtonian Flow in the Process Industries: Fundamentals and Engineering Applications* (Oxford: Butterworth-Heinemann) p 436

[2] Mezger T G 2006 *The Rheology Handbook: For User of Rotational and Oscillatory Rheometers. 2nd Revised Edition* (Hannover: Vincentz Network) p 298

[3] Mashelkar R A 1975 Hydrodynamic entrance-region flow of pseudoplastic fluids *J. Heat and Fluid Flow* **188** 683-90

[4] Kozhukhova E A, Kozhukhov N N and Ryazhskikh V I 2020 Experimental study of thermophysical and rheological characteristics of aqueous solution of surface-active substances *J. Polytechnic Bulletin Series: Engineering Studies* **3(51)** 19-24

[5] Collins M and Schowalter W R 1963 Behavior of Non- Newtonian Fluids in the Entry Region of a Pipe *J. AIChE* **9(6)** 804–9

Intensification of heat transfer during pool boiling of nitrogen on surfaces with capillary-porous coatings produced by 3D-printing

D V Kuznetsov and A N Pavlenko

Kutateladze Institute of Thermophysics, Siberian Branch, Russian Academy of Science, Russia, 630090 Novosibirsk pr. Akad. Lavrent'eva, 1

Kuznetsov_9308@mail.ru

Abstract. The study of heat transfer, critical heat fluxes (CHF) and evaporation dynamics at pool boiling of nitrogen at atmospheric and low pressures in a stationary heat generation regimes was performed. The two flat cooper heaters with porous coatings of various structural parameters obtained by additive 3D-printing as well as smooth one were used as working surfaces. According to obtained results such coatings significantly effect on pool boiling increasing up to six times the heat transfer coefficients (HTC) in comparison with uncoated sample. For all investigated heaters and pressures, visualization of boiling was performed using a video camera from which data on the bubble departure diameters and estimates of the active nucleation site density were obtained.

Introduction

As known, phase changes, including boiling, found their application in industry: in the nuclear power, in the chemical industry, in heat power engineering, at cooling of different electronic devices etc. However, the increasing of heat fluxes which must be removed from energy-intensive surfaces including by reducing the weight and size parameters of equipment stimulate the development of research in the field of heat transfer intensification and increasing the critical heat fluxes at pool boiling of various fluids. One of the most investigate methods in recent years aimed at solving the above problem are methods to creating modified heat releasing surfaces by structuring and creating porous coatings on the micro/nano scale. Today, there are a huge number of technologies for the production of these types of surfaces (mechanical processing, plasma spraying, sintering of particles, electrodeposition, lithography, etc.), which are constantly developing and each of which has its own advantages and disadvantages. In recent years, research has begun to actively develop for creation methods of using additive technology of 3D-printing (selective laser melting/sintering SLM/SLS) to produce modified surfaces which can be used in industry for increasing heat loads of devices. This method is interesting for researchers due to its flexibility in the creation of coatings. Namely it's able to create solid structures, porous media as well as combined areas from different materials with good reproducibility of microstructural characteristics [1], which allows to analyse the effect of individual coating parameters on the boiling process.

In recent years, the amount of research associated with the influence of individual microstructural parameters of the surfaces with different geometry and material, obtained by the SLM method, on HTC and CHF at nucleate boiling of fluids with different properties and variation of experimental conditions such as pressure, subcooling etc. is rapidly growing. For example, the authors of [2] studied the aluminium samples with micro-cavity and micro-fin structures during boiling of saturated FC-72 at atmospheric pressure. For all modified working sections, an increase of heat transfer coefficients and critical heat fluxes were observed relative to the smooth one, while the maximum values (70% for HTC and 76% for CHF) were achieved for a micro-ribbed surface with characteristic rib sizes (300-500) microns. In subsequent work [3], this team managed to achieve the high rates of HTC and CHF (up to 3 and 6 times higher respectively) using porous lattice structures relative to the uncoated heater. Similar research results for the same liquid were obtained in the work [4]. The authors used the 3D printing method to create combined structures on the surface of the samples, which are composite porous coatings (0.6 mm micro pores and 2 mm macro pores). It has been shown that composite coatings have the greatest increase in the critical heat flux, both in comparison with a smooth plate and a uniform coating. In this case, an important role is played the thickness of the coating, which significantly affects the value of the CHF. On the other hand the authors of [5] conducted the experiments using water as working liquid (saturated and subcooled) on surfaces with caverns of different geometry and characteristic size of about 1 mm and did not receive any heat transfer intensification. At the same time, the use of re-entrant microchannel structures obtained by the SLM method, as shown in [6], leads to a significant increase in HTC (up to 330%) at boiling of subcooled water. But the authors did not achieve CHF for the modified sample even with a 3-fold excess of the heat releasing power relative to the maximum (critical) heat load for the sample without modification. The influence of microstructural parameters of mesh coatings, obtained by this method, on HTC and CHF during boiling of saturated water was considered in [7]. The authors showed that the increasing of grid width in the rage from 0.5 mm to 1.3 mm significantly decreased the heat transfer coefficient and may cause to be even lower than HTC for plane surface. The maximum CHF values were obtained with a grid width of 1.1 mm (more than 3 times higher than CHF for the uncoated plate). Issues related to the intensification of heat transfer both during boiling and evaporation of n-dodecane were considered in the works [8, 9]. The experimental data were obtained at different pressures and thickness of working fluid layer. As shown in these studies, the use of porous coatings in the form of regular ridges on the surface of the heater, obtained by the SLM method, can lead to a 5-fold increase in the heat transfer coefficients relative to the uncoated one. The above studies show the promise of using the SLM method to create effective heat exchange surfaces for industrial applications, as well as the relevance of studying various forms and parameters of structures in relation to specific operating conditions (type of fluid, pressure, subcooling etc.) to identify optimal one.

This work is a continuation of the earlier research [10] and is aimed at studying heat transfer and critical heat fluxes on SLM porous coatings with varying their basic microstructural parameters during nucleate boiling of nitrogen at different pressures.

Experimental facility

Since this work is a direct continuation of earlier studies, a detailed description of all units of the experimental setup (the scheme of which is shown in the figure 1 for clarity), as well as the experimental techniques, is given in [10]. In this section, the main points related to the experimental conditions will be briefly noted.

Liquid nitrogen was used as a working fluid under conditions of saturation at atmospheric and reduced to 0.018 MPa pressures. To achieve subatmospheric pressures, the cryostat is equipped with a system for pumping out the internal volume using a vacuum pump and a fine-tuning valve that regulates the rate of pumping out nitrogen vapours. The current pressure values was directly monitored with precision of 0.01 kgf/cm^2 using an exemplary vacuum gauge installed in the upper part of the cryogenic tank and by implication according to the measurements of a temperature sensor located at a distance about 5 cm from the investigated surface in the working volume.

The working samples were square plates made of M1 copper with a size (22x22) mm and a thickness of 2.5 mm. The temperature sensor was installed in each sample at a distance of 0.5 mm from the heat exchange surface to minimize the temperature gradient along the heater thickness (the estimates in this configuration showed that the temperature difference did not exceed 0.1 K at maximum heat fluxes) during data recording. The working surface was heated with a constantan foil, pressed against the sample through a thin insulator and connected to a power supply. To maintain stationary conditions of heat release, after each change in electric power, several minutes were waited until the temperature of the sample surface reached stationary values. It should be noted that hysteresis of the boiling curve was found for all heaters; however, with a decrease in the heat flux density a good reproducibility of the experimental data was observed. In this connection, further analysis of the results was carried out with a decrease in heat load.

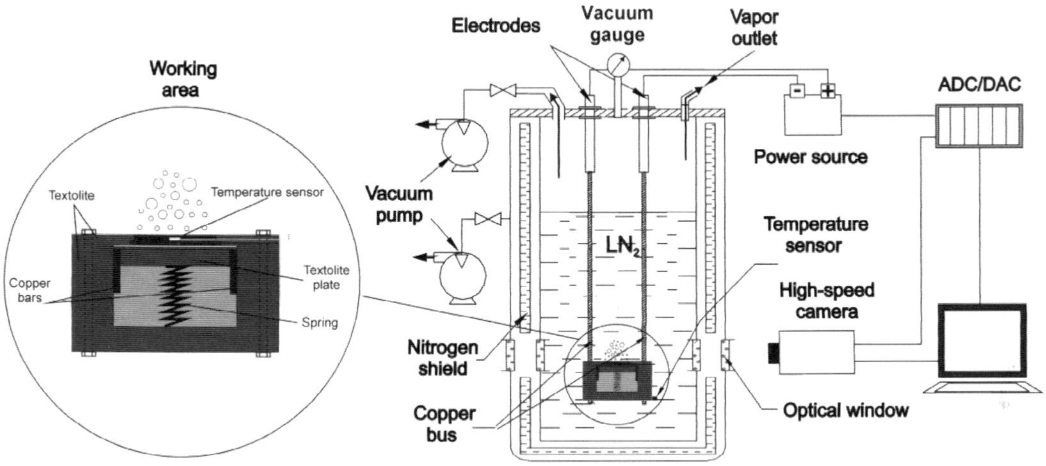

Figure 1. Scheme of experimental setup and working area.

To conduct the experiments two copper coatings with different microstructural parameters were created by additive method of selective laser melting/sintering [9] on the working surfaces. The morphology of the coatings was analysed using a BRUKER Contour GT-K1 optical microscope profilometer. 2D/3D profiles of them can be seen in figure 2. The main difference between the modified heater № 1 and № 2 is the significant difference in the width of the channels l and the distance between adjacent ridges λ. The porosity ε which was determined by mass method taking into account the shape of the profile of the coating as well as the thickness of uniform porous layer h have close values.

Table. Parameters of the samples.

Sample №	δ, μm	λ, μm	l, μm	h, μm	ε, %
1	480	2010	1140	70	30
2	430	3510	1820	50	34

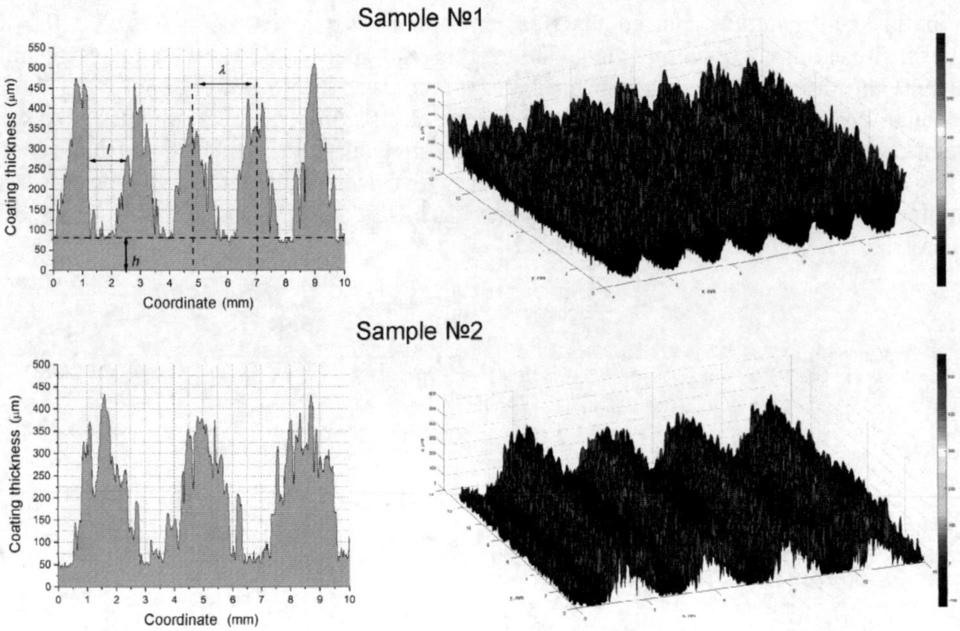

Figure 2. 2D/3D profiles of the samples.

Experimental results

The experimental data of the dependence of the heat flux on the overheating of the heat-transferring surface of a smooth and both modified samples with respect to the saturation temperature of liquid nitrogen at atmospheric pressure are shown in the figure 3a. The arrows on the figure indicate the critical heat fluxes for all heaters, which were determined by a sharp increase in temperature with an increase in the heat load and from the visualization of the boiling process. As it can be seen there is a characteristic break of the boiling curve for the smooth heater at a heat flux about 1-2 W/cm^2, which corresponds to the change of the regimes from the free convection to the dominating nucleate boiling. The value of the CHF for this working section was q_{CHF} = 16.9 W/cm^2 and is in good agreement with the well-known Kutateladze-Zuber model with a coefficient of 0.135. For both modified samples, there are no pronounced kinks in the heat transfer curve which is explained by the presence of boiling at very low heat fluxes until the complete shutdown of the heat release power. In addition, for sample № 1, which is characterized by a smaller distance between adjacent regular ridges, an increase in the CHF is observed by almost 30% relative to the smooth area. For heater № 2, the critical heat flux practically coincides with the q_{CHF} value for a smooth sample. This fact can be explained by the presence of wide zones (more than 1 mm) between the ridges with a quite thin coating layer in which, apparently, local film boiling occurs at high heat fluxes, leading to a heat transfer crisis.

The enhancement of heat transfer at atmospheric pressure, obtained as the ratio of the HTC on the heaters with coatings to the same value for the uncoated one at a fixed value of the heat flux, is presented on figure 3b. The intensity of heat transfer during pool boiling of liquid nitrogen on the heaters with coatings is much higher in all investigated range of heat releasing power up to the critical value on the smooth sample. The maximum increase in HTC (up to 6 times for the sample № 1 and up to 4 times for the sample № 2) is observed in the zone of low heat fluxes and decreases with increasing heat releasing power.

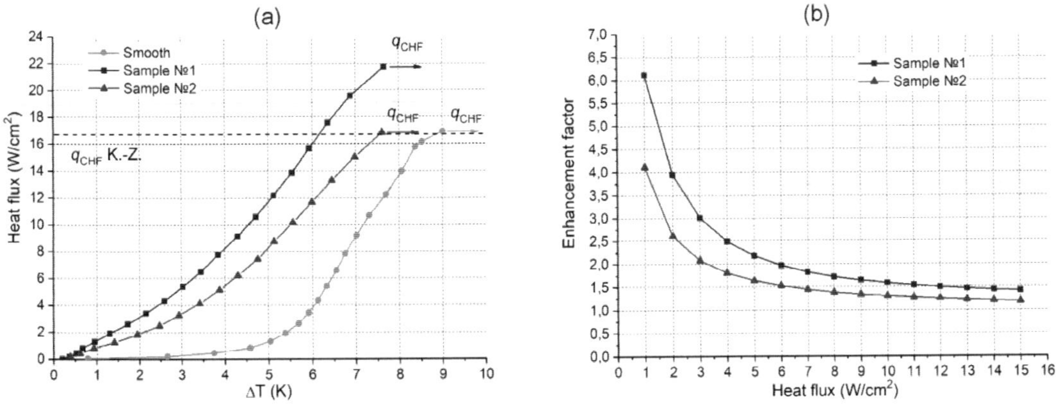

Figure 3. Atmospheric pressure: (a) boiling curves, (b) ratio of HTC.

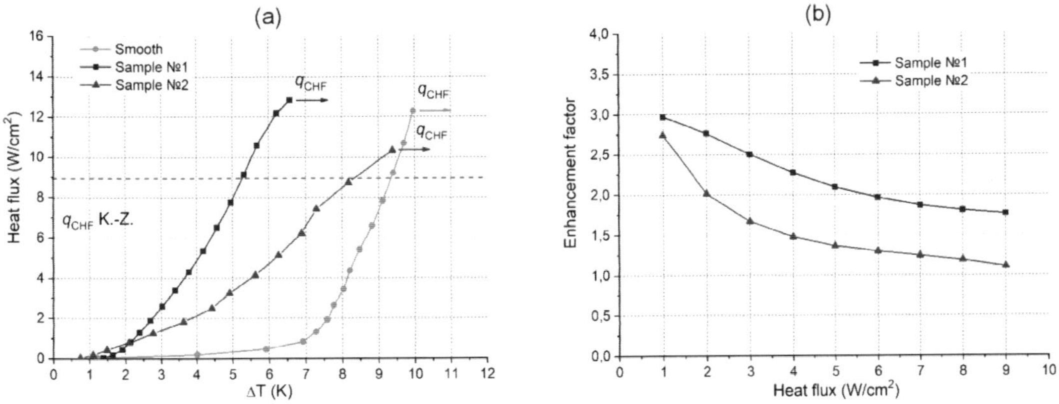

Figure 4. Low pressure – 0.018 MPa: (a) boiling curves, (b) ratio of HTC.

Decreasing pressure up to 0.018 MPa cause to the lower heat transfer rate both for the smooth and for the modified surfaces on the average 25% relative to the atmospheric pressure, as well as reducing the critical heat fluxes (see figure 4a). At the same time, the calculated Kutateladze-Zuber dependence in this case lies below q_{CHF} for all investigated samples. As in the case of atmospheric pressure, for both modified samples, there is a significant enhancement of heat transfer compare to the smooth area (figure 4b) in the investigated range heat flux density. The maximum increase in HTC up to 3 times was also obtained for sample № 1 with a smaller distance between adjacent ridges.

The frames of high-speed visualization of boiling for studied pressures at low heat fluxes when the individual bubbles can be determined are shown in figure 5. As shown by the qualitative analysis of the obtained data, at heat fluxes q <1 W/cm^2 the active nucleation site density (NSD) on the heaters with coatings is significantly higher than the similar value for the smooth one and boiling occurs in the pores of coating ridges both at atmospheric and at reduced pressures. This is clearly seen in the above figures. Also, the quantitative estimates of the bubble departure diameters for all studied samples at q < 2 W/cm^2 were carried out. Thus, at atmospheric pressure the departure diameters were (0.1-0.3) mm for the modified heaters (it should be noted that boiling was also observed at extremely low heat fluxes, however, the separation bubble diameters were less than 100 μm and could not be more accurately determined due to the insufficient resolution of the video camera), (0.4-0.7) mm for the smooth heater, and (0.5-1.5) mm, (1.2-4.5) mm at reduced pressure, respectively. An increase in NSD in the zone of low heat fluxes and the increasing of effective heat transfer area due to copper capillary-porous coatings, as well as a restructuring of a two-phase flow with a more ordered character of the removal of vapour bubbles and liquid supply to the surface are the main factors that can explain increased HTC during nitrogen boiling on such types of coatings obtained by 3D-printing.

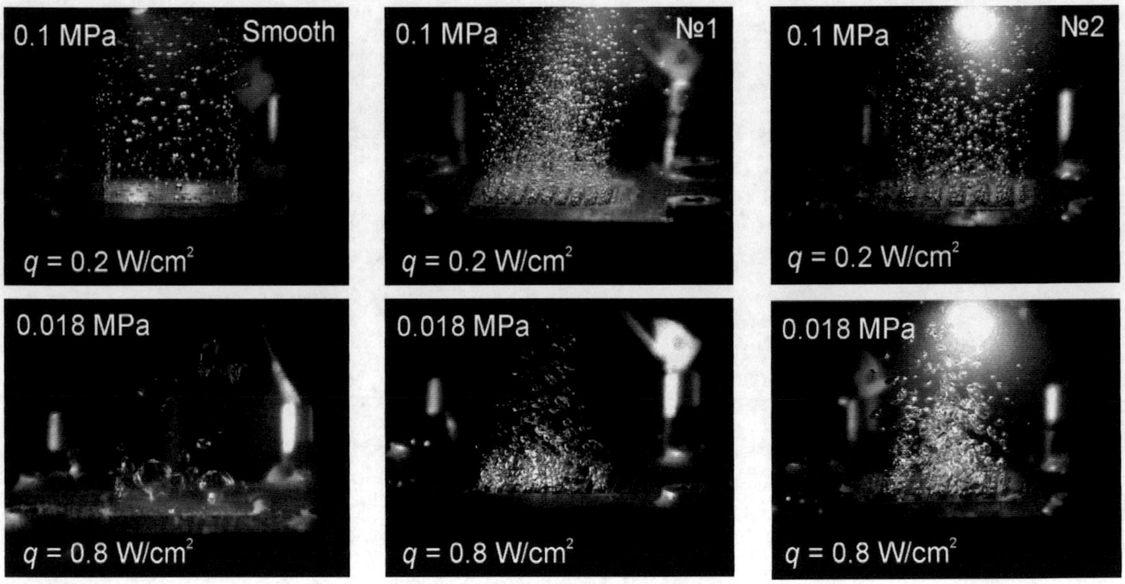

Figure 5. Boiling process at low heat fluxes.

Conclusion

In this paper, pool boiling of nitrogen on porous coatings, obtained by the method of additive 3D-printing, with different microstructural parameters was experimentally studied at atmospheric and reduced pressures. The following conclusions can be drawn from this work:

1) For both pressures the presence of such coatings leads to a significant enhancement of HTC (6 and 3-fold respectively) as compared to the plane surface as well as decreasing of bubble departure diameters. This increase of heat transfer coefficients is probably associated to the increased heat transfer surface area, increased NSD and effective liquid supply to the surface.

2) The microstructural parameters of the coatings play an important role in the value of critical heat fluxes. So for the sample №2 the wide zones (more than 1 mm) between the ridges with a quite thin coating layer are the reason for the lack of increase of CHF as compared to uncoated one at atmospheric pressure.

3) Pressure drop up to 0.018 MPa leads to the reduction of HTC and CHF for all investigated samples and higher value of bubble departure diameters.

Acknowledgments

This research was carried out at the Kutateladze Institute of Thermophysics SB RAS with the financial support of the Russian Science Foundation Grant No. 19-19-00180 (experimental study of the heat transfer and critical phenomena during boiling on modified surfaces produced by 3D-printing) and State Contract with IT SB RAS No. 121031800216-1 (debugging the technique for high-speed visualization of non-stationary boiling processes in cryogenic liquids).

References

[1] Read N, Wang W, Essa K and Attallah M M 2015 *Mater. Design (1980–2015)* **65** 417–424
[2] Ho J Y, Wong K K and Leong K C 2016 *Int. J. of Heat and Mass Transfer* **99** 107–121
[3] Wong K K and Leong K C 2018 *Int. J. of Heat and Mass Transfer* **121** 46–63
[4] Liu H, Zhang C, Wang J and Zhang L 2021 *Appl. Thermal Eng.* **197** 117396

[5] Kang Z and Wang L 2018 *Therm. and Fluid Sci.* **93** 165–170
[6] Pi G, Deng D, Chen L, Xu X and Zhao C 2020 *Int. J. of Heat and Mass Transf.* **156** 119920
[7] Zhang C, Zhang L, Xu H, Li P and Qian B 2019 *Int. J. of Heat and Mass Transf.* **128** 570–580
[8] Zhukov V, Pavlenko A and Bessmeltsev V 2019 *J. of Phys.: Conf. Ser.–IOP Publ.* **1369 (1)** 012007
[9] Bessmeltsev V P, Pavlenko A N and Zhukov V I 2019 *Optoelectronics, Instrumentation and Data Process.* **55 (6)** 554–563
[10] Pavlenko A N, Kuznetsov D V and Bessmeltsev V P 2021 *J. of Eng. Thermophys.* **30 (3)** 341–349

The influence of anionic surfactant in water on the quenching process

M A Lexin, A R Zabirov, I A Molotova, O N Kaban'kov and V V Yagov

Department of Thermophisic, National Research University «MPEI», Russia, 111250 Moscow, Krasnokazarmennaya, 14

Zabirov.arslan@gmail.com

Abstract. Understanding the patterns of heat transfer during quenching is important for many technical applications. Of particular interest is the boiling regime, which is characterized by high intensity and occurs at surface temperatures exceeding the temperature of attainable liquid superheat. This work is aimed at studying the effect of surfactants on the onset of intense heat transfer during quenching. For this, experiments were carried out on quenching spheres made of different metals (nickel, stainless steel and zirconium) in water with different concentrations of surfactants. The surfactant was alkylbenzene sulfonate, the concentration of which varied from 0.1 to 2%. The analysis of the obtained cooling thermograms revealed the influence of not only the surfactant concentration on the beginning of the intensive cooling mode, but also the state of the heat transfer surface.

1. Introduction

The high heat transfer rate associated with fine fragmentation of the liquids causes rapid vaporization of the cold liquid and the resulting high pressure mixture expands against the low-pressure surroundings. Since this process has the potential for releasing destructive mechanical energy and high pressure load to the system, it has been a safety concern in severe nuclear reactor accidents as well as many industrial accidents such as in metal foundries and paper mills. One way of studying the suppression of steam explosions has been the change of the physical properties of water; in particular, viscosity and surface tension. These properties were altered by dissolving small amount of additives such as polymers and surfactants. A number of studies have been reported on the effects of such additives on the suppression of vapor explosions. The studies of surface tension effects [1-4] indicate that dilute solutions of surfactants have a mitigating effect on both spontaneous and triggered vapor explosions, but there has been no observation of complete suppression of the explosion. Also it was reported that the surfactants rather increased the efficiency of explosions in a certain range of coolant temperature(~20 °C) [2].

In work [5] it was noted that density and viscosity of coolant liquid practically do not change with small additives of surfactants, while the surface tension changes significantly. In addition, the critical heat transfer is affected by the linear size of the sample. So, for thin wires, the critical heat flux increases with the increase in the concentration of surfactants, and for large bodies, for example, a 12 mm ball used in work [5], the critical heat flux decreases. Experiments have shown a decrease in critical heat flux when surfactants are added to water. In addition, there was a slight decrease in the minimum film boiling temperature, and an increase in the quenching time for surfactant solutions in comparison with pure water.

In work [6] comprehensive studies oriented at studying the effect of adding various substances to the water on the quenching process were carried out. Various salts, complex carbohydrates (sugar), surfactants, and ionic liquids were used as impurities. The authors found that the concentration of surfactants does not significantly affect the Leidenfrost temperature, which is due to the fact that the concentration in the volume of the solution differs from the concentration at the interface between liquid and vapor.

The influence of the addition of surfactants on the quenching process of vertical cylindrical samples (50x10 mm) in water was investigated at work [7, 8]. In comparison with pure water, cooling in aqueous surfactant solutions is characterized by lower values of minimum temperature of film boiling regime and critical heat flux, and, as a result, by longer cooling. The authors assume that the main reason for this effect is to reduce the surface tension and increase the stability of the liquid-vapor interface.

This work is aimed at the experimental study of the effect of surfactants on the temperature of the beginning of the intensive boiling regime (minimal film boiling temperature or Leidenfrost temperature), which in the text will be referred to as the transition temperature T_{tr}.

2. Experimental facility

Basing on the results of previous studies [9-14] main purpose of our experiments was to study the effect of changing of coolant surface tension in quenching process.

The experimental unit created in 2013 allows to conduct experiments in a wide range of coolants temperatures (from -200 °C to +200 °C) and pressure (up to 1.5 MPa). The experiment consists in heating the sample to a predetermined temperature and then immersing it into the test liquid.

In all our experiments spheres 40 mm in diameter from nickel, stainless steel and zirconium are used as the tested sample. Several thermocouples located at surface of ball at different polar (90, 135 and 180°) angle and one located in center. Surfactant solutions and pure distillated water were used as coolant liquid. All experiments were carried out at atmospheric pressure. Coolant temperature ranged from 50 to 90 °C. Sulfanol (alkylbenzene sulfonate) was used as anionic surfactant. The following mass concentrations were selected: 0.1, 0.25, 0.5, 1 and 2%. Figure 1 shows isotherm of surface tension of aqueous surfactant solution of sulfanol with different sulfanol mass concentration at border with air.

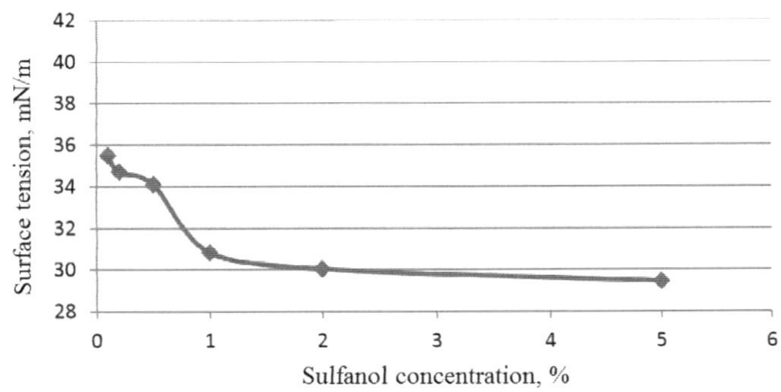

Figure 1. Isoterm of surface tension of aqueous solution anionic surfactant (sulfanol) at border with air.

3. Results and discussion

The main primary results of the experiments are thermograms of cooling. Typical thermogram of cooling process presented at figure 2 gives change in the temperature in the center of the tested sample and at several points of its surface over time of cooling process. This figure sows three typical cooling modes. At the beginning of cooling process (0-16 s) there is stable film boiling which is characterized by low heat transfer coefficient. Between 16 and 18 sec there is intensive cooling regime with high heat transfer coefficient. After 18 sec it was observed cooling by nucleate boiling mode and natural convection.

In many experiments the intensive cooling regime occurred almost simultaneously over the entire surface of the sphere. Therefore the value of transition temperature T_{tr} (transition between film boiling and intensive cooling regime) was taken as the average over three surface thermocouples.

At figures 3-5 presented transition temperature as dependence of liquid temperature and sulfanol concentration for different sphere materials. It can be seen from the figures that surfactant additions lead to a decrease in T_{tr}.

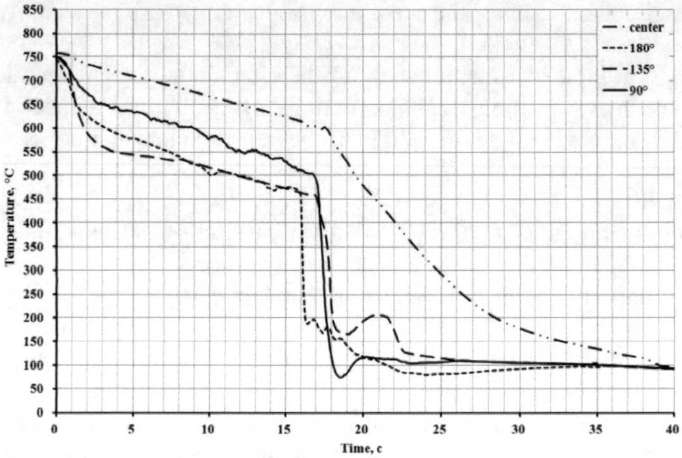

Figure 2. Thermogram of cooling (material of tested sample – nickel; diameter of tested sample – 40 mm; coolant – 0.25% water-sulfanol solution; coolant temperature – +60 °C).

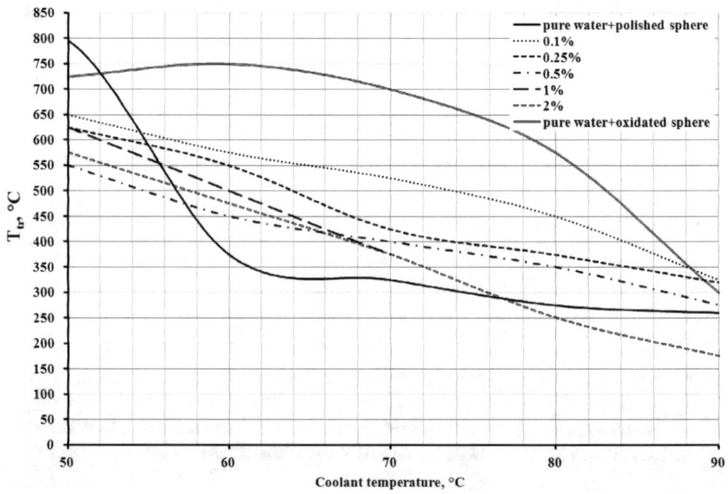

Figure 3. Transition temperature T_{tr} for different coolant temperature and surfactant concentration (material of tested sample – nickel; diameter of tested sample – 40 mm).

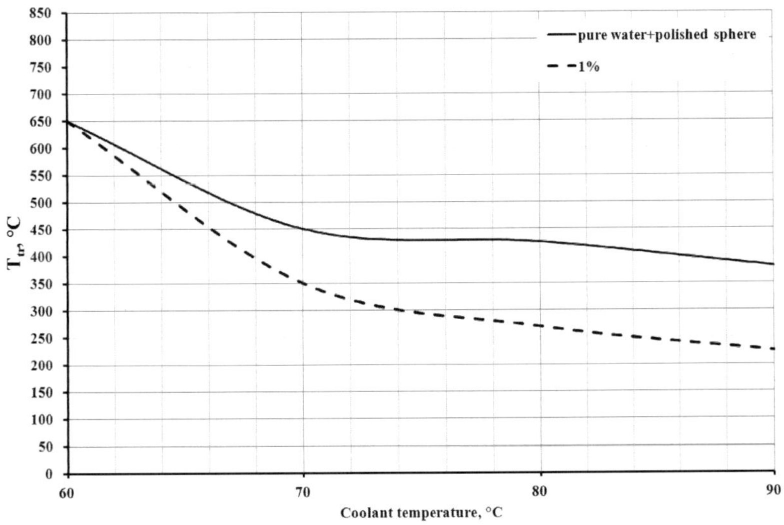

Figure 4. Transition temperature T_{tr} for different coolant temperature and surfactant concentration (material of tested sample – zirconium; diameter of tested sample – 40 mm).

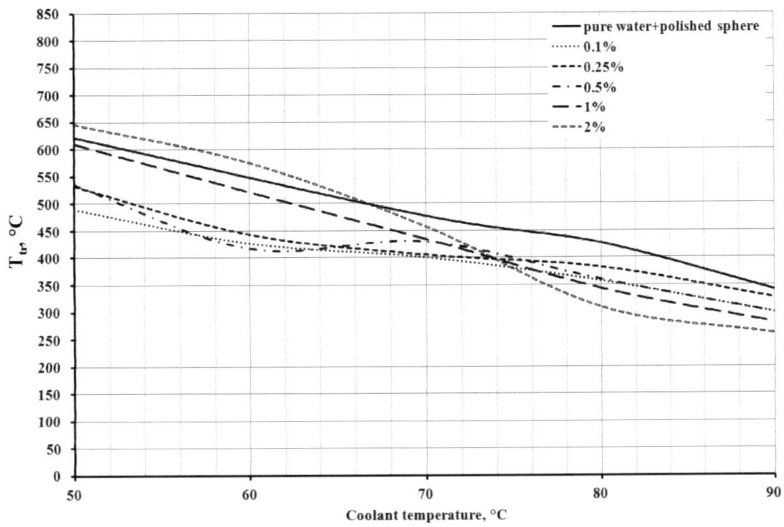

Figure 5. Transition temperature T_{tr} for different coolant temperature and surfactant concentration (material of tested sample – stainless steel; diameter of tested sample – 40 mm).

During experiments, when nickel and zirconium were used as material of tested samples, layer of oxide film was detected on its surface. In additional as tests were carried out contaminations precipitated on samples surface from solution. This fact affected on the intensive cooling regime in addition to sulfanol concentration.

The transition temperature was significantly lower in experiment with polished surface of tested sample than in experiment with oxidized surface. For purity of comparison distilled water without surfactant was used as coolant in both experiments. From data obtained in these experiments it can be concluded that there is an effect of cooling surface properties on the intensity of the cooling process.

Values of transition temperatures T_{tr} obtained in experiments with surfactant solutions as coolant are mainly between ones in experiments with pure water as coolant for the polished and oxidized surfaces of tested samples, except for the results at high subcooling level.

The growth of the oxide film was not so significant for stainless steel sphere. It can affect value of transition temperature at high subcooling level. As it is shown at figure 5 when subcooling level lower than 26 °C increase in surfactant concentration entails decrease in T_{tr} as well as samples from nickel and zirconium. At lower coolant temperatures on contrary increase of count of sulfanol in solution leads to increase in transition temperatures.

4. Conclusion

Experiments were carried out on samples of different materials (stainless steel, zirconium and nickel), on pure water and aqueous solutions of surfactants of different concentrations (0.1 - 2%) in a wide range of liquid subcooling (50>ΔT_{sub}> 0 K). An analysis of the experimental results made it possible to formulate the following conclusions:

Both during quenching in pure water and during quenching in surfactant solutions, an increase in T_{tr} with increasing subcooling is observed in all samples;

- There is a dependence of T_{tr} on the thermophysical properties of the metal. If we compare the T_{tr} obtained on “fresh” nickel, steel, and zirconium, then the minimum T_{tr} values are observed on the metal with the highest thermal effusivite coefficient, i.e. on nickel. The state of the surface of the body to be cooled (contamination, oxidation) has a strong effect on the quenching process;
- Almost always, when cooled in surfactant solutions, T_{tr} is lower than when quenched in pure water throughout the entire ΔT_{sub} range. This is most likely caused by the stabilization of the liquid-vapor interface and a decrease in the surface tension coefficient.

It should be noted that the experiments performed are still of a qualitative nature. To more accurately determine the effect of surfactants, experiments with minimal surface oxidation and contamination are required.

Acknowledgments
This work was supported by the Russian Science Foundation, Grant N 20-79-10363.

References

[1] Becker K M, Lindland K P 1990 KTH-NEL-50 *Royal Institute of technoligy* (Sweeden)
[2] Becker K M 1990 *M.S. Thesis* (University of Wisconsin-Madison)
[3] Kowal M G, Dowling M F, Abdel-Khalik S I 1993 *Nucl. Sci. & Eng.* **115** 185
[4] Bang K H, Kim M H, Jeun GD 1997 *Nuclear Engineering and Design* **177** pp 255-264
[5] Wu et al 1994 *International journal of heat and mass transfer* **37** pp 2377-2379
[6] Kumar V, Kumar N R S, Rishi R 2020 *Soft matter* **16** pp 6145-6154.
[7] Kang D G, Jae H K, Chi Y L 2018 *Journal of Mechanical Science and Technology* **32** pp 5919-5926
[8] Kang D G, et al 2020 *Journal of Mechanical Science and Technology* **34** pp 4753-4761
[9] Yagov V V, Zabirov A R, Lexin M A 2015 *Thermal Engineering* №11 pp 833-842
[10] Yagov V V, Zabirov A R, Kanin P K 2018 *International Journal of Heat and Mass Transfer,* Vol **126** pp 823-830
[11] Yagov V V, Lexin M A, Zabirov A R, Kaban’kov O N 2016 *International Journal of Heat and Mass Transfer*, Vol **100** pp 908-917
[12] Yagov V V, Lexin M A, Zabirov A R, Denisov M A 2016 *International Journal of Heat and Mass Transfer* Vol 100 pp 918-926
[13] Zabirov A R, Yagov V V, Kaban’kov O N, Lexin M A, Kanin P K 2016 *Journal of Engineering Physics and Thermophysics* Vol **89** pp 1466–1475
[14] Yagov V V, Zabirov A R, Kaban’kov O N, Minko M V 2017 *International Journal of Heat and Mass Transfer* Vol **110** pp 219–230

Heat transfer enhancement in superheated hydrocarbon fluids due to traces of water

K V Lukianov, A N Kotov and A A Starostin

Institute of Thermal Physics, Ural Branch of the Russian Academy of Sciences, Russia, 620016 Yekaterinburg, st. Amundsen, 107a

kireeshka90@mail.ru

Abstract
For the specific mode of pulse heating of a wire probe immersed in a test liquid (saturated hydrocarbon), the effect of enhancement of the heat transfer through the probe surface has been revealed. The characteristic heating time is from 10 to 20 ms. The objects of research are *n*-hexane, *n*-decane, and *n*-hexadecane. It is shown that the addition of 0.003% water increases the heat transfer to 10% in the course of approaching the boiling-up temperature of liquid. Simulation of the experimental conditions shows the possibility of microconvection in the most heated boundary layer of a liquid about 10 μm thick.

1. Introduction

The known studies on the water addition effect on the hydrocarbons' thermophysical properties revealed monotonic properties depending on the impurity concentration [1-3]. In the observed articles on the effect of water additives on the thermophysical properties of hydrocarbons, the monotonic dependences of the properties on the impurity concentration were obtained. For example, in [1], experimental results of a study of a binary mixture of water-hexane at various concentrations of these components near the critical point of water are presented, and critical lines and phase equilibria for the water-hexane system are calculated.

Moisture can negatively affect the properties of the process fluid, forming polar bonds with the components of the system, thereby changing the thermophysical properties of fuels, transformer and turbine oils and other process fluids [4, 5]. Water can be present in these process fluids in various states, such as dispersed, dissolved, and others. A change in the operating modes of technological units or the movement of process fluid from one section of a technological unit to another (change in temperature, pressure) can lead to a transition of water from a "bound" state to a "free" state.

Devices for monitoring the level of moisture in oil products were developed at our research group. These devices used the technique of pulsed thermal control [5, 6] which based on the general method of controlled pulse heating of the wire probe [7]. This method differs from the well-known THW-method [8] due to the relatively high level of heat generation in the probe. The idea of the technique [5, 6] is based on the results of studying the phenomenon of attainable superheating of mixtures (with respect of liquid-vapor equilibrium temperature) having limited mutual components' solubility, in the processes of their pulsed heating [9, 10]. Among the methods of physical control, the method of pulse heating of the wire probe is distinguished for its high sensitivity, ease of implementation, and suitability for organizing monitoring of the moisture content of hydrocarbon liquids in the laboratory and industrial conditions.

The study of the water addition effect on the thermophysical properties of hydrocarbons is usually carried out on laboratory equipment using stationary methods. However, in real technological installations, thermal processes often occur under non-stationary regimes and the addition of small amounts of water can have unexpectedly strong effects on the thermophysical properties of hydrocarbon liquids [5].

The methods of wire probe's pulsed heating are sufficiently sensitive to qualitatively detect heat transfer changes in superheated hydrocarbon liquids. This fact allows respond to the presence of moisture impurities in thousandths of a percent. However, the total measurement uncertainty in previously developed devices and techniques is large for quantitative assessment of heat transfer from the probe to samples with similar properties. The quality of measurements was improved by creating close temperature-time conditions for heating the probe in samples with different moisture content. Software and hardware have been developed for fine digital adjustment of the probe heating mode in order to minimize the methodological uncertainty in setting the heating mode. The purpose of the article is to develop a technique for reliably resolving the effect of moisture traces on heat transfer in pulse-heated hydrocarbon liquids. The objects of research are *n*-hexane, *n*-decane, and *n*-hexadecane with water addition form 5 ppm to 35 ppm.

2. Compensatory technique

The use of the two-pulse heating method previously revealed the effect of the increased influence of small moisture addition on the intensity of heat transfer from the probe's surface in liquid hydrocarbons near the boiling-up point [5, 6]. The development of the two-pulse heating technique has shown that the signal from the first (powerful) pulse is not very informative, since impurities practically do not affect it. Therefore, the main research was aimed at analyzing the signal on the second pulse.

The similarity of the time dependences on the thermograms made it possible to assume the appearance of an additional heat loss by the convection mechanism with an increase in the heat transfer coefficient of the probe to the moistened sample. The use of the method of controlled heating on the second pulse allows, by changing the preset value of the heating current, to "compensate" the observed changes in the thermogram. Compensating part of heating power $\Delta P=\delta I^2R(T-T_0)$ for relatively small temperature changes on the second pulse will repeat the dependence pattern of the compensated heat flux on temperature changes for the convection component of heat transfer $q=\alpha(T-T_0)$, in the first approximation.

The developed technique is used to select the current on the second pulse for one of the two samples, which differ only in the impurity content, in order to compensate for the difference in the heating thermograms (figure 1).

On the second current pulse I_2, there is a difference between the samples: the dried sample is heated to the temperature T_1 in the time t_0-t_c, the moistened sample - in the time t_0-$t_в$. In this case, it is possible to select a current $I_{2в}$, at which $(t_0-t_c)\approx(t_0-t_в)$. Thus, the thermograms for both samples will be the same. The similarity of the thermograms makes it possible to compare the intensity of heat loss from the probe surface in different samples under similar temperature-time conditions.

The proposed method makes it possible to introduce a relative quantitative characteristic of the effect of the addition of small amounts of moisture under non-stationary conditions of heating the wire probe. The proposed technique will be further referred to as the "compensation heating technique".

The compensation heating technique is similar to the differential scanning calorimeter technique. Comparison of the heating curves makes it possible to reveal the difference between the samples in their thermal response to a given effect.

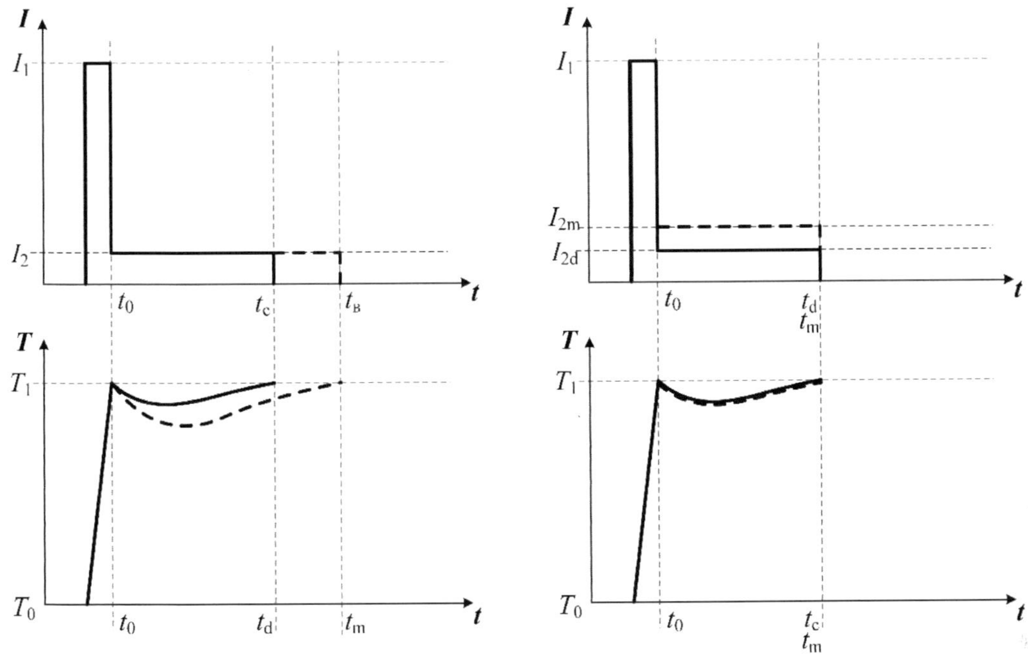

Figure 1. Compensation heating technique.

3. Measuring cell and probe

The scheme of the measuring cell is shown in figure 2. The measuring cell is made of inert material - fluoroplastic. The walls of the cell are thin and easily deformable; from the outside (from the outer part), they come into contact with the hydraulic fluid, and from the inside (from the inner part), they contact with the test fluid. Inside the cell, altogether with the liquid under study, there is a probe with a thin platinum wire and a sensor for the average temperature of the liquid volume. The top of the cell is closed by a sealed cover with fastening bolts and current leads. This design allows the application of pressure up to 10 MPa. The piston regulates the pressure in the cell. The pressure is monitored using a pressure gauge.

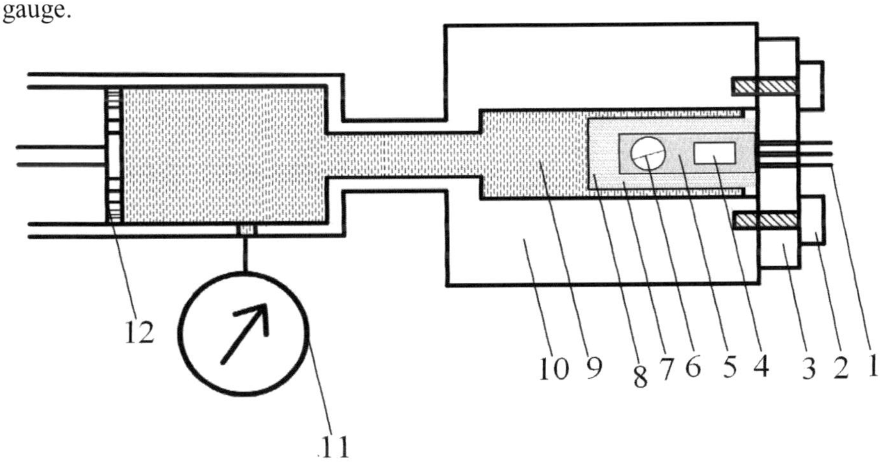

Figure 2. Measuring cell. 1 - current leads, 2 - mounting bolts, 3 - pressure chamber cover, 4 - liquid volume temperature sensor, 5 - probe board, 6 - wire probe, 7 - test fluid, 8 - fluoroplastic cup, 9 - hydraulic fluid, 10 - pressure chamber, 11 - pressure gauge, 12 - piston.

The measuring probe is installed in a textolite board, which is located inside a fluoroplastic cup and immersed in the test liquid (figure 2). This board also contains a temperature sensor that measures the temperature of the volume of the test liquid.

4. Results

The compensation technique is carried out automatically using specially developed software. The algorithm of the program is based on the cyclic loading of the experiment parameters from the table of the experiment parameters into the device. This table is generated automatically before the beginning of the experiment based on the data loaded by the experimenter. The data from this table are cyclically uploaded into the device, thus changing the parameters of each heating. After loading the parameters, the device performs heating and measurement. Automatic selection of the current allows quite accurately (0.3%) to select close temperature-time heating conditions.

The ratio of the heat transfer coefficients of the probe (K_{hw}/K_h) to the moistened (35ppm) and dried (5ppm) samples after the compensation procedure is practically independent of time (figure 3). This effect was observed for all studied samples (*n*-hexane, *n*-decane, *n*-hexadecane). Thus, the relative characteristic of the non-stationary heat transfer process at the second heating pulse of the probe, which does not depend on time, has been experimentally revealed. Should the average ratio of the heat transfer coefficients of a humidified hydrocarbon to a dried hydrocarbon be taken into consideration, it can be assumed that there is such a temperature for a given amount of moisture at which moisture begins to influence heat transfer.

The change in pressure significantly affects the average value of the ratio of the heat transfer coefficient of the watered (K_{hw}) and the original samples (K_h), which is clearly seen in figure 4 on the *n*-hexadecane example. It can be also seen that the effect of moisture on heat transfer practically disappears at a pressure of about 700 kPa. Similar effects are also observed with *n*-hexane and *n*-decane [11]. According to the relative increment of heat transfer from a moistened hydrocarbon to a dried one, it is possible to determine the pressure at which the effects of the influence of moisture become invisible under given heating conditions.

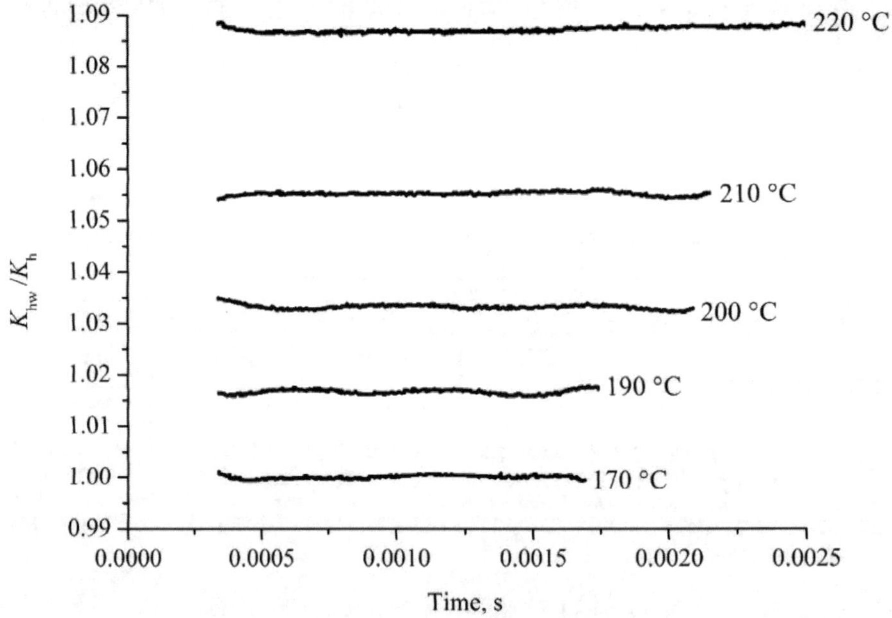

Figure 3. The ratio of the coefficients of unsteady "instantaneous" heat transfer from the probe to the moistened and dried samples of n-decane is stable over time at different temperatures of the probe heating. The weak dependence on time for a temperature of 260 ° C is explained by compensation errors.

5. Conclusions

A comparative method for measuring the temperature dependence of the heat transfer from a pulse-heated probe to dielectric liquids with impurities has been developed and tested on samples of hydrocarbon liquids with traces of water. A main feature of the technique developed is the ability to create similar heating conditions for the wire probe in various samples under discussion. It has been shown experimentally that the ratio of the heat transfer coefficients from the probe to the moistened and dried samples after compensation for the difference in heating can serve as a stationary characteristic of a non-stationary heat transfer process. The dependences of the relative heat transfer coefficient on the probe's temperature and external pressure are obtained. It is shown that the addition of 0.003% water increases the heat transfer to 10% in the course of approaching the boiling-up temperature of *n*-hexane, *n*-decane, and *n*-hexadecane. However, the application of an external pressure of several units of bar practically neutralizes this unexpected effect. The observed intensification of heat transfer manifests itself in pulsed experiments with a wire probe. Simulation of the experimental conditions shows that the effect of a heat transfer mechanism similar to microconvection in the most heated liquid layer with a thickness of about 10 μm is highly possible. The approximate values of the temperature gradient and heat flux in this thin layer are 10^7 K/m and 10^6 W/m^2, respectively. Probably, the addition of water has a destabilizing effect on the interface between the solid surface of the probe and the hydrocarbon liquid directly in the course of approaching the boiling-up temperature and promotes the development of microconvection in the boundary layer.

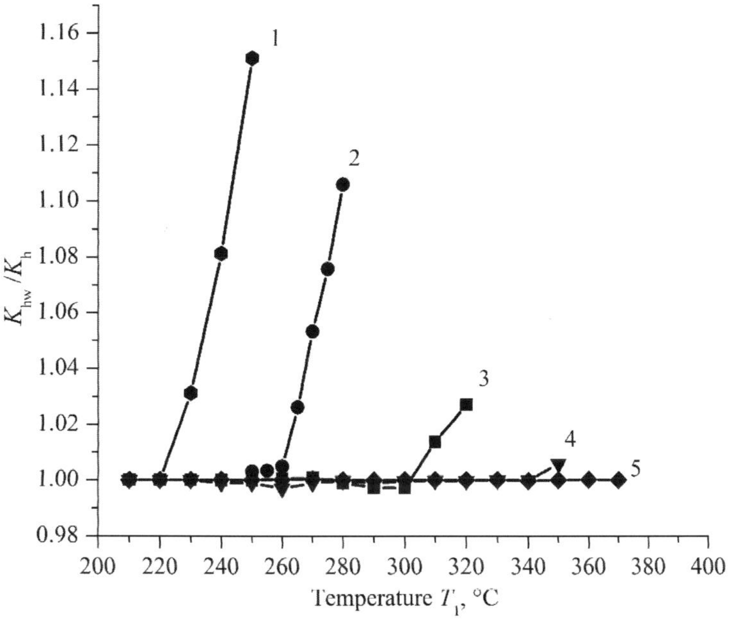

Figure 4. Temperature dependence of the K_{hw} / K_h ratio of the watered and dried *n*-hexadecane samples under specified temperature-time conditions at various pressures (1 - 0.084 MPa, 2 - 0.1 MPa, 3 - 0.212 MPa, 4 - 0.372 MPa, 5 - 0.692 MPa).

Acknowledgments
This work was supported by the Russian Science Foundation, grant no. 19-19-00115.

References

[1] Rasulov S M, Orakova S M and Abdulagatov I M 2015 *High Temperature* **53(4)** 497-501

[2] Bazaev E A, Bazaev A R, Osmanova B K and Dzhapparov T A G 2020 *J. Phys.: Conf. Ser.* **1683** 032031
[3] Koji O, Masaki M, Mototaka S, Hideo M 2003 *Electrical Engineering in Japan* **145 (2)** 21-27
[4] Orakova S M, Rasulov S M and Abdulagatov I M 2013 *Journal of Molecular Liquids* **187** 7-19
[5] Skripov P V, Demin V A, Shangin V V and Starostin A A 2016 *Thermal Eng.* **63(7)** 516–521.
[6] Lukynov K V, Starostin A A and Skripov P V 2017 *Int. J. Heat Mass Transfer* **106** 657–665
[7] Skripov P V, Smotritskii A A, Starostin A A and Shishkin A V 2007 *Journal of Engineering Thermophysics* **16(3)** 155-163
[8] Assael M J, Antoniadis K and Wakeham W A 2010 *Int. J. Thermophys*, **31 (6)** 1051-1072
[9] Igolnikov A A, Rutin S B and Skripov P V 2021 *Thermochimica Acta* **695** 178815.
[10] Kozulin I A, Bartkus G V and Kuznetsov V V 2020 *Journal of Physics: Conference Series* **1677(1)** 012065.
[11] Lukianov K V, Kotov A N, Starostin A A and Skripov P V 2019 *Interfacial Phenomena and Heat Transfer* **7(3)** 283–294

Calculation of thermodynamic properties of mixtures in two-phase equilibrium

Taiming Luo and A Yu Chirkov

Bauman Moscow State Technical University,
Russia, 105005 Moscow, 2-nd Baumanskaya, 5-1

luotaiming@yandex.ru

Abstract. Thermodynamic properties of mixtures in vapor-liquid equilibrium (VLE) were studied. Thermodynamic properties of the methane-ethane mixtures in VLE were calculated with highly accurate Helmholtz free energy equation of state GERG-2008, simplified GERG-2008 and common cubic PR equation of state (EOS). Results show that GERG-2008 has high accuracy in VLE calculations. However, simplified GERG-2008 and PR-EOS both work unsatisfactorily in VLE calculations.

1. Introduction

Calculation of the thermodynamic properties of mixtures in phase equilibrium is an important element in the modeling of chemical processes such as distillation and extraction. In this article, the stability analysis of a mixture at given temperature, pressure and overall composition was carried out before the phase equilibrium calculations, because the generated initial estimates by the stability analysis can be used for phase equilibrium calculations [1].

The stability analysis is based on the principle that the total Gibbs energy of a mixture at given temperature, pressure and overall composition has to be at the global minimum. Based on this principle, Baker et al. [2] proposed the Gibbs' tangent plane criterion to test thermodynamic stability, and Michelsen [3] proposed a computational approach based on this tangent plane criterion.

Successive substitution is a commonly used approach in phase equilibrium calculations, but sometimes it doesn't converge quickly. Therefore, Orbach and Crowe [4] proposed the dominant eigenvalue method, and Crowe and Nishio [5] proposed the general dominant eigenvalue method, in order to improve the convergence efficiency of successive substitution.

In the vicinity of critical point, the convergence rate of successive substitution is very low. Because of this, Michelsen and Mollerup [6] proposed Gibbs energy minimization by the second order method. In this method, the mostly used models are cubic equations of state, but multi-parameter Helmholtz free energy equations of state are rarely used as models. In this work, PR-EOS and GERG-2008, which is a highly accurate Helmholtz energy-based equation of state, were used to calculate thermodynamic properties of the methane-ethane mixtures in VLE. Besides, the influence of the deviation function (see Eq.(3)), which is used to model mixture behavior with higher accuracy, on accuracy of GERG-2008 was also investigated in phase equilibrium calculations.

2. Equation of state for mixtures

The GERG-2008 equation of state is the dimensionless form of the Helmholtz free energy which includes an ideal gas part and an empirically residual part [7]:

$$\alpha(\delta,\tau,\bar{x}) = \alpha^0(\rho,T,\bar{x}) + \alpha^r(\delta,\tau,\bar{x}), \tag{1}$$

where $\bar{x}$ is the molar composition in a mixture, δ is the reduced mixture density and τ is the inverse reduced mixture temperature.

The ideal-gas part α^0 is given by

$$\alpha^0(\rho,T,\bar{x}) = \sum_{i=1}^{N} x_i[\alpha^0_{oi}(\rho,T) + \ln x_i], \tag{2}$$

where α^0_{oi} is the reduced Helmholtz free energy in the ideal-gas state of the pure component i.

The residual part α^r is

$$\alpha^r(\delta,\tau,\bar{x}) = \sum_{i=1}^{N} x_i \alpha^r_{oi}(\delta,\tau) + \Delta\alpha^r\left(\delta,\tau,\bar{x}\right), \tag{3}$$

where α^r_{oi} is the residual part of the reduced Helmholtz free energy of the pure component i and $\Delta\alpha^r$ is a deviation function used to model complex mixture behavior.

The simplified GERG-2008, i.e. GERG-2008 without the deviation function (DF), means that Eq. (3) doesn't include the deviation function $\Delta\alpha^r$.

The PR-EOS for mixtures is described in [8], and the binary interaction coefficient is obtained from [9].

3. Phase equilibrium calculation

For a given n-component mixture of composition $\bar{z}(z_1, z_2, ..., z_n)$ at specified temperature T and pressure p, the necessary and sufficient condition for stability of this mixture is that the reduced tangent plane distance is non-negative for any trial phase composition $\bar{w}(w_1, w_2, ..., w_n)$ [3]:

$$\text{tpd}(\bar{w}) = \sum_i w_i\left(\ln w_i + \ln\varphi_i\left(\bar{w}\right) - \ln z_i - \ln\varphi_i\left(\bar{z}\right)\right) = \sum_i w_i\left(\ln w_i + \ln\varphi_i\left(\bar{w}\right) - d_i\right), \tag{4}$$

where φ_i is the fugacity coefficient of component i and $d_i = \ln z_i + \ln\varphi_i(\bar{z})$.

The generated initial estimates by the tangent plane criterion can be used for phase equilibrium calculations. The successive substitution is a commonly used method in vapour-liquid equilibrium calculations. However, if the efficiency of the successive substitution decreases, it is recommended to use the Gibbs energy minimization by the second order method [6, 10]. At specified T, p and $\bar{z}$, the VLE calculations can be formulated as

$$\min G(T, p, \bar{v}, \bar{z} - \bar{v}), \tag{5}$$

where G is the Gibbs energy and $\bar{v}$ is vapor amounts (per unit feed).

The algorithm for calculating the VLE is described in [11]. In the stability analysis and phase equilibrium calculations, fugacity coefficients and their derivatives are needed. For PR-EOS, the method in [12] is used to calculate these properties. For GERG-2008, all thermodynamic properties of a mixture can be obtained by combining various derivatives of Eq. (1), which is described in detail in [7, 10].

Thermodynamic properties of the methane-ethane mixture at methane mole fraction of 15% in VLE were calculated with the help of PR-EOS, GERG-2008 without DF and GERG-2008. Figure 1 shows that GERG-2008 has high accuracy in predicting molar density of the methane-ethane mixture in VLE. The average relative deviation of molar density predicted by GERG-2008 is much smaller than the average relative deviation of molar density predicted by GERG-2008 without DF and PR-

EOS. Besides, GERG-2008 without DF works better than PR-EOS in prediction of molar density in VLE.

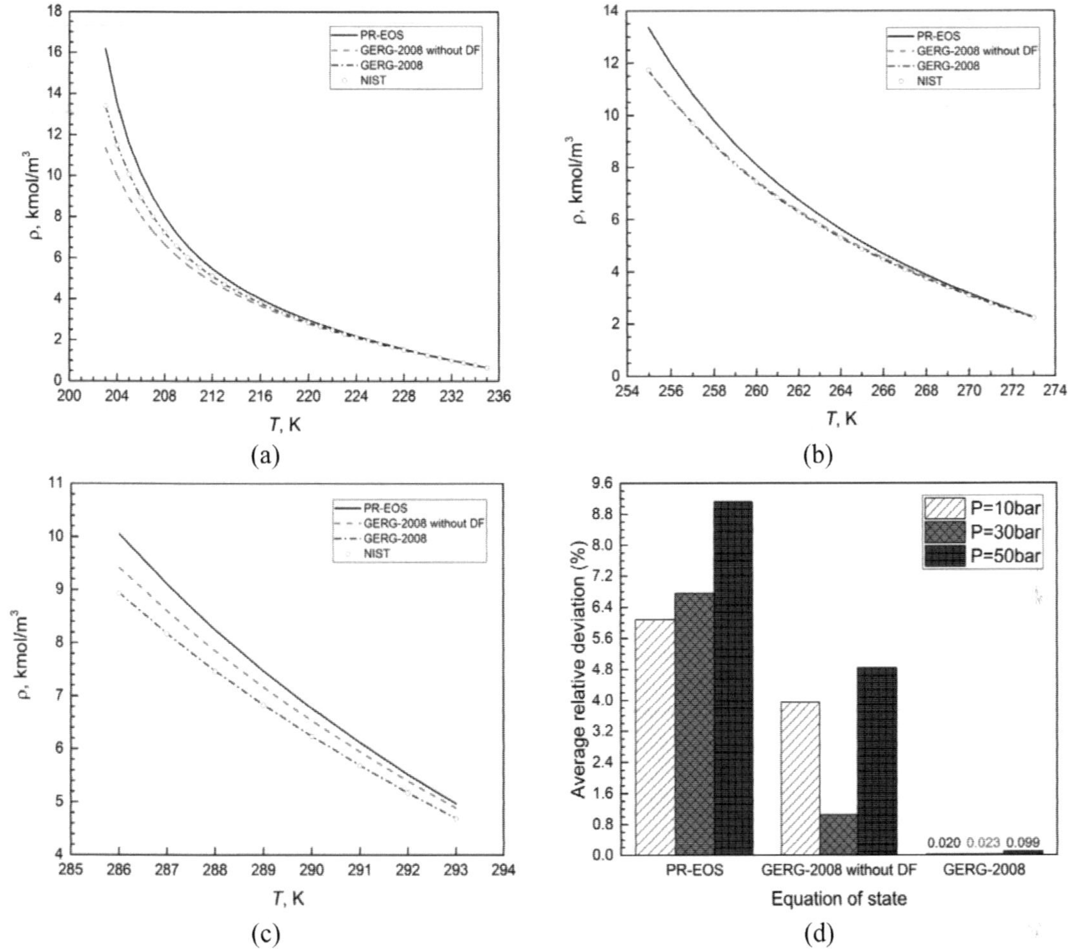

Figure 1. Density versus temperature and its average relative deviation from reference data for the methane-ethane mixture at methane mole fraction of 15% in VLE with the help of PR-EOS, GERG-2008 without DF and GERG-2008 at different pressures: (a) 10 bar, (b) 30 bar and (c) 50 bar.

Figure 2 shows that PR-EOS works worse than GERG-2008 and GERG-2008 without DF in predicting enthalpy difference of the methane-ethane mixture in VLE. The average relative deviation of predicted enthalpy difference by GERG-2008 is smaller than the average relative deviation of enthalpy difference predicted by GERG-2008 without DF and PR-EOS.

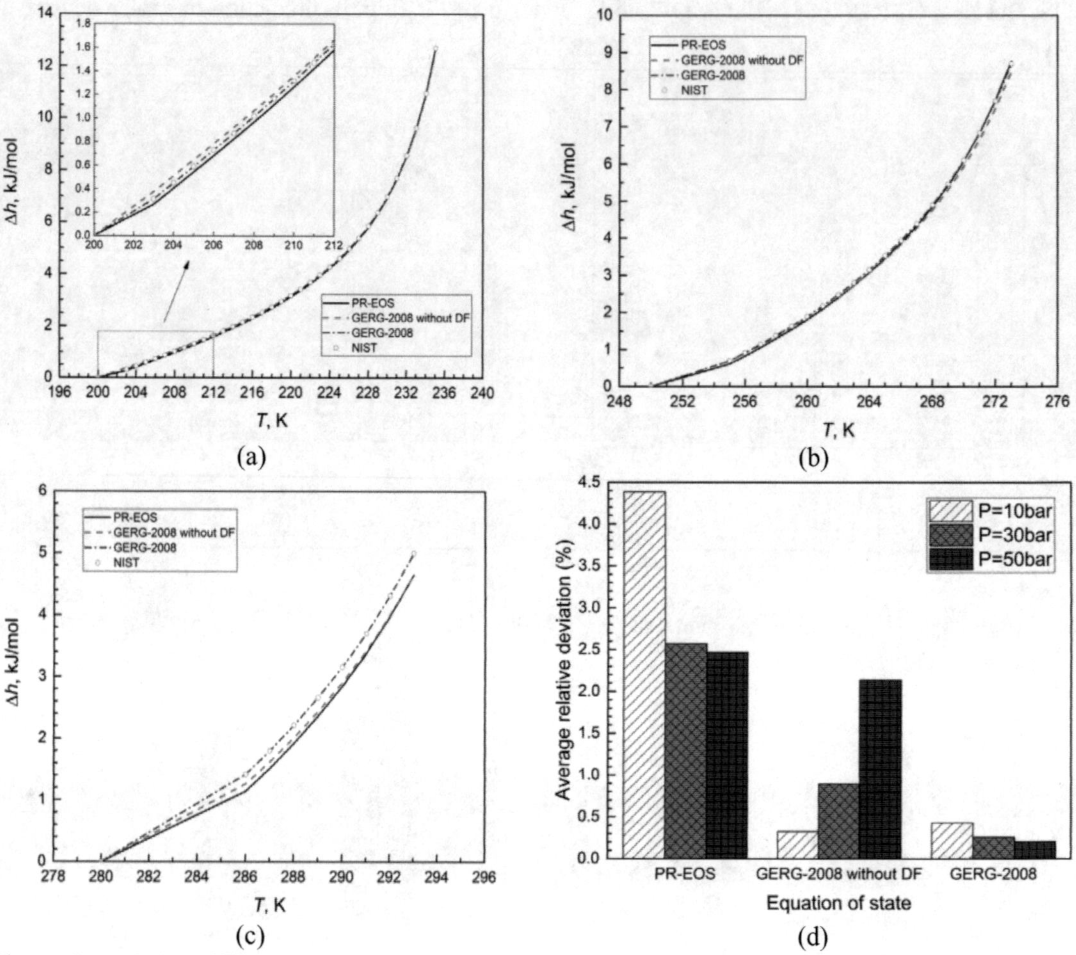

Figure 2. Enthalpy difference versus temperature and its average relative deviation from reference data for the methane-ethane mixture at methane mole fraction of 15% in VLE with the help of PR-EOS, GERG-2008 without DF and GERG-2008 at different pressures: (a) 10 bar, (b) 30 bar and (c) 50 bar.

4. Conclusions

In VLE calculations, GERG-2008 is superior to PR-EOS and GERG-2008 without DF. GERG-2008 without DF and PR-EOS both work unsatisfactorily in VLE calculations, but in general GERG-2008 without DF works better than PR-EOS. The deviation function in GERG-2008 (see Eq. (3)) can significantly affect the accuracy of GERG-2008 in VLE calculations.

References

[1] Michelsen M L 1982 *Fluid Phase Equilibria* **9** 21
[2] Baker L E, Pierce A C, Luks K D 1982 *Society of Petroleum Engineers Journal* **22** 731
[3] Michelsen M L 1982 *Fluid Phase Equilibria* **9** 1
[4] Orbach O, Crowe C M 1971 *Canadian Journal of Chemical Engineering* **49** 509
[5] Crowe C M, Nishio M 1975 *AIChE Journal* **21** 528
[6] Michelsen M L, Mollerup J M 2004 *Thermodynamic models: Fundamentals & Computational Aspects* (Holte: Tie-Line Publications)
[7] Kunz O, Wagner W 2012 *Journal of Chemical & Engineering Data* **57** 3032
[8] Peng D Y, Robinson D B 1976 *Industrial & Engineering Chemistry Fundamentals* **15** 59

[9] Nishiumi H, Arai T, Takeuchi K 1988 *Fluid Phase Equilibria* **42** 43
[10] Kunz O, Klimeck R, Wagner W, et al. 2007 *The GERG-2004 Wide-Range Equation of State for Natural Gases and Other Mixtures* (Düsseldorf: VDI Verlag GmbH)
[11] Luo T, Chirkov A Y 2020 *Int. J. of Eng. Systems Modelling and Simulation* **11** 138
[12] Mollerup J M, Michelsen M L 1992 *Fluid Phase Equilibria* **74** 1

Study of the Leidenfrost Effect on Heterogeneous Surfaces of Complex Structure

P G Makarov, A V Artamonov and A S Dmitriev

National Research University "Moscow Power Engineering Institute",
Russia, 111250 Moscow, Krasnokazarmennaya, 14

MakarovPG@mpei.ru

Abstract. This work is devoted to a research of water droplets that are put in-between two parallel metal strings the distance between which is comparable to linear size of the droplet. Strings are heated by Joule heating to temperatures that exceed critical temperatures of nucleate and film boiling. Different configurations of strings` side surface have been tested: smooth and with winding made of the same material (intermittent and uninterrupted). Experiments have shown that droplets on these types of surface do not boil away quickly or fall down. Instead they displayed behavior that can be described as floating, either stable or with directed motion, depending on surface structure or relief. Multiple experiments have shown that it is quite similar to Leidenfrost effect demonstrated on a flat overheated surface by liquids.

1. Introduction

The Leidenfrost effect is known since the middle of XVIII century. It consists in, when liquid is placed on a surface, heated to temperatures greatly exceeding boiling point, immediate evaporation of the lower part of droplet, being the closest region to surface. The resulting vapour supports the remainder of the droplet above it, preventing further direct contact between liquid and hot object. It was shown, that vapor, that supports droplet in levitating position, is generated most intensively in center of droplet [1-2].

Among other features Leidenfrost effect can also cause a direct or erratic movement of a droplet along overheated surface. For the last couple of decades, it grew into discrete branch if researches [3].

Previously it was assumed that, in most cases, gravity is the cause of the droplets' movement. In addition, the unsteady behavior of a drop can be caused by the Marangoni effect [4]. The essence of the effect is as follows: drops on the surface of liquid with a non-uniform distribution of the surface tension coefficient begin to shift to the region where this value has a maximum.

In [3] a self-propelled transportation of a droplet on a ratchet-structured brass heated surface was studied. Droplets of R-134a freon of millimeter diameter began to accelerate directionally with an acceleration of 1–2 m/s^2, after some time acquiring a constant speed of 5 cm/s. Later scientists from the Ecole Polytechnique and the Laboratory of Physics and Mechanics of Inhomogeneous Media in Paris investigated the phenomenon described above for the ethanol droplets in more detail [5].

A congregate classification of main reasons that can cause displacement was formulated:

1) the Laplace pressure gradient caused by the inhomogeneity of the radius of curvature of the drop;
2) processes of matter transfer from the back of the droplet to the front (waves), generating the motion of the droplet;

3) spontaneous droplet oscillations due to the inhomogeneous thickness of the vapor layer are capable to be transformed into kinetic energy of directed motion;
4) the Marangoni effect: the surface tension coefficient has a temperature dependence, and therefore a possible uneven temperature can provoke its inhomogeneous distribution, and hence droplet displacement.
5) while the surface on which the Leidenfrost drop is located is smooth, vapor flows out from under the drop uniformly and isotropically (equally in all directions). If the surface is corrugated, the vapor flow becomes anisotropic, acquiring a certain direction.

It should be noted that the characteristics of the droplet motion also depend on geometric parameters of the substrate relief. According to [5] if radius of drop does not exceed distance between teeth of the corrugated surface, the Leidenfrost effect is not able to move drop. At a larger size, as follows from the theory developed by the authors of the article, droplet's velocity is proportional to the square root of the ratio of distance between surface teeth and its height.

A self-centering movement and transportation of droplets with low friction using a herringbone structure heated to the Leidenfrost temperature was investigated by Linzi E. Dodd, Prashant Agrawal, Matthew T. Parnell in 2019 [6]. Within the framework of this work, scientists for the first time managed to obtain a directed self-centering motion of a drop with the Leidenfrost effect. They used a surface structure resembling surface used in Linke's work, but the difference was it was inhomogeneous forming a herringbone-shaped mirror structure.

The study of droplet dynamics is also of great practical importance: its results are used in lab on a chip technology, inkjet printing, spray cooling and even in such a mundane process as painting.

Transporting droplets in a controlled and energy efficient manner can also play an important role in some engineering applications such as low resistance fluid transport, water basins, and advanced microfluidic devices.

2. Methods and materials

Investigation of Leidenfrost effect was held on experimental stand (figure 1) that was setup particularly for that occasion.

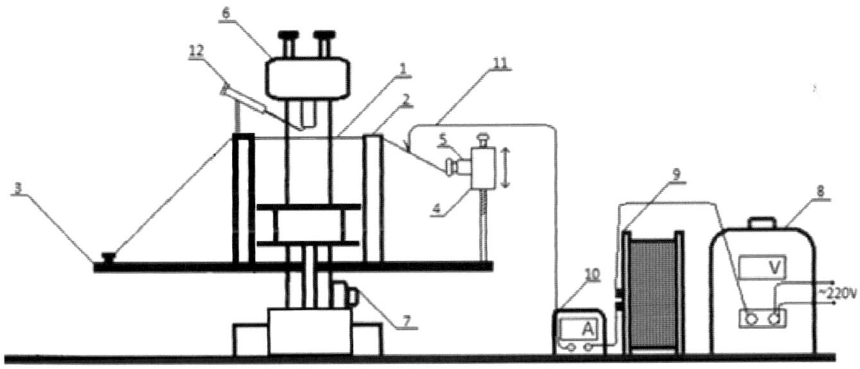

Figure 1. Experimental setup scheme.

2.1. Experimental stand

Two parallel strings (1) are stretched between two vertically aligned ceramic tiles (2) acting as supports. Strings lay in grooves that are chiselled on upper sides of tiles. The distance between centers of grooves is 2 mm. Strings are made of nichrome – an alloy, that is characterized by high operating temperatures (up to 1100 °C) and mechanical strength (tensile strength 0.65-0.70 GPa). We chose ceramic tiles as supports for stretching strings because of its great heat and electrical insulation properties and availability. Tiles are set vertically on a platform made of composite epoxy material (CEM) (3). There is a tensioning system (4) fixed on a platform aside from tiles. It provides ability to stretch strings with tension, because strings can sag when heated due to nichrome's high coefficient of linear thermal expansion (18×10-6K-1). The system consists of a bar that is mobile controllably in vertical dimension, on which two ceramic insulators (5) are nailed down. Strings are rigidly fixed on these insulators. Platform 3 is installed under optical microscope (6), thereby its vertical position could be regulated by microscope`s adjusting wheel (7), helping us to get clear view of droplets from above.

Strings are heated by potential difference. Its value is controlled by a set of transformers, variable autotransformer (8) in particular, connected to 220 V electrical network. Step-down transformer (9) is also in circuit. Amperage is controlled by ammeter (10) connected in series. Next step in the circuit – metal strings, connected by pin clips (11).

Distilled water, dosed to 4 μl droplets, is supplied onto strings via stated on a rack dispenser (12).

2.2. Studied surfaces

Metal strings had different configurations: smooth nichrome wire, diameter 0.4 mm (figure 2a); wire with intermittent winding (figure 2b); wire with uninterrupted winding (figure 2c):

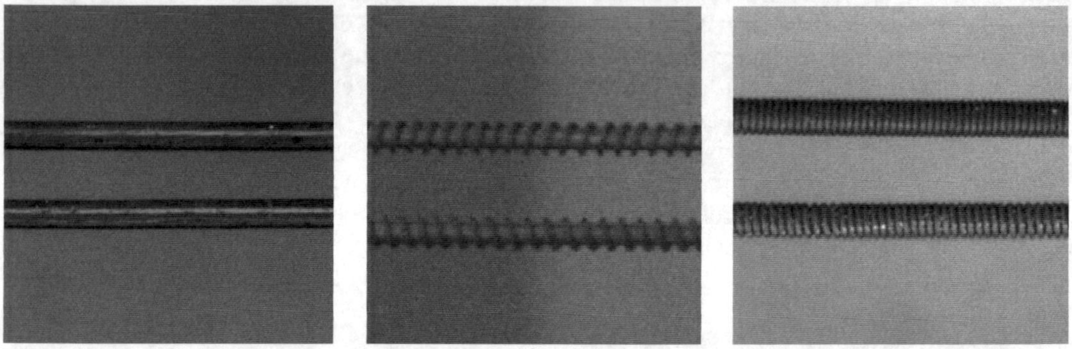

a b c

Figure 2. Nichrome strings: a) smooth; b) with intermittent winding; c) with uninterrupted winding.

In case of a winded strings second wire of 0.1 mm diameter is wound around the first one. Both wires are made of nichrome. Winding was done using electric motor: its shaft has been mounted coaxially with a main string attached to the shaft. Constant thread of 0.1 mm was ensured by clamp and fixed rotation speed of the shaft. Direction of the winding on two strings was oppositely directed, so one string with a winding was a mirror image of the other, so that the projection in the top view was a "herringbone". For uninterrupted winding a set of components is the same as in previous configuration. The difference is that in this case there is no thread – every other turn of winding lays back to back to previous one.

3. Experiments

Voltage was applied to strings according to the scheme shown in Figure 1. Due to Joule heating, because of the finite resistance of the conductor, strings began to heat up. Introduction of winding reduces heating rate of strings, i.e. to reach the same temperature, strings with a winding required higher voltage than strings without it. One of the reasons is the increase in the heat dissipating area. Thanks to this property, a finer temperature setting can be achieved (figure 3).

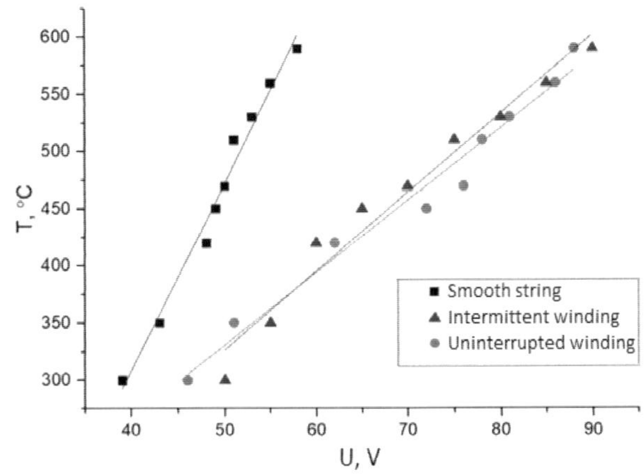

Figure 3. Temperature of strings depends on the voltage applied to them and their configuration.

3.1. Smooth strings

Putting droplets on different configurations of strings gave different results. In particular, on smooth nichrome wires the profile of droplet looked as shown on Figure 4a. Droplet was either still or wobbled around stable position. It stayed upon strings until its diameter became smaller than distance between strings (figure 4b).

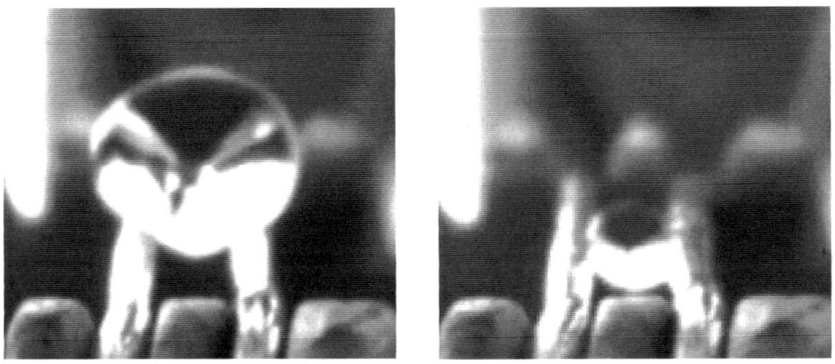

a b

Figure 4. Droplet on smooth strings at the initial (a) and final (b) moments of its lifetime.

During first series of experiments voltage was adjusted to set of values from 40 V to 60 V, which corresponded to temperatures from 300 °C to 600 °C (Figure 3). For every temperature a lifetime of a droplet on strings was measured. Based on results obtained, the following graph was constructed for the dependence of lifetime of droplets from surface temperature of strings (figure 5):

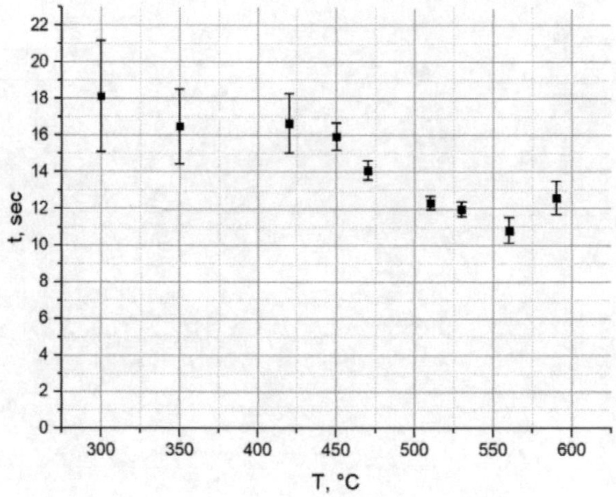

Figure 5. Temperature of smooth strings affect lifetime of water droplets.

According to this dependence, we can conclude that the higher surface temperature of strings was set, the shorter average lifetime of droplet became. It was also noticed that at lower temperatures periods of chaotic oscillations of droplets were observably shorter than the whole lifetime. But starting from 470 °C almost all the time that droplets spent on strings they wobbled. It is worth noticing that droplets explosively evaporate away just as they touch ceramic walls of supporting tiles.

It can be assumed that the lifetime decreases due to the fact that with increasing temperature of the superheated surface, the intensity of vaporization increases, which leads to a more rapid evaporation of the droplet.

3.2. Strings with constant-stepped winding

The second part of research consisted in experiments with nichrome strings of a 0.4 mm diameter with intermittent winding (0.1 mm spacing between coils). At temperatures of 300 °C and 350 °C, droplets evaporate immediately when in contact with strings. When the temperature of strings was increased up to 420 °C, the Leidenfrost effect occurs in nearly half of all of cases, drops kept stationary (figure 6).

a b c

Figure 6. Droplet staying on strings with constant-stepped winding: a) floating on them, front and b) side view; c) hanging under strings.

As droplets stayed on strings they kept evaporating. The volume of liquid, that was held above the strings, decreased and droplets` center of gravity shifted lower until the drop ceased to be held above the strings. However, unlike in the case of ordinary strings, droplets did not fall down, but they hung on strings from below, continuing to evaporate (Figure 6c).

Also, at this temperature, droplets were seen to explode or jump back onto dispenser needle once touching heated strings. In the remaining 50% of cases, the drop evaporated.

With temperature increasing up to 450 °C, in 10% of cases a directed motion of a drop along the surface of strings was observed. In the remaining 90%, droplets remained stationary above strings. With further increase of strings' temperature, the ratio of number of moving droplets to stationary increases. For example, at 590 °C, 100% of the droplets move. Possible explanation: due to ribbed structure of strings some amount of vapor flows away along winding near the droplet, thus decreasing effective vapor "cushion" that holds droplet above strings. Hence difference of graphs on Figure 6 and Figure 7. The higher temperature of superheated strings with constant-stepped winding is, the more vapor is being exhaled from under the droplet, which helps to compensate vapor escaping through winding. Because of the geometry of the surface droplets tend to move. Thus, the higher the surface temperature is, the less ratio of vapor escapes, the thicker the vapor layer becomes and the greater reactive forces, acting from the superheated surface perpendicular to the drop center, are (Figure 7):

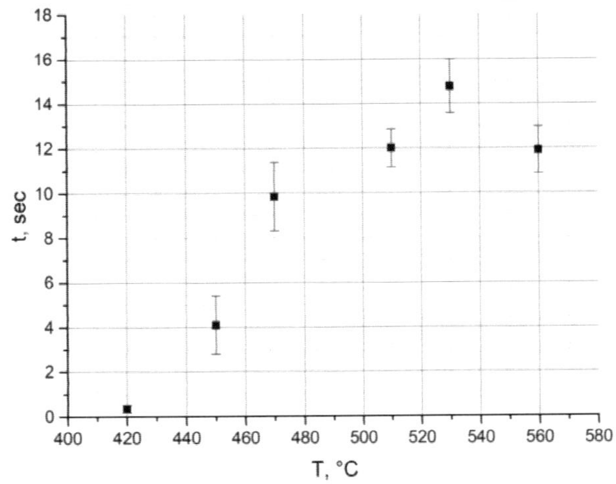

Figure 7. Temperature of strings with intermittent winding affect lifetime of water droplets.

3.3. Strings without spacing between coils

For strings with uninterrupted winding it was observed that for interval of temperatures from 300 °C to 350 °C, the Leidenfrost effect is not present – the drop evaporates instantly once touching strings. But with further increasing temperature up to 420 °C and higher, the Leidenfrost effect appears (figure 8).

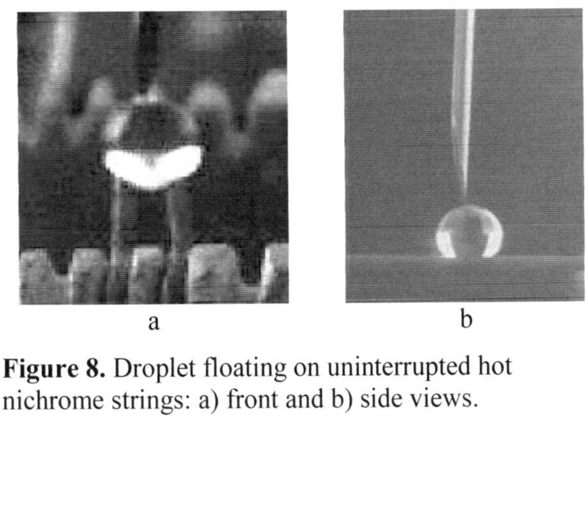

Figure 8. Droplet floating on uninterrupted hot nichrome strings: a) front and b) side views.

According to results of observations, the drop "moves" along the direction of the strings nearly in every two out of three cases of contact with strings at 420 °C. In the remaining one third of all cases, droplets dripped down from strings. With further increase of the temperature of strings number of "moving" drops increases. For example, at 560 °C, almost all of the droplets "move".

In contrast with two previous parts of the experiment, droplets on uninterrupted winding of strings did not rest steady. Therefore, in this part of the research, there was no possibility to measure lifetime of droplets.

4. Conclusion

Experiments have shown that droplets on two heated parallel metal strings, the distance between which is comparable to linear size of the droplet, do not boil away quickly or fall down. Instead they displayed behavior that can be described as floating, either stable or with directed motion, depending on surface structure or relief. Multiple experiments have shown that it is quite similar to Leidenfrost effect demonstrated on a flat overheated surface by liquids.

On smooth nichrome strings without winding droplets either float still or wobble around stable position and fall off only when droplet's diameter becomes smaller than distance between strings. The higher surface temperature of strings was, the shorter average lifetime of droplet became.

On strings with intermittent winding a process observed depended on temperature quite strongly. On strings` with temperatures of 300 °C and 350 °C, droplets evaporate immediately. On strings heated up to 420 °C nearly half of droplets falls off, another half floats stationary. At 450 °C 9 out 10 droplets slowly evaporate away steady, 1 out of 10 begin to move. This proportion changes with temperature`s increase. At 590 °C 10 out of 10 droplets began to move.

On strings with uninterrupted winding situation with temperature dependence was almost the same except that no stationary floating of droplets was observed. They either drip down or display constant movement.

References

[1] David Quéré 2013 Leidenfrost Dynamics *Annual Review of Fluid Mechanics* vol.45 pp 197–215

[2] Sobac B, Rednikov A, Dorbolo S, Colinet P 2014 Leidenfrost effect: Accurate drop shape modeling and refined scaling laws *Phys. Rev*. vol 90 issue 5 pp 053011-1–053011-12

[3] Linke H, Alemán B, et. al. 2006 Self-Propelled Leidenfrost Droplets *Physical review letters* vol 96 issue 15 pp 154502-1–154502-4

[4] Chen-li. Sun, Van P Carey 2004 Marangoni effects on the boiling of 2-propanol/water mixtures in a confined space *International Journal of Heat and Mass Transfer* vol 47 issue 25 pp 5417–5426

[5] Guillaume Lagubeau, Marie Le Merrer, Christophe Clanet, David Quéré 2011 Leidenfrost on a ratchet *Nature Physics* vol 7 pp 395–398

[6] Linzi Dodd, Prashantet Agrawal, et al. 2019 Low-Friction Self-Centering Droplet Propulsion and Transport Using a Leidenfrost Herringbone-Ratchet Structure *Physical Review Applied* vol 11 issue 3 pp 034063-1–034063-11

Modern approaches to the intensification of heat and mass exchange in multi-temperature condensation filters

M V Malevany and D A Konovalov

Voronezh State Technical University "VSTU",
Russia, 394006 Voronezh, 20-letiya Oktybrya, 84

dmikonovalov@yandex.ru

Abstract. The article considers the problems and features of heat and mass exchange on developed surfaces in the conditions of both single-phase and vapour-liquid flow during its condensation. We give a brief analytical review of studies of hydrodynamics and heat exchange in such systems. We analyzed the efficiency of the working channel of the condensation filter and identified problematic points. We offer possible methods for intensifying heat and mass transfer on working surfaces.

1. Introduction

Currently, in connection with the active development of the oil and gas industry, the issue of cleaning gases from mechanical and chemical impurities is quite acute. Capturing aerosols is necessary to recover valuable substances and prevent environmental pollution. Gas purification can be carried out quite successfully using a multi-temperature filter [1]. The impure gas is fed into the working channel, the walls of which have different temperatures (figure 1).

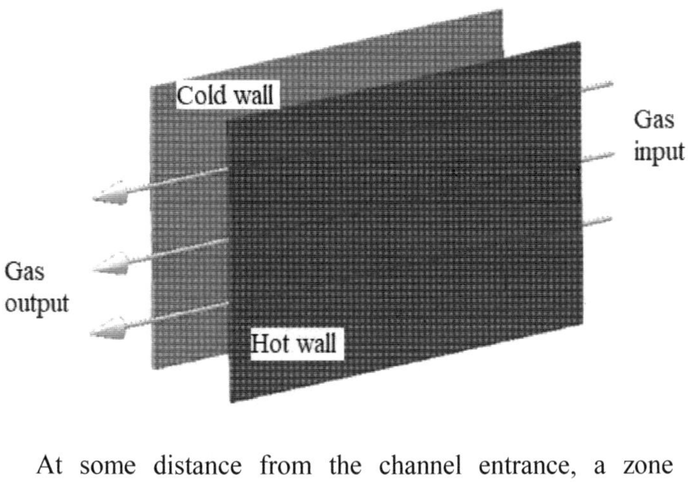

Figure 1. Schematic diagram of a working channel for a multi-temperature condensation filter.

At some distance from the channel entrance, a zone of stable supersaturation is formed, characterized by the formation of moisture drops with a further increase in their size and subsequent condensation on gas ions, mechanical inclusions, and various impurities. Moisture vapours independently trap impurities, and, increasing to the size of drops, condense on the cold wall.

Studies [2, 3] showed that the efficiency of heat and mass exchange processes in a different-temperature channel is influenced by the intensity of heat exchange between the channel wall and the gas flow, an increase in the flow rate, as well as the intensity of moisture evaporation on a hot surface, followed by condensation on a cold wall.

The capture coefficient of the liquid aerosol phase in this case substantially depends on the gas flow rate and has a pronounced nonlinear character (figure 2).

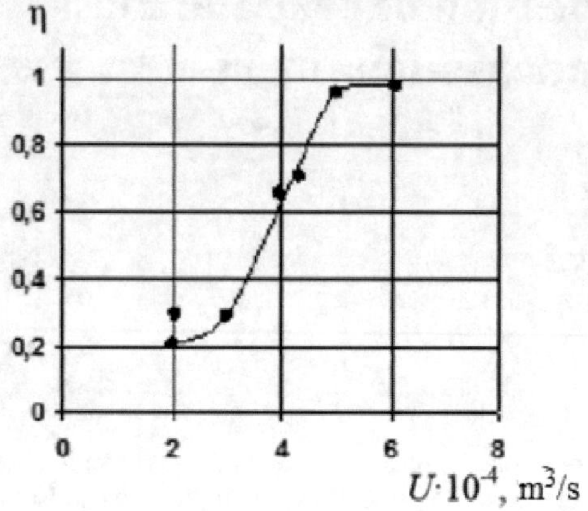

Figure 2. Efficiency of capturing the liquid phase depending on the gas flow rate.

Thus, the task is to increase the efficiency of work, which consists in intensifying the processes of heat and mass exchange, including phase transitions in a small range of gas flow rates. This task can be successfully solved due to the formation of developed surfaces of heat and mass exchange in the working area.

2. Heat exchange on large-scale developed surfaces

One of the ways to intensify heat exchange is to create developed surfaces, which have confirmed their high efficiency [4, 5]. The simplest technical solutions include the use of plates of various shapes [6] (figure 3), sandwich panels filled with metal gratings [7] (figure 4), surfaces with dimples of various configurations [8] (figure 5).

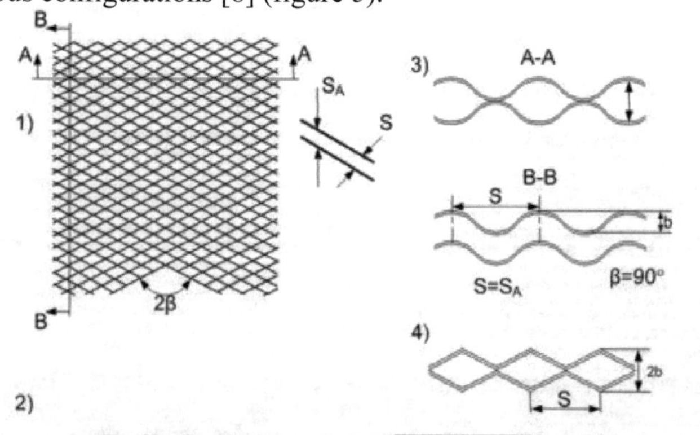

Figure 3. Scheme of Plate Heat Exchange channel corrugated field with various forms of plates corrugations: 1, 2 – two neighbour plates intersection; 3 – channels cross-sections with sinusoidal corrugations; 4 – channels cross-sections with triangular corrugations.

Surfaces with various types of ribbing are widely used, for example [9]. In [10], the authors presented a comparative analysis of surfaces with lamellar-ribbed, mesh-wire and needlelike ribbing under conditions of forced convection. Thus, the aerodynamic efficiency of a surface with transverse ribbing is 1.15-1.65 and 1.8-2.0 times higher than that of a standard plate and a needlelike one, respectively, and also surpasses the mesh structure by 3.0-3.2 times. It is shown that partial cutting of smooth plates perpendicular to the base of the surface leads to an increase in thermal efficiency. In this case, greater efficiency has been achieved for wire mesh ribs, and the use of needlelike ribs leads to a 40% increase in efficiency compared to wire mesh surfaces. Similar experiments were carried out in [11, 12], where they obtained similar qualitative results and established the effect of the anisotropy of mesh materials on hydrodynamics and heat transfer.

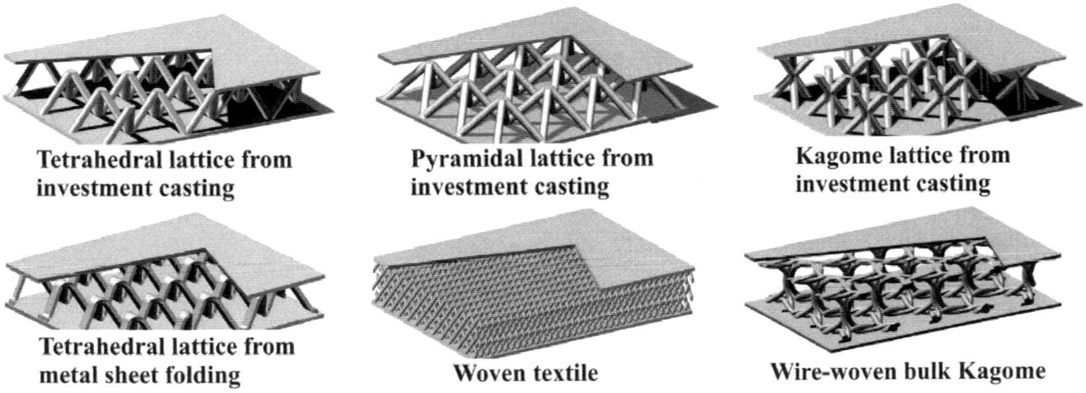

Figure 4. The sandwich panels cored with the metallic lattices fabricated with different methods to bear mechanical and thermal load concurrently.

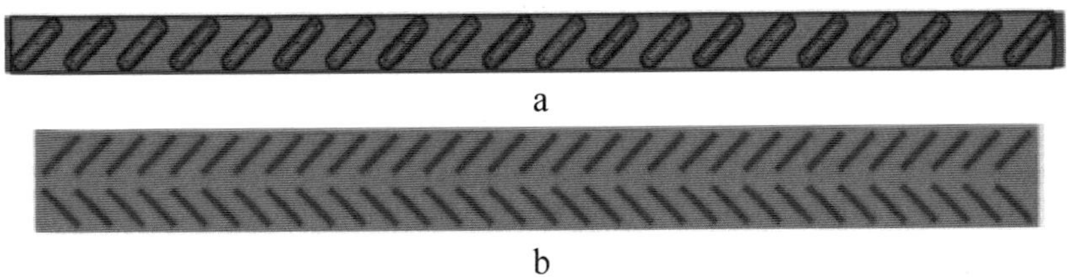

Figure 5. Packs of sparse single (a) and zigzag (b) corridor packages of oval-trench holes.

The methods for intensifying heat exchange proposed above are typical, as a rule, for macroscales. In this case, we observe a pronounced separation of the flow into the main core and the near-wall region, where heat transfer will directly occur. In the context of increasing the compactness and efficiency of the devices being developed, this may not be sufficient. To further increase efficiency, it is necessary to use microchannel and porous media.

3. Heat exchange in microchannels and porous media

The use of microchannels and porous media makes it possible to significantly intensify heat exchange, which is a prerequisite for the creation of compact heat exchange devices. Mesh materials and classical porous structures with interchannel transpiration of the heat transfer agent have become widespread [13, 14]. In modern realities, the most promising is the use of highly porous [15] and/or anisotropic media [16, 17]. The use of such technologies makes it possible to make the heat exchange process in different-temperature filters as efficient as possible. The efficiency of heat exchange [4] is characterized by the ratio of hydrodynamic efforts for pumping the heat transfer agent when the heat transfer coefficient exceeds the one at the common section.

Anisotropic media deserve special discussion since they allow a high heat transfer coefficient in combination with relatively low efforts for transporting the heat transfer agent. Figure 6 offers various types of anisotropic media [18, 19]. It is shown that the use of media with a similar geometry makes it possible to significantly intensify heat exchange.

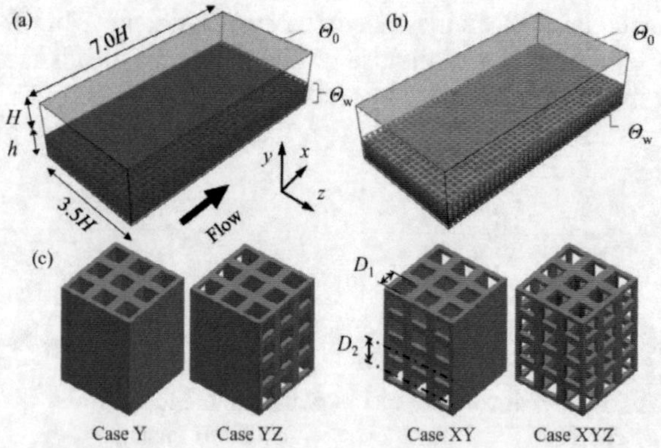

Figure 6. Computational domain and geometry of the channel: (a) isothermal and (b) conjugate wall heat transfer conditions and (c) porous media: cases Y, YZ, XY, and XYZ.

4. Influence of the structure and properties of developed surfaces on heat and mass exchange

Cleaning in multi-temperature filters is also associated with the mass exchange and the surface state of the phase transition. As in the case of heat exchange, the mass transfer takes place most intensively on developed surfaces. Thus, in [20], the authors investigated the effect of surface roughness and inhomogeneity on vapour condensation. Similar studies were also carried out in [21]. They showed the effect of the surface structure, including the wetting angle, on the heat transfer coefficient in smooth microchannels. Similarly, in [22], they studied the behaviour of drops and heat transfer during condensation on a microrelief surface made of copper by mechanical processing. Comparison of indicators for a smooth and hydrophobic surface showed that microchannels contribute to an increase in the removed heat flow by 55-102% but, at the same time, the growth of droplets is significantly limited.

In [23], the authors carried out a study of condensation on microstructured surfaces (figure 7). They assessed the effect of surface wettability on heat exchange. In this case, a hybrid surface was considered: the wettability of the base and pillars changed from superhydrophobic to hydrophilic. They showed that the heat exchange in the hybrid case of wettability increased by 28% in comparison with the superhydrophobic case.

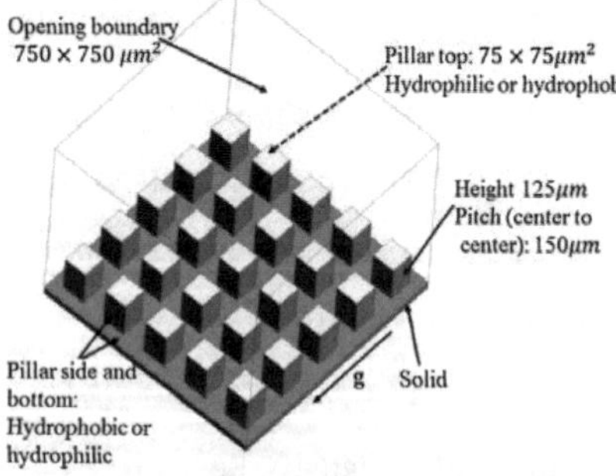

Figure 7. Computational domain.

Similar studies were carried out in [24, 25] for microstructured surfaces made of aluminium within the framework of a free convective flow of a coolant. The authors found that a structured hydrophilic surface demonstrates an increase in convection by 10-12%, and the maximum value of heat transfer corresponds to a superhydrophilic wall. Similar experiments for condensation were carried out in [26]. Also, in [27 - 29], combined use of surfaces with variable wettability is proposed to improve the thermohydraulic characteristics and improve the removal of condensate during drop condensation. These developments have found their application in the creation of heat pipe condensers.

5. Conclusions of the study

The analysis of the conducted studies showed that the intensification of heat transfer is achieved through the use of developed surfaces of various configurations, and the highest efficiency is achieved for highly porous microchannel heat exchange elements. Their use in combination with hybrid properties of wettability makes it possible to form a developed system of centers of drop formation and condensation, as well as to intensify the process as a whole. Since hybrid surfaces is a complex material science problem, further work is to produce a developed heat and mass exchange surface with desired properties. As a possible option, the authors considered a highly porous surface or a microchannel one based on silicon single crystals. [13, 14].

References

[1] Ryazhskikh V I, Voronov G G, Solzhenikin P A, Stogney V G 2014 Improvement of the engineering methodology for calculating the multi-temperature condensation filter *Bulletin of VSTU* vol 10 no 1 pp 107-109

[2] Ting Wang, En Lin, Yun-Lei Peng, Yao Chen, Peng Cheng, Zhenjie Zhang 2020 Rational design and synthesis of ultramicroporous metal-organic frameworks for gas separation *Coordination Chemistry Reviews* vol 423 15 Article 213485

[3] Ryazhskih V I, Keller A V, Ryazhskih A V, Nikolenko A V, Dakhin S V 2020 Mathematical model of the acceleration laminar flow of a Newtonian fluid in an anisotropic porous channel of rectangular cross section *Bulletin of the South Ural State University, Series: Mathematical Modelling Programming and Computer Software* vol 13 iss 3 pp. 17-28

[4] Dzyubenko B V, Kuzma-Kichta Yu A, Leont'ev A I, Fedik I I, Kholpanov L P 2008 Intensification of heat and mass transfer at macro-, micro- and nanoscale / B V Dzyubenko Monograph Moscow: FSUE "TSNIIATOMINFORM" 532 p

[5] Nield D A, Bejan A 2017 Convection in porous media – New York: *Springer* 998 p

[6] Kapustenko P O, Klemes J J, Arsenyeva O P, Kusakov S K, Tovazhnyanskyy L L 2020 The influence of plate corrugations geometry scale factor on performance of plate heat exchanger as condenser of vapour from its mixture with noncondensing gas *Energy* vol 201 15 June 117661

[7] Xin Jin, Beibei Shen, Hongbin Yan, Bengt Sunden, Gongnan Xie 2018 Comparative evaluations of thermofluidic characteristics of sandwichpanels with X-lattice and Pyramidal-lattice cores*International Journal of Heat and Mass Transfer* Volume 127 Part B December Pages 268-28

[8] Isaev S A, Leontiev A I, Milman O O, Nikushchenko D V and Popov I A 2020 Energy Efficient Surfaces with Multi-Row Inclined Oval-Trench Dimples for Air Condensers *News of RAS Power Engineering* no 4 pp 3–10

[9] Popov I A 2007 Hydrodynamics and heat transfer in porous heat-exchange elements and apparatuses Kazan: Center for Information Technologies 240 p

[10] Rohachev V A, Terekh O M, Baranyuk A V, Nikolaenko Yu E, Zhukova Yu V and Rudenko A I 2020 Heataerodynamic efficiency of small size heat transfer surfaces for cooling thermally loaded electronic components *Thermal Science and Engineering Progress* vol 20 1 December 100726

[11] Pelevin F V 2016 Hydraulic resistance of porous metals *Izvestiya VUZov Mechanical engineering series* no 2 pp 42 - 52

[12] Pelevin F V 2018 Heat transfer in metal mesh materials during interchannel transpiration and two-dimensional intermesh movement of the coolant *Thermophysics of high temperatures* vol 56 no 2 pp 219-228

[13] Konovalov D A, Ryazhskikh V I, Lazarenko I N and Kozhukhov N N 2019 Model of Cooling of Compact Surfaces by Microchannel Recuperative Heat Exchangers with a Matrix of Filamentary Silicon Single Crystals *Journal of Engineering Physics and Thermophysics* March vol 92 2 pp 355–364

[14] Konovalov D A, Ryaszhkih V I, Drozdov I G, Shmatov D P and Kozhukhov N N 2020 Experimental study on heat exchange in a compact porous aluminium foam heat exchanger *Journal of Physics: Conference Series* vol 1565 iss 1 28 July 01207

[15] Soloveva O V, Soloveva S A and Popkova O S 2018 Modeling of the Three-Dimensional Structure of Open Cell Foam and Analysis of the Model Quality Using the Example of Pressure Drop Calculation *Uchenye Zapiski Kazanskogo Universiteta Seriya Fiziko-Matematicheskie Nauki (Proceedings of Kazan University Physics and Mathematics Series)* vol 160 no 4 pp 681–694

[16] Ying J, Lu L, Tian L, Yan X and Chen B 2018 Anisotropic Porous Structure Modeling for 3D Printed Objects *Computers&Graphics* vol 70 no 2 pp 157–164

[17] Ryazhskikh V I, Konovalov D A, Kozhukhov N N, Ryazhskikh A V, Nikolenko A V and Portnov V V 2020 Pressure filtration of the Newtonian fluid in the Darsi-Brinkman approximation through the horizontal porous rectangular channel with orthotropic anisotropy *Journal of Physics: Conference Series* The Third Conference "Problems of Thermal Physics and Power Engineering" 1683 022005

[18] Yudai Nishiyama, Yusuke Kuwata, Kazuhiko Suga 2020 Direct numerical simulation of turbulent heat transfer over fully resolved anisotropic porous *structures International Journal of Heat and Fluid Flow* vol 81 108515

[19] Yang Hu, Decai Li, Shi Shu, Xiaodong Niu 2017 A multiple-relaxation-time lattice Boltzmann model for the flow and heat transfer in a hydrodynamically and thermally anisotropic porous medium *International Journal of Heat and Mass Transfer* vol 104 pp 544–558

[20] Sahand Etemad, Arash Behrang, Peyman Mohammadmoradi, S Hossein Hejazi, Apostolos Kantzas Effects of surface roughness and mineral heterogeneity on pore-scale steam Condensation *Journal of Petroleum Science and Engineering* S0920-4105(17)30752-0

[21] Yong Ding, Li Jia, Yongxin Zhang, Zhoujian An 2019 Investigation on R141b convective condensation in microchannel with lowsurface energy coating and hierarchical nanostructures surface *Applied Thermal Engineering* vol 155 5 pp 480–488

[22] Qi Peng, Li Jia, Chao Dang, Zhoujian An, Yongxin Zhang, Liaofei Yin 2019 Analysis of droplet dynamic behavior and condensation heat transfer characteristics on rectangular microgrooved surface with CuO nanostructures *International Journal of Heat and Mass Transfer* vol 130 pp 1096–1107

[23] Zhaoqing Ke, Junxiang Shi, Bo Zhang, Chung-Lung Chen 2017 Numerical investigation of condensation on microstructured surface with wettability patterns *International Journal of Heat and Mass Transfer* vol 115 part B pp 1161–1172

[24] MisyuraS Y, Kuznetsov G V, Feoktistov D V, Volkov R S, Morozov V S, Orlova E G 2019 The influence of the surface microtexture on wettability properties and drop evaporation *Surface and Coatings Technology* vol 375 15 pp 458–467

[25] Misyura S Y 2020 Dependence of wettability of microtextured wall on the heat and mass transfer: simple estimates for convection and heat transfer *International Journal of Mechanical Sciences* Volume 170 15 105353

[26] Yeonghwan Kim, Jin Sub Kim, Dong Hwan Shin, Jin Hyeuk Seo, Seung M You, Jungho Lee Effects of hydrophobic and superhydrophobic coatings of a condenser on the thermal performance of a two-phase closed thermosyphon *International Journal of Heat and Mass Transfer* vol 144 118706

[27] Rongfu Wen, Xingdong Zhou, Benli Peng, Zhong Lan, Ronggui Yang, Xuehu Ma 2019 Falling-droplet-enhanced filmwise condensation in the presence of non-condensable gas *International Journal of Heat and Mass Transfer* vol 140 pp 173–186

[28] Atsushi Tokunaga, Takaharu Tsuruta 2020 Enhancement of condensation heat transfer on a microstructured surface with wettability gradient *International Journal of Heat and Mass Transfer* vol 156 119839

[29] Koukoravas T P, George Damoulakis G, Constantine M 2020 Megaridis experimental investigation of a vapor chamber featuring wettability-patterned surfaces *Applied Thermal Engineering* vol 178 115522

CCGT unit with a carbonation system for CO_2 capture

G E Maslennikov, V V Nazarova, I P Lazebny and A F Ryzhkov

Ural Federal University, Russia, Yekaterinburg, Mira Street, 19

g26m12@gmail.com

Abstract. The paper considers the issues of creating energy-efficient and environmentally friendly resource-saving energy generation technologies based on fossil fuels, solid waste and materials. A natural gas CCGT with a MSW ash carbonation unit was chosen as the research object. The prototype of the CCGT unit is the CCGT "Akademicheskaya" with an installed capacity of 230 MW. For the carbonation of MSW ash, a direct semi-dry route is used. In the calculation of the CCGT, the power and efficiency of the cycle are determined, as well as the parameters of the flue gas, which are transferred to the calculation of carbonation. As a result of calculations, it was found that this specific CO_2 emissions from the power plant using carbonation can be reduced by almost 20% from 403 to 329 g kWh-1, however, a large amount of required MSW (2.6-6.7 kg per kg of flue gases) causes certain difficulties for the implementation of the project.

Introduction

The problems of tightening the standards of emissions of harmful substances and the assumed obligations to reduce greenhouse gas emissions require the development of environmentally and economically viable technologies for capturing CO_2 with subsequent utilization [1, 2]. According to global environmental programs, the scenario of reducing CO_2 emissions will inevitably be realized for all industries, but only the energy industry needs to switch to emission-free low-carbon technologies in 2050-2060 with a reduction in CO_2 emissions by 80-95% in relation to 1990, and emissions of others harmful substances by 100% [3].

One of the options for solving this problem is the CCS method (Carbon Capture and Storage) [4], including the capturing CO_2 from industrial and energy source flue gases, its transportation to the storage site and long-term isolation from the atmosphere. A more promising method is CO_2 mineralization by carbonation of materials with an increased content of basic oxides: rocks, ash and slag waste (ASW), construction industry wastes, etc. For the first time, the idea of carbonation of minerals for the sequestration of CO_2 in a stable solid phase was expressed in 1990 in the work [5].

Currently, the one-stage process of ex-situ carbonation of solid materials with CO_2 capture, called direct accelerated mineral carbonation – AMC, is recognized as the most effective [6, 7]. The main problem of direct accelerated carbonation is the need to intensify reactions and rational use of heat release (thermal effects) [8]. The direct method means that CO_2 is fed into dry or humidified chemically untreated alkaline waste such as fly ash or aqueous fly ash slurry. In this process, fly ash cations such as Ca^{2+} and Mg^{2+} are leached into solution and then precipitated as carbonates on the surface of the solids. Direct carbonation products are suitable, for example, for use as cement additives.

To start solving this problem, it is necessary to consider the technological options for the implementation of the chemical problem. In the review [4] the process of accelerated direct carbonation is considered in three ways by liquid to solid ratio: (1) L/S=0 – dry process; (2) L/S=0.03-1 – semi-dry process; (3) (1) L/S>0 – aqueous process.

Studies of the mineral carbonation of CO_2 using fly ash have focused primarily on the direct aqueous route, which involves the direct reaction of fly ash with CO_2 in a single reactor with water as the reaction medium. Direct carbonation can also be carried out in a semi-dry manner. The low water content makes the product easier to dry. This semi-dry processing route also allows control of the grain size and chemical composition of the carbonated material produced, which will help find applications for these materials, for example, in the construction industry. The third option for direct carbonation is the dry route. Dananjayan compared dry and aqueous carbonation and showed that dry is a slower process than wet carbonation with water. [9].

This paper considers the issues of creating energy-efficient and environmentally friendly resource-saving energy generation technologies based on fossil fuels, solid waste, and materials. A natural gas CCGT with a MSW ash carbonation unit was chosen as the research object. The prototype of the CCGT unit is the CCGT "Akademicheskaya" with an installed capacity of 230 MW. The direct semi-dry route described in [7] is used for the carbonation of MSW ash.

When calculating the CCGT, the power and efficiency of the cycle are determined, as well as the parameters of the flue gas, which are transferred to the calculation of carbonation. The purpose of calculating the carbonation plant is to assess the size of MSW ash consumption and the impact on the efficiency of the entire power plant.

Calculation methods

Modern technologies provide a wide range of software packages that can significantly reduce the time and improve the accuracy of mathematical calculations. Currently, there are many widely used commercial software such as "AspenPlus", which is used to simulate chemical and thermodynamic processes. Among open source software, the alternatives are "COCO Simulator" and "DWSIM". The programs allow performing stationary thermodynamic calculations of technological schemes, determining phase equilibria of multicomponent media, and simulating chemical processes. Based on our own experience in these programs, we may say that "DWSIM" has a better solver that can handle complex circuits with many circular threads. The advantage of the "COCO Simulator" is a wide functionality for setting new substances that are absent in the standard library. However, these products are still inferior to "AspenPlus" in terms of substance bases, thermodynamic models, and functional elements.

To calculate the thermal scheme of the CCGT unit, the "DWSIM" package was chosen, and for the calculation of the carbonation unit – "COCO Simulator". The property package based on the IAPWS-IF97 formulation is used to determine the thermodynamic properties of pure water flows. Gas streams (air, fuel, combustion products) are considered as real gases, and their enthalpy, entropy and heat capacity are calculated using Peng-Robinson equations of state for real gas [10].

CCGT calculation

The thermal schematic of the CCGT-230 "Akademicheskaya" power unit was taken as a prototype of the design scheme. The CCGT unit includes: one GT13E2 gas turbine (GT) with a rated electrical power (according to ISO conditions) of 166.6 MW and an efficiency of 36.5%, one P-86 heat recovery steam generator (HRSG) and one T-63-7.0 steam turbine (ST). The first three stages of GT have air cooling, which must be considered for a more correct calculation of the exhaust gas temperature. Detailed data on this system are closed by the manufacturer, so the approximate values were taken from [11] and corrected further, so 7% of the inlet air consumption goes to cooling the turbine, of which 72.4% falls on the 1st stage cooling (supplied to the combustion chamber element), 19, 7% and 7.9% – at the 2nd and 3rd steps, respectively. Double-circuit HRSG has heating installation for domestic hot water needs. The deaerator is fed with steam taken from the LP drum.

The calculation is carried out with the ambient parameters: temperature 15 °C, pressure 1.013 bar, humidity 60%, and the nominal operating mode of the CCGT equipment. The steam turbine is assumed to operate in condensing mode. Other main assumptions and schematic for calculating the CCGT are presented in appendix.

To verify the calculation, the data on the operation of the CCGT unit at the "Akademicheskaya" are used. The verification data and the data obtained in the calculation are presented in table 1.

Table 1. Main results of the CCGT.

Parameters		Measurement units	Calculated values	Verification values	Relative deviations, %
GT outlet temperature		°C	505.8	507.8	0.39
Combustion products composition		%			
	N_2		75,0	75,0	0.00
	O_2		14.15	14.06	0.64
	H_2O		6.93	6.98	0.72
	CO_2		3.02	3.07	1.63
	Ar		0.89	0.89	0.00
HP steam capacity		kg s^{-1}	58.22	59.64	2.38
LP steam capacity		kg s^{-1}	13.4	13.06	2.60
HRSG flue gas temperature		°C	91.3	91.9	0.65
GT electric power (gross)		MW	163.87	166.6	1.64
ST electric power (gross)		MW	67.51	65	3.86
Fuel compressor electric power		MW	-2.39	–	–
Pump electric power		MW	-0.97	–	–
CCGT electric power (gross)		MW	231.38	–	–
CCGT electric power (net)		MW	228.01	–	–
Thermal input		MW	455.87	–	–
GT efficiency (gross)		%	35.95	–	–
HRSG efficiency		%	72.63	–	–
ST efficiency (gross)		%	27.2	–	–
CCGT efficiency (gross)		%	50.76	–	–
CCGT efficiency (net)		%	50.02	–	–
Specific CO_2 emissions		g kWh^{-1}	403.7	–	–

The relative deviations of the calculated values from the verified ones are less than 4%. Thus, the computational model of the CCGT-230 thermal scheme can be considered complete and satisfactory. At the next stage, the flue gas parameters is used for calculating the carbonation of MSW ash.

Calculation of carbonation

1. Verification

The AMC (Accelerated Mineral Carbonation) process proposed in [7] was chosen as the design scheme and includes dryer, blower fan, humidifier, heater, and fluidized bed reactor (FBR) (figure 1). The flue gas is removed from the funnel and fed to the dryer. Condensed moisture is trapped to protect the blower. The humidifier and heater allow to control the humidity and temperature of the flue gases.

The upper inlet of the reactor is connected to the ash hopper to supply the required amount of fresh material to the reactor. The fly ash particles are fluidized by the flue gas stream, thereby ensuring proper mixing and good contact between the particles and the gas. Solid lines 1-8, 18, 31 represent material flows. Dotted lines 21, 22, 33, 34 are information streams providing the adjustment of ash and water supply.

According to the conditions [7], the flue gas consists of CO_2 – 13%, H_2O – 10%, N_2 – 66%, O_2 – 11%, its consumption is 10.38 kg min^{-1}. The composition of fresh ash is presented in table 2. The pressure in the reactor is 115.1 kPa. The isentropic efficiency of the compressor is 75.0%. Before entering the reactor, the gas is humidified to a water content of 16%. The gas temperature in front of the reactor is 60 °C.

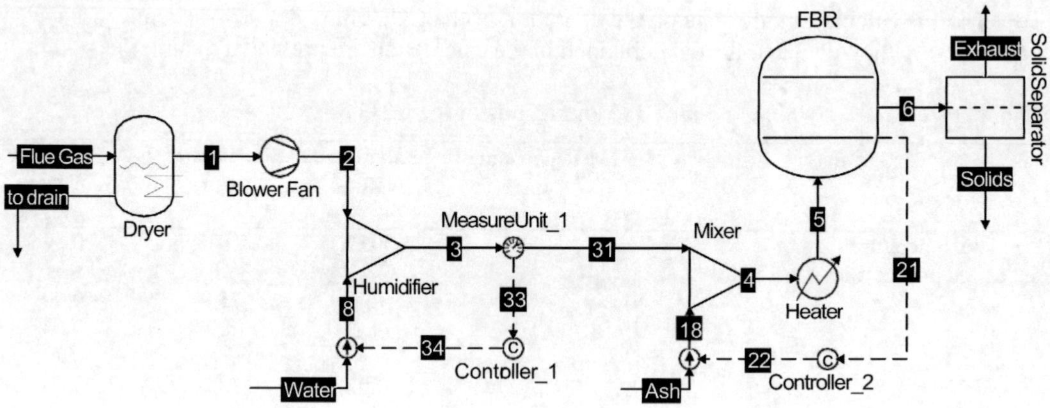

Figure 1. The AMC process diagram for the direct capture of CO_2 from flue gases.

Table 2. Elemental composition (in terms of oxides) of fresh ash for verification (% wt.) [7].

SiO_2	Al_2O_3	Fe_2O_3	CaO	MgO	Na_2O	K_2O	$CaCO_3$	Water
58.61	19.06	5.37	7.5	3.85	0.7	0.94	<0.1	0.042

The calculation considers the following four reactions of interaction of CO_2 with metal oxides:

- $CaO+CO_2=CaCO_3$;
- $MgO+CO_2=MgCO_3$;
- $K_2O+CO_2=K_2CO_3$;
- $Na_2O+CO_2=Na_2CO_3$.

The author does not give the value of the carbonation efficiency; therefore, the completeness of the reactions was taken as 30% as for a semi-dry process in [9].

Verification is carried out by the concentration of CO_2 in the gases after the reactor and by the content of $CaCO_3$ in the carbonized ash. The results of the verification of the calculation according to the data [7] are presented in table 3, it speaks about the adequacy of the model.

Table 3. AMC model verification.

Flue gases (% mol.)				**Ash (% wt)**			
Component	Before carbonation [7]	After carbonization		**Component**	Before carbonation [7]	After carbonization	
		[7]	Calculation			[7]	Calculation
CO_2	13	9.6	9.6	**$CaCO_3$**	0.10	3.5-4	3.98

The calculation results show that the degree of CO_2 conversion is 22.6% and 0.033 kg of CO_2 per kg of ash is captured.

2. *Calculation of MSW ash carbonation*

Own calculation of the MSW ash carbonation is made for two ash compositions, the composition of which is taken from [12]. In The work author investigated the solid residues that are formed during the thermal processing of MSW using the two most common combustion technologies – combustion grate (CG) and a swirl fluidized bed (SFB) furnace. The composition of these ashes is given in table 4.

The carbonation parameters obtained at the verification stage and included in the main calculation are as follows: the degree of conversion of carbonated oxides is 30%, the degree of CO_2 conversion is 22.6%. The flue gas of the CCGT unit consists of CO_2 – 3.02%, H_2O – 6.93%, N_2 – 75.0%, O_2 – 14.14%, its consumption is 547.6 kg s^{-1}. The calculation results for two evils are presented in table 5.

Table 4. Elemental composition (in terms of oxides) of MSW ashes from two incineration plants,% (by weight) [12].

Oxide	CG ash				SFB ash			
	min	max	average	taken in calculation	min	max	average	taken in calculation
SiO_2	29	43.2	36.1	45.1	54.9	77	65.95	68.15
TiO_2	0.9	2.6	1.75	0	0.6	0.8	0.7	0
Al_2O_3	7.2	13.6	10.4	10.4	4.5	11.1	7.8	7.8
Fe_2O_3	1.5	5.9	3.7	3.7	0.7	2.4	1.55	1.55
CaO	13.8	37	25.4	25.4	10.8	19	14.9	14.9
MgO	1.5	2.5	2	2	0.5	1.3	0.9	0.9
K_2O	0.5	4.5	2.5	2.5	0.9	1.4	1.15	1.15
Na_2O	2.2	6.5	4.35	4.35	1.3	2.1	1.7	1.7
SO_3	4.5	8.6	6.55	6.55	1.2	6.5	3.85	3.85
P_2O_5	2.1	6.5	4.3	0	0.5	2.1	1.3	0
Cl	0.8	2.7	1.75	0	0.3	1.4	0.85	0

Table 5. Results of calculating the MSW ash carbonation.

Parameter	**Measurement units**	**CG ash**	**SFB ash**
Ash consumption	kg s^{-1}	73.0	133.6
Specific ash consumption	kg kg^{-1} of flue gas	0.133	0.207
Specific yield of fly ash [12]	kg kg^{-1} of MSW	0.02	0.08
The amount of MSW consumed	kg kg^{-1} of flue gas	6.65	2.59
CO_2 concentration in exhaust gas	% mol.	2.11	
Concentration of carbonates in finished ash	% wt.		
$CaCO_3$		9.21	7.94
$MgCO_3$		1.01	0.56
K_2CO_3		0.54	0.50
Na_2CO_3		1.43	0.87
Flue gas compression power required	MW	10.04	

As a result of the calculation, we find that for capturing 22.6% of CO_2 from flue gases of the natural gas CCGT 0.1-0.2 kg of MSW ash per kg of flue gases will be required. Energy consumption for the injection of flue gases into the carbonization unit reduces the net efficiency of the power unit from 50.02% to 47.81%. At the same time, the specific CO_2 emissions of the entire power plant will decrease by almost 20% from 403 to 329 g kWh^{-1}. However, in this work, the system of humidification and heating of flue gas in the installation has not yet been thought out, therefore, the heat costs of the carbonization unit are not estimated.

Conclusions

Simulation of the scheme of natural gas CCGT unit with the installation of MSW ash carbonation has been carried out. The prototype of the CCGT unit is the CCGT "Akademicheskaya" with an installed capacity of 230 MW. The direct semi-dry route described in [7] is used for the carbonation.

The calculation of the CCGT was carried out in the open software for modeling chemical processes "DWSIM". The relative deviations of the calculated values from the declared ones are less than 4%. Calculation of the carbonization process of MSW ash of two compositions (CaO content is 25.4–14.9% wt.) was carried out in the free software package "COCO Simulator". The carbonization parameters obtained at the verification stage and included in the main calculation are as follows: the degree of conversion of carbonated oxides is 30%, the degree of conversion of CO_2 is 22.6%.

As a result of the calculation, we find that to capture a given amount of CO_2 from the flue gases of a CCGT fired with natural gas 0.1-0.2 kg of MSW ash per kg of flue gases will be required. Energy consumption for the injection of flue gases into the carbonization unit reduces the net efficiency of the power unit by 2.2% absolute. At the same time, the specific CO_2 emissions of the entire power plant will decrease by almost 20% from 403 to 329 g kWh^{-1}. However, in this work, the system of humidification and heating of flue gas in the installation has not yet been thought out, therefore, the heat costs of the carbonization unit are not estimated. A large amount of required MSW (2.6-6.7 kg per kg of flue gases) causes certain difficulties for the implementation of the project.

Appendix

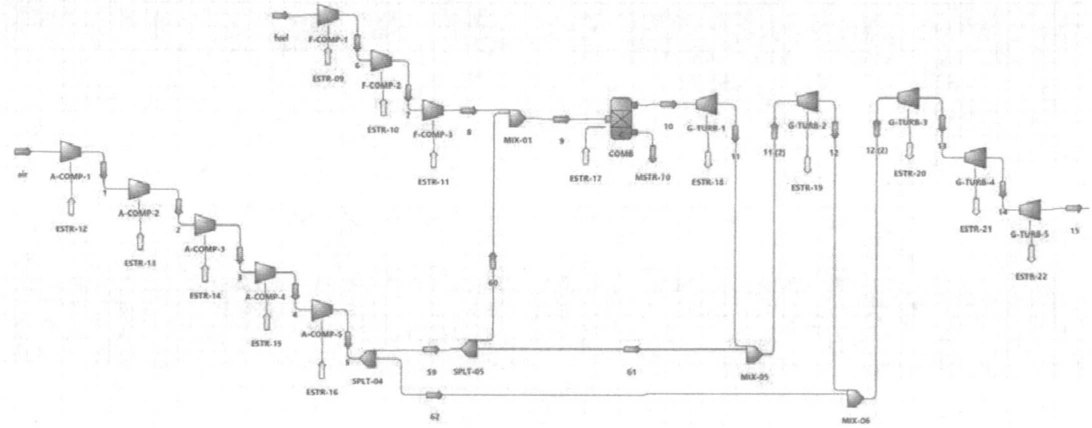

Figure 2. Schematic of gas turbine part.

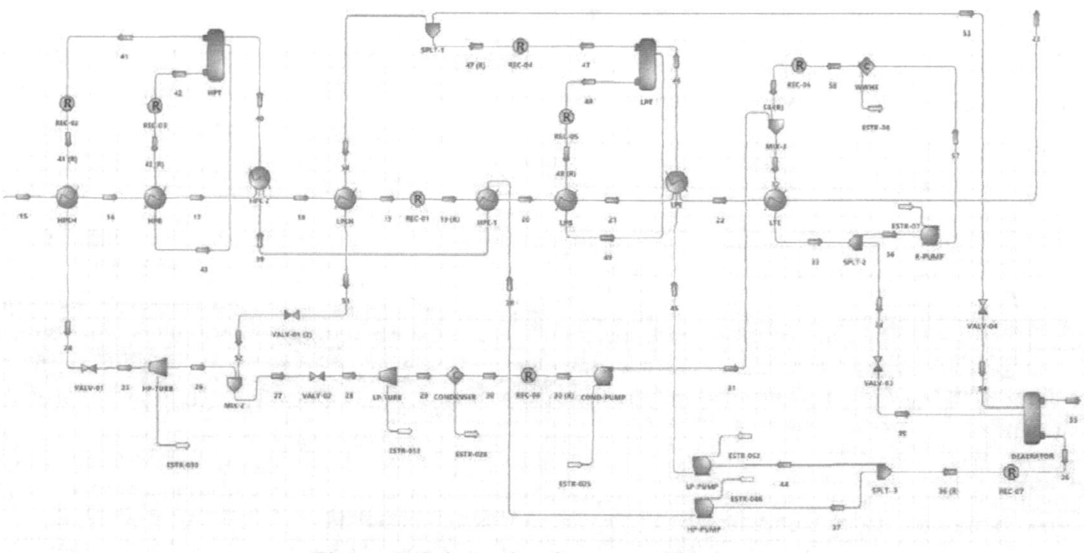

Figure 3. Schematic of steam turbine part.

Table 6. Main assumptions for the calculations of CCGT.

Parameter	Dimension	Value
Fuel composition: CH_4, C_2H_6, C_3H_8, N_2	%	97.69, 0.82, 0.33, 1.16
Air/fuel compressor pressure ratio		16.9/3.9
Isentropic efficiency of compressors/turbine	%	92.0/84.0
Turbine inlet temperature	°C	1095
Flue gas mass flow rate	kg s^{-1}	547.6
Heat loss at combustor	% of fuel LHV	0.9
Pressure loss at combustor	%	3
Mechanical efficiency of the gas turbine/compressors	%	99.8
HRSG gas side pressure loss	kPa	3
Live steam temperature (HP/LP)	°C	483.6/276.7
Live steam pressure (HP / LP)	bar	79.8/11.9
Deaerator pressure	bar	6
Condenser pressure	kPa	5
Condensate temperature before LTE	°C	60
Thermal power of WWHE	Gcal h^{-1}	22.9
HRSG heat losses from ambient cooling	%	0.7
Pinch point (HP / LP)	°C	10
Condensate underheating to saturation temperature at the deaerator	°C	7
Feed water underheating to saturation temperature in drums	°C	5
Hydraulic losses in HPSH/LPSH/live steam lines/stop valves	%	5/5/8.5/3
Isoentropic efficiency of HPT/LPT/pumps	%	83.0/78.0/75.0
Mechanical efficiency of the steam turbine	%	99.5
Generator efficiency	%	98.7

References

[1] IEA 2019 *Putting CO_2 to use* (Paris: IEA)
[2] Ho H-J, Iizuka A and Shibata E 2019 Carbon capture and utilization technology without carbon-dioxide purification and pressurization: a review on its necessity and available technologies Ind. Eng. Chem. Res. **58(21)** 8941–54
[3] European Comission 2011 *Energy Road Map 2050, impact assessment and scenario analysis*
[4] Dindi A, Quang D V, Vega L F, Nashef E and Abu-Zahra M R M 2019 Applications of fly ash for CO_2 capture, utilization, and storage *J. CO_2 Util.* **29** 82—102.
[5] Seifritz W 1990 CO_2 disposal by means of silicates *Nature* **345(6275)** 486–486
[6] Fernández Bertos M, Simons S J, Hills C D and Carey P J 2004 A review of accelerated carbonation technology in the treatment of cement-based materials and sequestration of CO_2. *J. Hazard. Mater.* **112(3)** 193–205
[7] Reddy K J, Weber H, Bhattacharyya P, Morris A, Taylor D, Christensen M, Foulke T and Fahlsing P 2010 Instantaneous capture and mineralization of flue gas carbon dioxide: pilot scale study *Nat. Prec.*
[8] Chiang P C and Pan S-Y 2017 *Carbon Dioxide Mineralization and Utilization* (Singapore: Springer) p 456

[9] Dananjayan R R T, Kandasamy P and Andimuthu R 2016 Direct mineral carbonation of coal fly ash for CO_2 sequestration *J. Clean. Prod.* **112** 4173–82
[10] Peng D Y and Robinson D B 1976 A new two-constant equation of state *Ind. Eng. Chem. Fundam*. **15** 59–64
[11] Chiesa P and Macchi E 2004 A thermodynamic analysis of different options to break 60% electric efficiency in combined cycle power plants *ASME J. Eng. Gas Turbines Power* **126(4)** 770–785
[12] Tugov A N 2013 Prospects for the use of municipal solid wastes as secondary energy resources in Russia *Therm. Eng. (Engl. Trans. Teploenergetika)* **60(9)** 663–668

Effect of dustiness in atmospheric air on the formation and characteristics of condensation aerosol

V O Mayorov[1,2], A K Yastrebov [2] and V Yu Levashov [1]

[1] Lomonosov Moscow State University, Institute of Mechanics, Russia, 119192 Moscow, Michurinsky Pr. 1
[2] Research University Moscow Power Engineering Institute, Russia, 111250 Moscow, Krasnokazarmennaya, 14

MayorovVO@imec.msu.ru

Abstract. In this work, the simulation of the formation process of condensation aerosol in atmospheric air is carried out. The influence of the initial supersaturation ratio and dustiness content on the formation mechanism and macroscopic characteristics of the aerosol is investigated.

1. Introduction

Condensation aerosol is a two-phase medium containing discrete particles of the liquid phase (droplets). This system can form if the vapor pressure exceeds the saturation pressure at the ambient temperature. This metastable state is called supersaturated vapor. Supersaturated vapor can condense using two mechanisms: the droplet formation due to homogeneous nucleation and heterogeneous condensation on the surface of solid particles. The simplest example of such a process is the formation of fog, in this case the supersaturated state is achieved due to isobaric cooling of atmospheric air. Also, condensation aerosol can be formed during adiabatic expansion of steam in the nozzle [1].

With intense evaporation near the interface, the vapor can be supersaturated [2] therefore bulk condensation is also possible. An analysis of the aerosol formation process under such conditions can be important for the investigation of droplet clusters [3], which is fundamental for understanding the mechanisms of propagation of harmful impurities and viruses.

In this paper, the effect of dustiness in atmospheric air on the formation and characteristics of aerosol is considered.

2. Formulation of the problem and mathematical description

The zero-dimensional problem of bulk condensation of steam is considered. At the initial time, steam is supersaturated, its partial pressure is higher than the saturation pressure corresponding to the ambient temperature $T_0 = 300$ K, the pressure of the vapor-gas mixture p_0 is equal to atmospheric one. The mass of solid particles per unit volume is constant and is given by the value of Φ. The initial size distribution of particles is normal:

$$f_0 = \frac{1}{\sqrt{2\pi D}}\exp\left[-\frac{r-M^{\,2}}{2D}\right], \qquad (1)$$

where D is the variance and M is the expected value. Normal distribution parameters are selected based on the regulations governing the content of dust in the air [4].

There can be a large number of droplets with different sizes and formation mechanism in considering system. Therefore, two droplet size distribution functions are used: F for homogenous droplets and f for heterogeneous droplets. The products $F\mathrm{d}r$ and $f\mathrm{d}r$ determinate the number of droplets per volume unit, which size are in the range from r to r + dr. The evolution of these distribution functions is described by the kinetic equations [5]:

$$\frac{\partial F}{\partial t}+\frac{\partial(\dot{r}F)}{\partial r}=I\delta(r-r_{cr})\,; \tag{2}$$

$$\frac{\partial f}{\partial t}+\frac{\partial(\dot{r}f)}{\partial r}=0\,. \tag{3}$$

Here t is time, r is the droplet radius, $\dot{r}$ is the droplet growth rate, I is the frequency of formation of critical droplets per volume unit (nucleation rate), r_{cr} is the droplet critical radius, and δ is the Dirac delta function. The nucleation rate is calculated using classical nucleation theory [6]. Equations (2) and (3) are very similar, the only difference is on the right parts. In equation (3), the right part is equal to zero, because the number of drops formed as a result of condensation on solid particles in the system is constant. To solve these equations, the method of direct numerical solution presented in [7] is used.

The system includes four components: non-condensable gas, vapor, liquid and solid particles. Mass fraction of gas Y_{gas} and solid particles Y_{dust} do not change throughout the entire condensation process:

$$\frac{\mathrm{d}Y_{gas}}{\mathrm{d}t}=0\,; \tag{4}$$

$$\frac{\mathrm{d}Y_{dust}}{\mathrm{d}t}=0\,. \tag{5}$$

The mass fraction of liquid Y_{liq} increases due to the formation of a new phase nuclei and its growth, as well as due to heterogeneous condensation of vapor on solid particles. The increase in the mass of the liquid is calculated as follow:

$$\frac{\mathrm{d}Y_{liq}}{\mathrm{d}t}=\frac{\mathrm{d}}{\mathrm{d}t}\left[\frac{\rho_{liq}}{\rho_{\Sigma}}\left(\frac{4}{3}\pi\int_{r_{cr}}^{\infty}(F+f)r^{3}\mathrm{d}r-\frac{1}{\rho_{dust}}\Phi\right)\right], \tag{6}$$

where ρ_{liq}, ρ_{Σ} and ρ_{dust} are the density of the liquid, the multiphase mixture and the solid particles, respectively. The mass fraction of water vapor is determined from the equation of all system components:

$$\sum_{i}Y_{i}=1\,. \tag{7}$$

In this work, it is assumed that all droplets and solid particles have the same temperature, which changes due to release of the latent heat of the phase transition and interphase heat transfer:

$$(\rho_{dust}Y_{dust}c_{p,dust}+\rho_{liq}Y_{liq}c_{p,liq})\frac{\mathrm{d}T_{d}}{\mathrm{d}t}=\left[L-c_{p,vap}(T_{d}-T_{m})\right]\rho_{liq}\dot{r}-q\ S_{d}\,, \tag{8}$$

where $c_{p,\ dust}$, $c_{p,\ liq}$ and $c_{p,\ vap}$ are the heat capacities of the solid particles the liquid and vapor, respectively, T_d and T_m is the temperature of the droplets and the environment, L is the heat of vaporization, q is the interfacial heat flux, S_d is the surface area of the droplets per volume unit, which is calculated as the second moment of the distribution function:

$$S_d = 4\pi \int_{r_{cr}}^{\infty} (F + f) r^2 \mathrm{d}r \,. \tag{9}$$

In the general case, the temperature of solid particles differs from the temperature of the liquid, but as a first approximation it is assumed that they are equal. Equation (8) takes into account that for condensation the vapor should be heated to the droplet temperature.

The change in the temperature of the mixture is described by the following energy equation:

$$\rho_m c_{p,m} \frac{\mathrm{d}T_m}{\mathrm{d}t} = q S_d \,. \tag{10}$$

Here ρ_m and $c_{p,\,m}$ are the density and heat capacity of the gaseous phase.

The droplet growth rate and interfacial heat flux are calculated using expressions for the free molecular regime [8]:

$$\dot{r} = \frac{1}{\rho_{liq}} \left[\frac{p_{vap}}{(2\pi R/\mu_{vap} T_m)^{1/2}} - \frac{p_s(T_d)}{(2\pi R/\mu_{vap} T_d)^{1/2}} \right]; \tag{11}$$

$$q = 2p_{gas} \left(\frac{RT_m}{2\pi\mu_{gas}} \right)^{1/2} \frac{T_d - T_m}{T_m} + 2p_s(T_d) \left(\frac{RT_d}{2\pi\mu_{vap}} \right)^{1/2} - 2p_{vap} \left(\frac{RT_m}{2\pi\mu_{vap}} \right)^{1/2} + \frac{5}{4} w (p_s(T_d) + p_{vap}); \tag{12}$$

$$w = -\frac{\rho_{liq} \dot{r} R}{\mu_{vap} (p_{vap}/T_m + p_s(T_d)/T_d)} \,. \tag{13}$$

Here w is the transport speed of the vapor.

It is assumed that water vapor and non-condensable gas are ideal gases; therefore, the Mendeleev-Clapeyron equation of state is used to close the system of equations (2) - (13).

Results and discussion

Calculations are carried out for supersaturated steam in atmospheric air, which also contains solid particles. The supersaturated ratio of water vapor $S_0 = p_{vap} / p_s(T_m)$ at the initial time is varied in the range from 1.25 to 4.5, and the content of solid particles Φ is from 0.01 to 1 mg/m^3. The calculation is continuing while the relative temperature difference between gaseous phase and droplets is greater than 0.1%. The evolution of the homogenous droplets size distribution function is shown in figure 1(a). At the initial stage, the nuclei of a new phase are formed (curves 1 and 2 in figure 1(a)), then, as the supersaturation decreases, the nucleation rate decreases, and the distribution function changes mainly due to the condensed particles growth (curves 3 and 4 in figure 1(a)). By the time moment corresponding to equilibrium between the liquid and vapor phases, the distribution function takes its final form (curve 5 in figure 1(a). The number of drops formed as a result of heterogeneous condensation on solid particles does not change, therefore the distribution function f moves to the region of large radius throughout the entire process of condensation relaxation (figure 1(b)).

At the beginning of bulk condensation, the droplet temperature T_d rapidly increases due to active vapor condensation on their surface (figure 2). The growth rate decreases over time (figure 3) due to decrease of supersaturation, and the temperatures of the droplets T_d and the gaseous phase T_m equalize due to interphase heat transfer (figure 2).

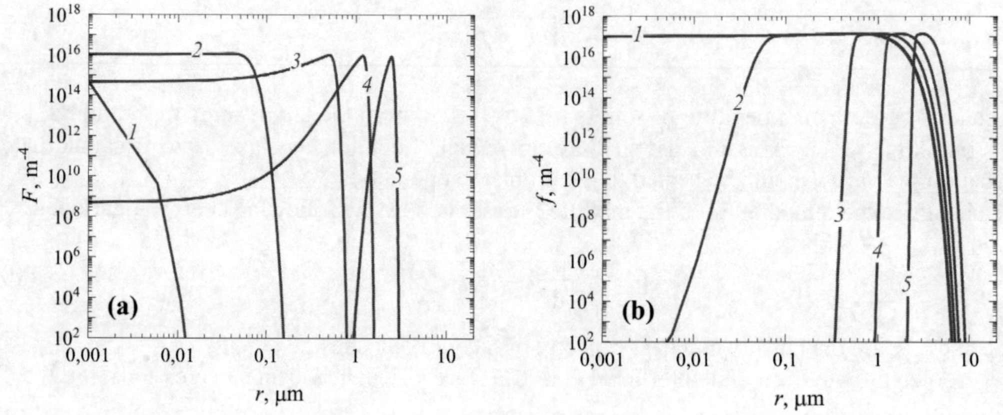

Figure 1. The evolution of size distribution functions F (a) and f (b) at $S_0 = 2.75$ and $\Phi = 0.1$ mg/m^3. *1* – 0.02 μs, *2* – 1 μs, *3* –10 μs, *4* – 25 μs, *5* – 4746 μs.

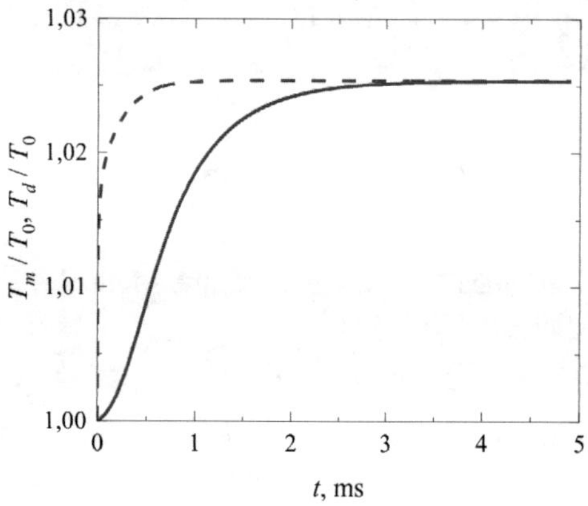

Figure 2. Time depends of droplet temperature (dash curve) and gas phase temperature (solid curve) at $S_0 = 2.75$ and $\Phi = 0.1$ mg/m^3.

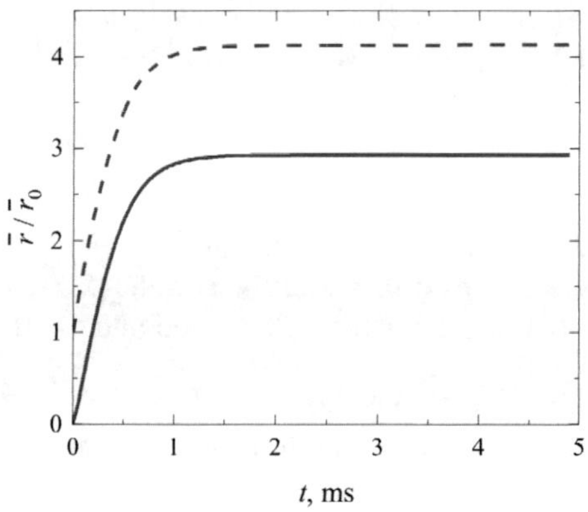

Figure 3. Time depends of homogenous (dash curve) and heterogeneous (solid curve) average relative radius at $S_0 = 2.75$ and $\Phi = 0.1$ mg/m^3.

For clarity of the results, the value Ω is introduced, which is the ratio of the number of heterogeneous droplet to the total number of drops:

$$\Omega = \int_{r_{cr}}^{\infty} f \mathrm{d}r \left(\int_{r_{cr}}^{\infty} (F + f) \mathrm{d}r \right)^{-1} . \quad (14)$$

With an increase in the supersaturation ratio, the number of droplets formed by homogeneous nucleation increases greatly due to increase of nucleation rate. However, at S_0 <2.5, the number of such droplets is negligible in comparison with the number of drops are formed by heterogeneous condensation on solid particles (figure 4). With an increase of the dust concentration in the air, the total surface for heterogeneous condensation increases, therefore the supersaturation ratio decreases faster and number of homogeneous droplets decreases.

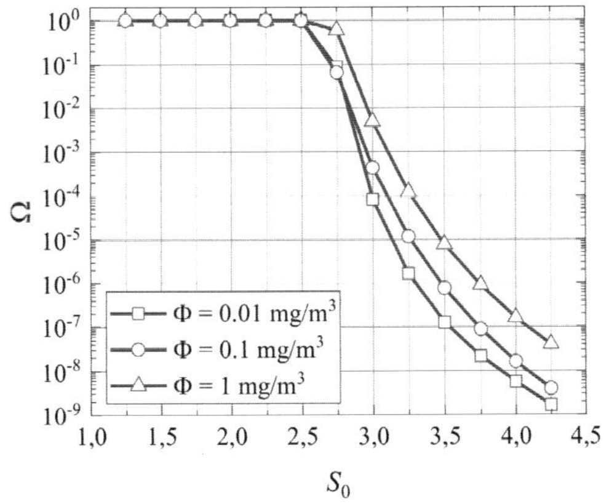

Figure 4. Dependences of the ratio heterogeneous droplet to the total number of droplets on the initial supersaturation for different contents of solid particles.

The droplet size at the end of the condensation process is determined by two factors: the duration of the process and the intensity of condensation on the surface. Accordingly, with an increase in the initial supersaturation ratio, the growth rate of drops increases, and it takes for the system to reach thermodynamic equilibrium decreases. At low supersaturation $S_0 < 3$, the intensity of condensation on the surface increases faster than the duration of process decreases; therefore, the droplets reach large sizes (figure 5). With a subsequent increase in water vapor supersaturation, the duration of the condensation process rapidly decreases due to intense homogeneous nucleation (figure 5). An increase in the concentration of solid particles also slightly reduces the volumetric condensation time, since the total surface area for heterogeneous condensation increases. However, at high supersaturation, this effect is practically leveled out (figure 5). The average radius of heterogeneous droplets at the end of the process is somewhat larger than that of homogeneous ones.

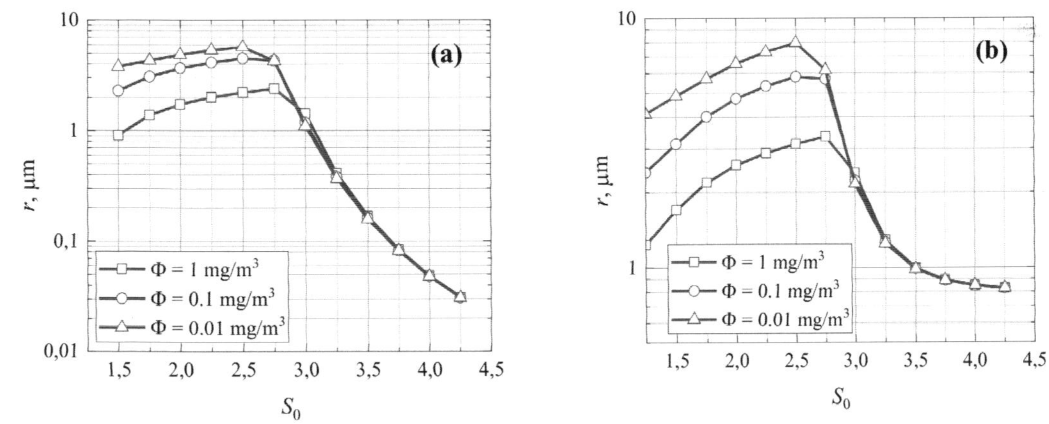

Figure 5. Dependences of the average radius of homogeneous (a) and heterogeneous droplets (b) on the initial supersaturation for different contents of solid particles.

Conclusion

The characteristics of droplet condensation aerosol formed in atmospheric air are investigated. It is shown that, at a low initial supersaturation ratio $S_0 < 2.5$, the droplets are mostly formed due to heterogeneous vapor condensation on the surface of solid particles contained in the air. On the other hand, at $S_0 > 3$, the number of droplets formed as a result of homogeneous nucleation significantly exceeds the number of heterogeneous droplets. The content of solid particles in the air has little effect on the size of homogeneous and heterogeneous droplets at high supersaturation ratio.

Acknowledgments

The work was supported by Russian Foundation for Basic Research (project 20-38-90042)

References

[1] Hou D, Jiang W, Zhao W, Bian J, Liu Y and Lai X 2018 *Chem. Eng. Process.* **133** 128-136
[2] Anisimov S I , Imas Ya A, Romanov G S and Khodyko Yu V 1970 *Deistvie izlucheniya bol'shoi moshchnosti na metally* (The Effect of High Power Radiation onto Metals) (Moscow: Nauka)
[3] Fedorets A 2004 *JETP Letters* **79** 372-374
[4] Hygienic standards GN 2.1.6.2604-10 2010 Maximum permissible concentrations (MPC) of pollutants in the atmospheric air of populated areas
[5] Hulburt H M and Katz S M 1964 Some problems in particle technology: A statistical mechanical formulation *Chem. Eng. Sci.*19555
[6] Kashchiev D 2000 *Nucleation. Basic Theory with Applications* (Oxford: Butterworth-Heinemann)
[7] Kortsensteyn N M, Samuilov E V and Yastrebov A K 2009 *Int. J. Heat Mass Transfer* **52** 548
[8] Kortsenshteyn N M and Yastrebov A K 2011 *Int. J. Heat Mass Transfer* **55** 1133

Flow and heat transfer near a cylinders row

A V Mityakov, A A Gusakov, M A Grekov and V V Seroshtanov*

Peter the Great St. Petersburg Polytechnic University,
Russia, 195251 St. Petersburg, Politekhnicheskaya ul., 29

vladvik1992@gmail.com

Abstract. The paper aims to investigate the dependence of heat transfer classification on the Reynolds number (Re) during flow around circular heated cylinders row. The investigated range of Re number varies from 4.5×10^3 up to 42×10^3. The distance between cylinders *S* was changed from 0.5*d* to 4*d* (where *d* is the cylinders dia). Cylinders surface temperature was kept constant. For each Re number, the case when the cylinders were mounted one after the other was investigated. To measure heat transfer and flow parameters (velocity, heat flux and heat transfer coefficient) near and at the cylinders surface, two experimental methods were used: gradient heatmetry and PIV. Heat flux and velocity fields were obtained from gradient heatmetry and PIV results, based on which the flow mode could be determined and compared with heat transfer mode. As a result, it was found that heat transfer is influenced by both the Reynolds number and the distance between the cylinders. The observed features are associated with influence on characteristics such as separation point location, boundary layer thickness, change in flow between the cylinders and vortices formation.

1. Introduction

Experimental study of heat transfer and hydrodynamic between multiple surfaces used in heat exchangers is of both fundamental and practical importance. Flow in and behind the cylinders row is by far more complicated than that behind a single cylinder. In this configuration, in addition to the interrelationship between flow and heat transfer, described by the similarity equation Nu = Re (Nu), it is important to evaluate the effect of distance *S* between cylinders (Re = Wd/ν, where *W* is free flow velocity, *d* is cylinder dia, and ν is kinematic viscosity of liquid; Nu = hd/k, where *h* is heat transfer coefficient and *k* is heat conductivity of liquid). The flow classification in terms of Re and *S* is crucial for understanding of heat transfer and has received considerable attention.

Lately, tubes rows and bundles of various profiles and configurations have been actively investigated. For example, Aiba and Yamazaki [1] carried out series of experiments to study flow near a tube bundle of different configurations. They investigate the flow and heat transfer characteristics. Based on the results, the authors argue that the most significant is contribution of vortex formed behind the first tube. The experiment was carried out for a narrow range of Reynolds numbers: from 10^4 up to 5×10^5. In other work, Aiba and others [2] studied heat transfer in a row of 7 tubes. It was concluded that Nusselt number varies slightly in the far tubes, starting from the second.

There is a lot of numerical studies conducted to determine cross flow around isothermal circular cylinders in arrangements such as single cylinder, inline and staggered arrays. So, in the paper [3] flow over the cylinder with low Reynolds number was investigated. The authors solved two-dimensional stationary problem for a viscous incompressible fluid in a channel. The main conclusion, the authors believe that heat transfer from a staggered array of cylinders is slightly higher than that from an inline array.

Another direction of such studies is associated with flow around cylinders of non-circular cross-section. It is known that cross-sectional shape can both improve and worsen the hydrodynamic characteristics. The study [4] esteems flow characteristics and associated instability of convective flow behind a semi-circular cylinder at incidence with a downstream circular cylinder. Unsteady computations are performed for different incidence angles and Reynolds numbers from 60 up to 160. Dependencies of coefficient of drag and lift coefficient on the angles of incidence are examined. The simulation results compared with the experimental ones [5] showed satisfactory agreement. The authors concluded that two separation points on a semicircular cylinder change its location with angles of incidence. However, flow topology behind a circular cylinder resembles flow situation behind a rotating circular cylinder. Vortex shedding frequency is nonlinearly.

The work [6] aims to investigate the dependence of Reynolds number for the wake of two staggered cylinders. Hotwires were used to measure fluctuating velocities in vortex streets behind two cylinders. Power spectral density functions were obtained, based on which flow structure pattern could be determined. According to the article, there is a departure in the border between different flow modes, which is partially revealed with refined measurement grids, and hence improved resolution in determining of flow mode.

2. Experimental details

2.1. Experimental model

Schematic of experimental arrangement is represented at figure 1. The model consists of single (*I*), two (*I* and *II*) or three (*I*, *II* and *III*) circular cylinders, which are mounted inline one by one at distance of *S*. The dia of all cylinders is of 66 mm, and the thickness of the cylinder wall is of 0,1 mm. Cylinders had a length of 600 mm. At the investigated cylinder (*I*, *II* or *III*) gradient heat flux sensor (GHFS) are mounted, and the cylinder could rotate around its axis at an angle φ from 0 up to 180 °C. Saturated water steam at atmospheric pressure was supplied inside all cylinders. This heating scheme made it possible to maintain constant temperature at the surface of the model.

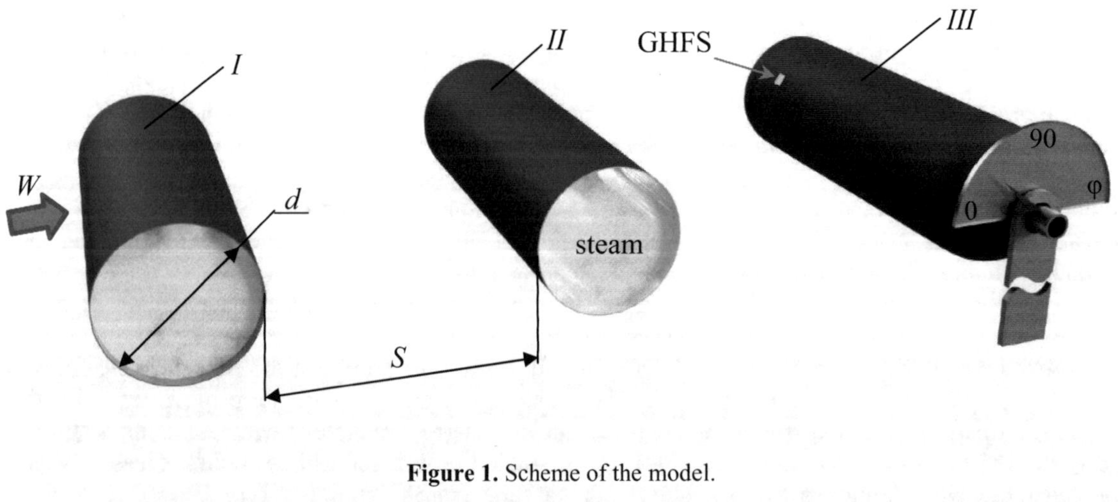

Figure 1. Scheme of the model.

Our experimental setup made it possible to change distance *S* between investigated cylinders. The experimental model was installed in the working section of a subsonic wind tunnel. The wind tunnel features associated with maintaining of constant temperature of free-stream airflow, velocities range and turbulence degree are described in detail in the paper [7]. The range of modes (Re = 500 ... 42000) is realized by change of airflow velocity in the tunnel.

2.2. Experimental methods

Two modern methods were used in the work, namely particle image velocimetry (PIV) and gradient heatmetry. The choice of methods depends on both the model used and the problem statement. Velocity measurements using PIV are described in detail in the work [7]. We used the POLIS system [8] in two-dimensional setting. The resolution of the camera used was 4MPx, and dry wood smoke was used as tracers media.

In study of heat transfer during flow around a cylinders row technique of gradient heatmetry developed in Peter the Great St. Petersburg Polytechnic University was used. Gradient heatmetry is based on use of gradient heat flux sensors. Physically, the action of these sensors is based on transverse Seebeck' effect. It is thermal to electric energy conversion, where direction of electric field is normal to direction of heat flux. The electric field $\boldsymbol{E}$ generated by transverse Seebeck' effect is given by $\boldsymbol{E}=\boldsymbol{S}\nabla \boldsymbol{T}$ (where **S** is Seebeck tensor and T is temperature). Moreover, the electric field $\boldsymbol{E}$ and the temperature gradient $\nabla \boldsymbol{T}$ are not always collinear. Unlike thermoelectric devices in which we use only the diagonal term $\boldsymbol{S}$, transverse thermoelectric devices use off-diagonal term **S** [9]. Tensor **S** is not diagonal in two cases. Either sensor's media is of anisotropic properties, or the main axes of media's anisotropy must be inclined with respect to directions of thermal input and electrical output. GHFSs used in our experiments are made of single-crystal bismuth of purity 0.9999. Since this material is anisotropic, it implements transverse Seebeck' effect described above. The GHFS is of 2×2×0.2 mm in size. Its volt-watt sensitivity was about 8 mV/W.

3. Results

Velocity fields in the vicinity of single, pair, and three cylinders row make it possible to say that the flow in the wake depends on the number of cylinders. For example, figure 1 shows the average velocity fields near cylinders for Reynolds number of Re = 9600.

It can be noted that for given distance of $S=d$, vortex street length depends on the number of cylinders. Average velocity fields make it possible to estimate the size as well as the average velocity distribution in the wake-zone. However, for a more complete characterization, it is necessary to analyze instantaneous velocity fields. Such analysis is given as an example in our paper [10].

In this paper, we will discuss relationship between heat transfer and system geometry in detail. Thus, figure 3 shows dependences of local Nusselt number ($Nu_\varphi = h_\varphi d/k$, where h_φ is local heat transfer coefficient) on the rotation angle φ for the second and third cylinders.

Distribution of Nusselt number shows that heat transfer at the second and third cylinders is different both in nature and in magnitude. The differences are more noticeable for larger Reynolds numbers: the local Nusselt number maximums differ by about two times. However, in the φ range of 150 ... 180° there are practically no differences. The difference between surface-averaged Nusselt numbers is of 10...50%. Moreover, small differences are typical for low Reynolds numbers (Re<5000). As the distance between the cylinders increases, the difference in the average Nusselt number decreases.

For a visual comparison of all three cylinders, figure 4 shows distributions of heat flux per unit area. This type of presentation of results does not allow to estimate values of parameters numerically.

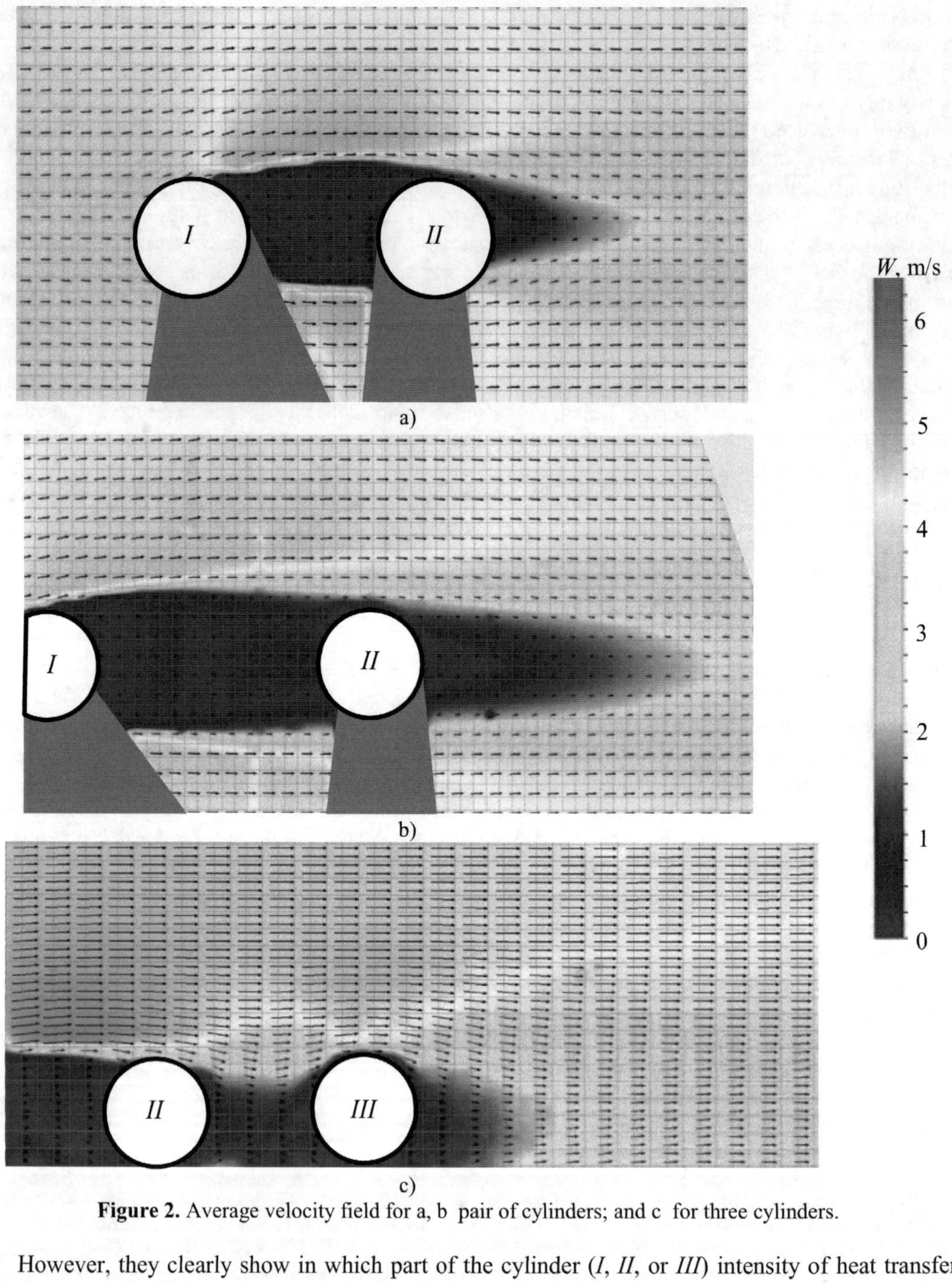

Figure 2. Average velocity field for a, b pair of cylinders; and c for three cylinders.

However, they clearly show in which part of the cylinder (*I*, *II*, or *III*) intensity of heat transfer is the lowest or the highest. For example, for the first and third cylinders there is a region where forced convective heat transfer is practically absent ($\varphi \approx 90^\circ$ for the first and $\varphi \approx 120^\circ$ near the third one).

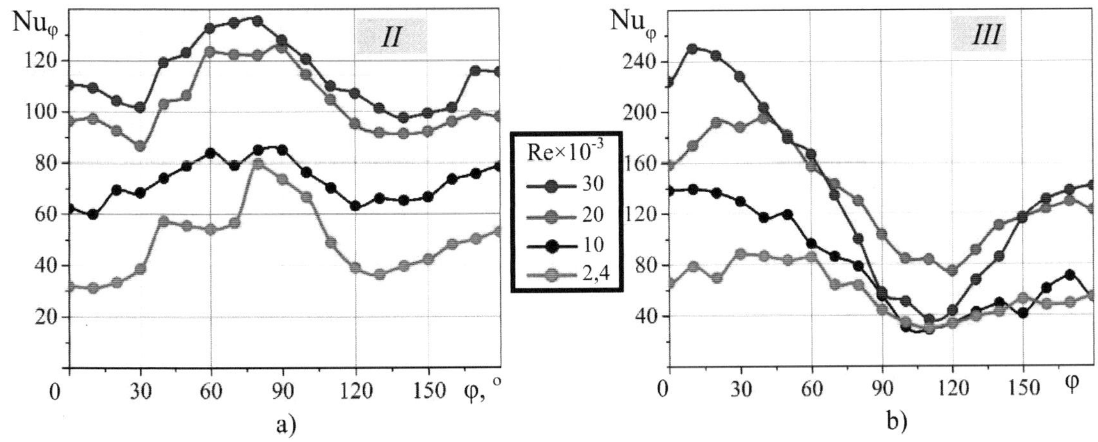

Figure 3. Dependence of local Nusselt number on the angle φ for the second (*II*) and the third (*III*) cylinders ($S = d$).

Heat transfer intensity at the second cylinder is uniform. This illustration allows us to see that heat transfer in the aft part of the second and third cylinders is weakly dependent on the distance *S*. Another advantage of such figures is that in terms of heat transfer enhancement, one can immediately determine its ways and, which is most important, geometry of intensifiers.

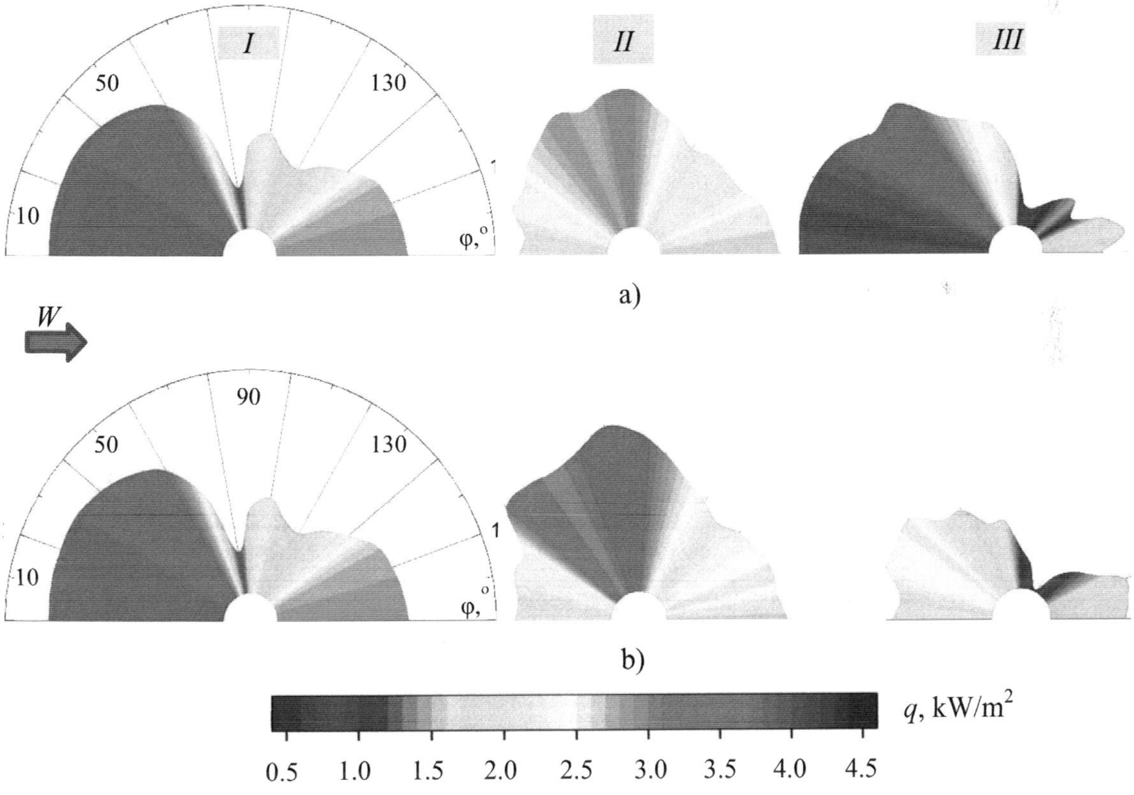

Figure 4. Heat flux at the cylinders surface: a) for $S = d$ and b) for $S = 2d$.

Conclusions

Simultaneously use of gradient heatmetry and PIV makes it possible to evaluate not only the qualitative, but also quantitative parameters of convective heat transfer. First of all, to derive the similarity equations for the flow around a cylinders row.

The results showed that the local heat transfer coefficient's distribution at the surface of the second and third cylinders depends largely on the distance S in the case of Re = 500… 5000. This is due to the action of rather large vortices that periodically descend from the first cylinder. In addition, at the second cylinder near the frontal generatrix area (φ = ±30°) the heat transfer intensity of depends on whether the second cylinder is in the zone of developed vortices (approximately (1.5 … 3)d from the first cylinder) or in the stagnation zone (less than 1d from the first cylinder). Experiments have shown that the velocity field behind the second cylinder is practically uniform, which affects heat transfer near the third cylinder as follows: the dependence of the heat transfer coefficient on the angle φ is qualitatively similar to the dependence for the first cylinder. There is also a displacement of the separation point to the φ = 120…130°. In addition, averaged heat transfer coefficient for the third and for the first cylinder are comparable.

The combined technique and the results obtained will serve as basis for study of flow and heat transfer in tube bundles and heat transfer enhancement while maintaining acceptable hydrodynamic parameters.

References

[1] Aiba S and Yamazaki Y 1976 An Experimental Investigation of Heat Transfer Around a Tube in a Bank *ASME. J. Heat Transfer.* **98(3)** pp 503–508

[2] Aiba S, Tsuchida H, Ota T 1982 Heat transfer around tubes in in-line tube banks *Bull. JSME* **25 (204)** pp 919–926

[3] Mia Abrahamsen Prsic, Muk Chen Ong, Bjørnar Pettersen, Dag Myrhaug 2014 Large Eddy Simulations of flow around a smooth circular cylinder in a uniform current in the subcritical flow regime *Ocean Engineering* **77** pp 61-73

[4] Haider Jibran, Noman, Sabbir, Khan Waqar, Mehdi Uzair, Abbasi Bilal 2010 Heat transfer and fluid flow over circular cylinders in cross flow. *NUST Journal of Engineering Sciences*. **3** 68

[5] Xiantao Fan, Zhongchen Wang, Yang Wang, Wei Tan 2021 The effect of vortices structures on the flow-induced vibration of three flexible tandem cylinders *International Journal of Mechanical Sciences* **192**

[6] Wong C W, Zhou Y, Md Mahbub Alam, Zhou T M 2014 Dependence of flow classification on the Reynolds number for a two-cylinder wake *Journal of Fluids and Structures* **49** pp. 485-497

[7] Gusakov A A, Kosolapov A S, Markovich D M, Mityakov A V, Mityakov V Y, Mozhayskiy S A, Nebuchinov A S, Sapozhnikov S Z 2014 Simultaneous PIV and gradient heat flux measurement of a circular cylinder in cross-flow *Applied Mechanics and Materials* **629** pp 444-449

[8] http://polis-instruments.ru/

[9] Sapozhnikov S Z, Mityakov V Yu, Mityakov A V 2020 H*eatmetry The Science and Practice of Heat Flux Measurement* (Springer International Publishing)

[10] Guzeev A S, Sapozhnikov S Z, Mitiakov V Yu, Seroshtanov V V, Suchok V V, Vlasov A S, Zhidkov N A 2020 Features of vortex formation and heat transfer during cross flow around two cylinders *Journal of Physics: Conference Series* **1683** 2

Study of the heat flux fluctuations intensity during flow around a circular cylinder

V Yu Mityakov, V V Seroshtanov*, V V Suchok, N A Zhidkov and A S Vlasov

Peter the Great St. Petersburg Polytechnic University,
Russia, 195251 St. Petersburg, Politekhnicheskaya ul., 29

vladvik1992@gmail.com

Abstract. An experimental study performed in an air channel using gradient heatmetry is presented, for a configuration of three heated circular cylinders with different distance between them. Cylinders were installed one by one. The main objective is to analyze the heat flux fluctuations, employing gradient heat flux sensor. The behavior of the flow in the wake behind the first cylinder under various regimes, for the specified configuration, changes the fluctuations level at the surface of the second and third cylinders. By visualizing the flow using PIV, it was possible to see areas of stagnation, separation points and other features of flow around the model. The results showed that the fluctuations level for the second cylinder is an order of magnitude lower than for the first. We can say that the first cylinder stabilizes the flow. However, at the third cylinder, this level is comparable and even higher than on the first. Use of gradient heatmetry thus made it possible for the first time to estimate the fluctuating nature of the flow and heat transfer around cylinders.

1. Introduction

Circular cylinders are a typical configuration that appears in many applications in industry and engineering, being a common geometry in pipelines, heat exchangers, transmission lines and nuclear reactor etc. This problem in terms of time-averaged heat transfer has been investigated both in academia and industry, given their theoretical and practical significance. If we talk about nonstationarity, then the Von-Karman street inherent in the flow around cylindrical surfaces has been thoroughly studied by previous researchers. The mechanisms of tandem cylinders especially under different scales of vortices structures are still undergoing investigation. Flow patterns around cylinders (for the two-dimensional case) were systematically generalized by Sumner [1] and Zhou [2].

The vortex shedding phenomenon behind a circular cylinder develops over a wide range of Reynolds number ($Re = Wd/\nu$, where d – is cylinders dia, m; ν – kinematic air viscosity, m^2/s), starting from $Re \sim 49$, above which the symmetry behind the body becomes unstable. This phenomenon is due to a wake instability associated with a rather sudden inception and growth in amplitude of wake fluctuations, as one increases the Reynolds number [3]. Vortex shedding is a periodic phenomenon characterized by frequency f. It is normalized with the free-stream velocity W and the cylinder dia d. Thus obtaining the Strouhal number $St = f\, d/W$, which is a function of the Reynolds number. The relationship between the Strouhal number and the Reynolds number for single circular cylinders has been reported by many authors in the literature [4].

Coplex flow and heat transfer in the turbulent wake were studied by both experimental and numerical techniques. An experimental study of variation of momentum and heat transfer associated with these vortices was carried out by Antonia et al. [5]. They concluded that, vortices transport the heat effectively than momentum in the far wake zone. Several researchers have investigated using numerical simulations, in order to understand this flow. The effect of vortex shedding frequency on forced convection heat transfer from bank of circular tubes, in heat exchangers, under cross flow was studied by Kumar and Jayavel [6].

Thus, we can say that the research topic is relevant, and the use of new research methods may open up new prospects.

2. Experimental model

Experimental model consists of three circular cylinders, which were installed one by one. The diameter of cylinders is 66 mm. Wall thickness of the cylinders is 0.1 mm. Cylinders could rotate around the axis at an angle φ. All cylinders were heated by saturated steam with temperature $T_s \approx 100$ °C. Heat transfer surface' temperature was constant during all experiments. The distance S between the cylinders could be changed. The model was installed into the subsonic wind tube of REC "Engineering Thermophysics". The velocity range developed in the wind tunnel provided a range of Re from 500 to 42000 (Re = Wd/ν, where d – is cylinders dia, m; ν – kinematic air viscosity, m^2/s). Heated hollow cylinder models and their installation is similar to that described in the work [7].

3. Experimental methods

Study of heat transfer during flow around a cylinders row, we used a gradient heatmetry. This method is based on the use of gradient heat flux sensors (GHFS). Operating of GHFS is based on the transverse Seebeck's effect [8]. In this study we used GHFS a single-crystal bismuth, its size is 2×2×0.2 mm. The volt-watt sensitivity of the sensor was about 8 mV/W.

Another most important characteristic of any sensors is their response time. It was found by calibration: the "hot" surface of the GHFS was irradiated by calibrate monochromatic pulsed laser ray (wavelength 635 nm, peak power 50–120 mJ, pulses frequency 1–10 Hz). There is a contact between "cold" surface of the GHFS and the aluminium plate. The GHFS signal is recorded by the Tektronix oscilloscope. Results of the experiment are shown on Fig. 1. The peak value is late (concerning the front of a pulse) on 10 ns [8].

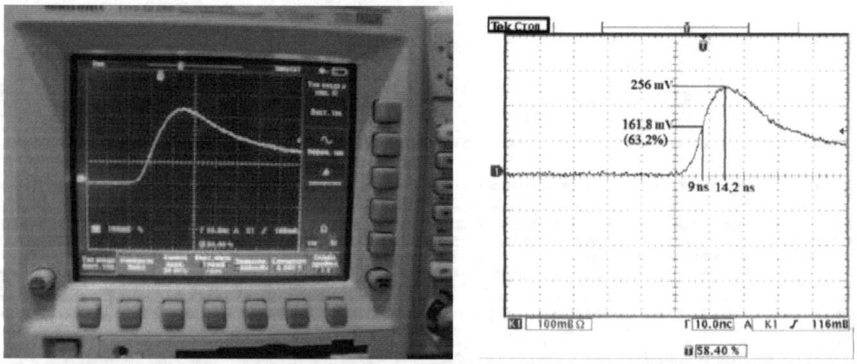

Figure 1. GHFS based on bismuth response time calibration results.

According to Fourier theory the "working" layer of semi-boundary media is about 0.2…0.6 μm, which is enough for realization of Seebeck's effects. So, the GHFS signal is proportional to temperature gradient in "working" (heated) layer. That is why we have suggested response time order of 10^{-9} s. If we take into account that, according to the literature [4-6], the velocity fluctuations during flowing around a cylinder for our sizes and regimes does not exceed tens of Hz, then the use of the GHFS to estimate these fluctuations is legitimate.

It is also important to note that PIV was used to identify the most interesting points. Features of our PIV system by POLIS [9] are also described in [8].

4. Results

The instantaneous velocity fields which was obtained using PIV help us to divide the cylinder into zones where the cylinder flows more actively and areas where there is stagnation. For example, figure 2 shows the velocity fields near the second and third cylinders for the Reynolds number Re = 9600. The off-design area under the cylinders is marked in gray. Since the laser sheet shines from top to bottom this area is in the shadow. It can be seen, that velocity near the second cylinder in the aft section practically does not differ, while the separation on the third cylinder is clearly noticeable. A further vortex street can be observed far from the model.

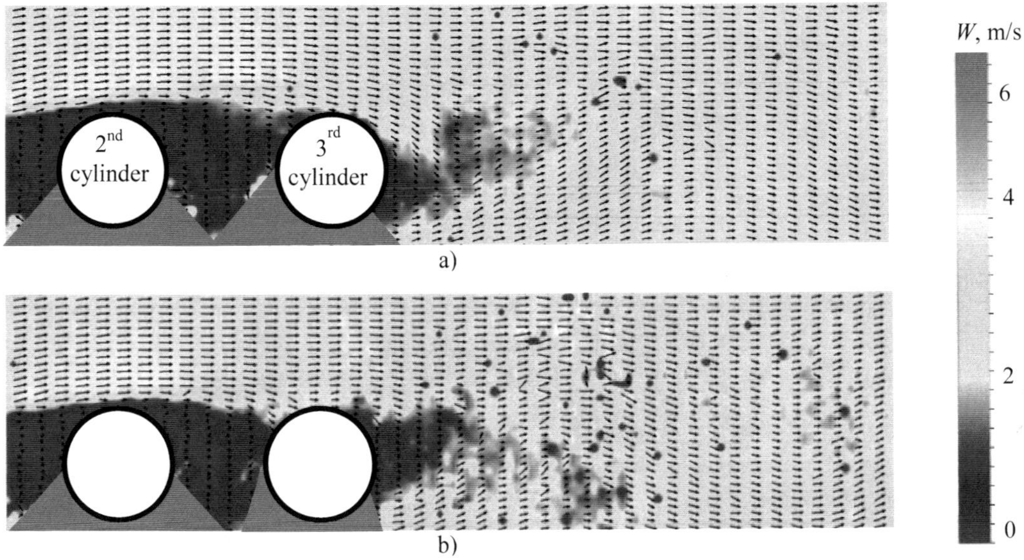

Figure 2. Instantaneous velocity fields near second and the third cylinder at different times: for a – τ =14 s, for b – τ =17 s (Re = 9600, $S = d$).

Flow visualization near the cylinders made it possible to observe unsteady flow processes, which, in turn, should affect the heat transfer parameters. Figure 3 shows heat flux fluctuations intensity on the rotation angle φ for all cylinders.

Because investigated cylinder could rotate around the axis, and the GHFS mounted on its surface, thus, fixed the heat flux at each angle φ. How the angle φ was determined is also added in figure 3 as a diagram. The GHFS' signal was measured with a frequency of 100 Hz by DAC NI 9238.

The intensity value was determined by the formula

$$\eta = \frac{\sqrt{q'^{\,2}_{\varphi}}}{\bar{q}_{\varphi}},$$

where $\bar{q}_{\varphi}$ and $\sqrt{q'^{\,2}_{\varphi}}$ (W/m^2) are the time-average heat flux per unit area and the root mean square (RMS) of the heat flux fluctuations at a fixed cylinder rotation angle φ, respectively.

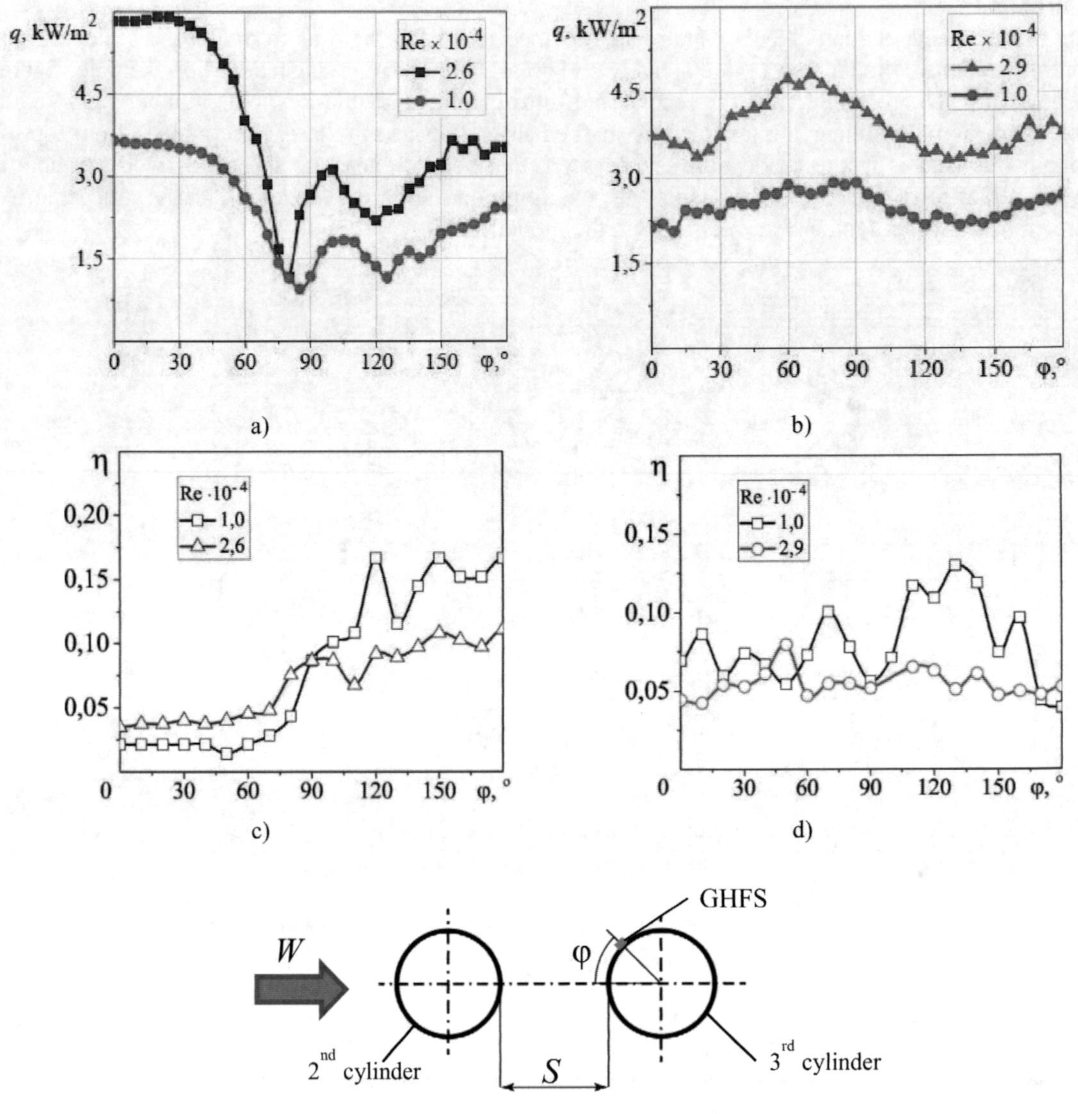

Figure 3. Time-averaged heat flux (a, b) and heat flux fluctuations intensity (c, d) over : a, c -first cylinder; b, d - the second cylinder ($S = 2d$).

At each angle φ, the heat flux was measured for 10 s. In studies for a pair of cylinders, the distance S between the cylinders varied from $0.5d$ to $4d$. The maximum intensity for the second cylinder is lower, by almost a third, and the dependence differs depending on the flow regime.

The distributions for the single and second cylinder differ in both level and shape. For a single cylinder, the fluctuation intensity increases in the range φ = 70...180 °.

The influence of the regime in the investigated range of Reynolds numbers for the first cylinder is expressed in the region behind the separation point of the boundary layer. With an increase in velocity the fluctuation level of pulsations decreases since the flow is turbulent. For the second cylinder the dependence of the fluctuation intensity on the angle φ is less pronounced and does not exceed 15 % in the investigated range of regimes. Heat transfer near the second cylinder depends on which zone of the wake formed by the first cylinder is the second cylinder located. With an increase in the distance S between the cylinders, the intensity φ distribution becomes similar to the first cylinder.

Conclusions

In the future, it is planned to obtain the distributions of the power spectrum of the heat flux fluctuations at each corner for cylinders. Having obtained the dependence of the power spectrum on frequency for the heat flux, it will be possible to compare them with the energy spectrum of the velocity, which will make it possible to compare the flow and heat transfer when flowing around cylinders at the level of velocity and heat flux fluctuations.

References

[1] Sumner D 2010 Two circular cylinders in cross flow: a review *Journal of Fluids and Structures* **26** pp 849-899
[2] Zhou Y, Yiu M W 206 Flow structure, momentum and heat transport in a two-tandem–cylinder wake. *J Fluid Mech* **548** pp 17–48
[3] Williamson C H 1996 Vortex dynamics in the cylinder wake. *Annu. Rev. Fluid Mech.* **28** (1) pp 477–539
[4] Norberg C 2003 Fluctuating lift on a circular cylinder: review and new measurements. *J. Fluids Struct.* **17** pp 57–96
[5] Antonia R A, Zhou Y, Matsumura M 1993 Spectral characteristics of momentum and heat transfer in the turbulent wake of a circular cylinder, *Exp. Therm. Fluid Sci.* **6** pp 371–375
[6] Kumar R S, Jayavel S 2017 Influence of flow shedding frequency on convection heat transfer from bank of circular tubes in heat exchangers under cross flow *Int. J. Heat Mass Transf.* **105** pp 376–393
[7] Sapozhnikov S Z et al 2019 *J. Phys.: Conf. Ser.* **1421** 012064 http://dx.doi.org/10.1088/1742-6596/1421/1/012064TI
[8] Sapozhnikov S Z, Mityakov V Yu, Mityakov A V 2020 *Heatmetry The Science and Practice of Heat Flux Measurement* (Springer International Publishing)
[9] http://polis-instruments.ru/

Development of an automated system for experimental investigation of thermal processes in a hydrogen-oxygen steam generator

I M Molotov[1], A I Schastlivtsev[1], L V Yamshchikova[1] and I A Molotova[2]

[1] Joint Institute of High Temperature,
Russia, 125412 Moscow, Izhorskaya, 13, Bd. 2
[2] National Research University "MPEI",
Russia, 111250 Moscow, Krasnokazarmennaya, 14

h2lab@mail.ru

Abstract. The paper presents the results of the development and creation of an automated system of scientific research (ASSR). It provides experimental studies of heat and mass transfer processes in a hydrogen-oxygen steam generator (HOSG). The most relevant fields of application of hydrogen-oxygen steam generators are considered. The paper discusses the most relevant areas of application of hydrogen-oxygen steam generators, scientific and technical barriers to the introduction of technology and the features of the construction of ASSR for experimental research. The schematic diagram of the primary measuring transducers and the control mechanisms of the experimental stand are described. The choice of the configuration of the automated control and measurement system is justified from the point of view of completeness and reliability of the obtained data.

1. Introduction

Common knowledge that hydrogen-oxygen steam generators (HOSG) can be widely used in the energy industry [1], associated with increased maneuverability and reliability of nuclear power plants (NPPs), alignment of energy consumption schedules in the daily and weekly intervals, etc. [2], since they have high thermodynamic parameters of the generated steam, compactness, environmental friendliness, and fast start-up [3]. The joint use of a steam generator as part of a hydrogen storage complex integrated with a nuclear power plant can be the most effective [4, 5]. The main idea of this technology is that during the hours of minimum load, hydrogen and oxygen are produced by electrolysis of water, sent to storage, and during the hours of maximum load are used to generate energy by burning hydrogen in oxygen in a HOSG, increasing the power of the steam turbine, including for power supply to the NPP's own needs in case of emergencies [6].

In previously published works, the efficiency of experimental samples of HOSG was shown and experimental data were obtained, which showed high dynamic characteristics and other technical parameters of promising HOSG [7]. However, for the introduction of HOSG in the energy sector, it is necessary to solve a number of important scientific and technical problems, namely:

- achieving more complete combustion of hydrogen in an oxygen environment [8, 9];
- providing more reliable thermal protection of the most heat-stressed sections of the combustion chamber of a HOSG;

- studies of the hardening effect.

The design features of the new steam generator model are considered in [11]. The principal difference between the new sample is that the steam is cooled through the wall along the entire length, and the cooling zone has an increased length, compared to the previous samples of HOSG [8], sufficient to ensure a uniform decrease in the temperature of the combustion products, thus minimizing the tempering effect.

Common knowledge that the tempering effect [12] consists in a sharp decrease in the rate of chemical reactions during rapid cooling. Therefore, hydrogen, which is in a dissociated state at high temperatures, does not have time to recombine and form steam as a result of rapid cooling. Thus, during the operation of the installation, hydrogen gas can accumulate, which creates risks of an explosion and fire situation.

In this regard, the automated system for the study of processes on a new sample of the HOSG has the following features:

- input module with a channel polling rate of at least 200 ms for temperature and pressure measurement;
- digital input module with an input sampling frequency of at least 150 kHz for receiving data from turbine flow meters;
- a software package for dispatching control and data acquisition (SCADA), designed for operator control of experiments;
- remote secure installation management.

2. Experimental setup

A schematic diagram of the HOSG stand, including diagnostic and control elements, is shown in figure 1.

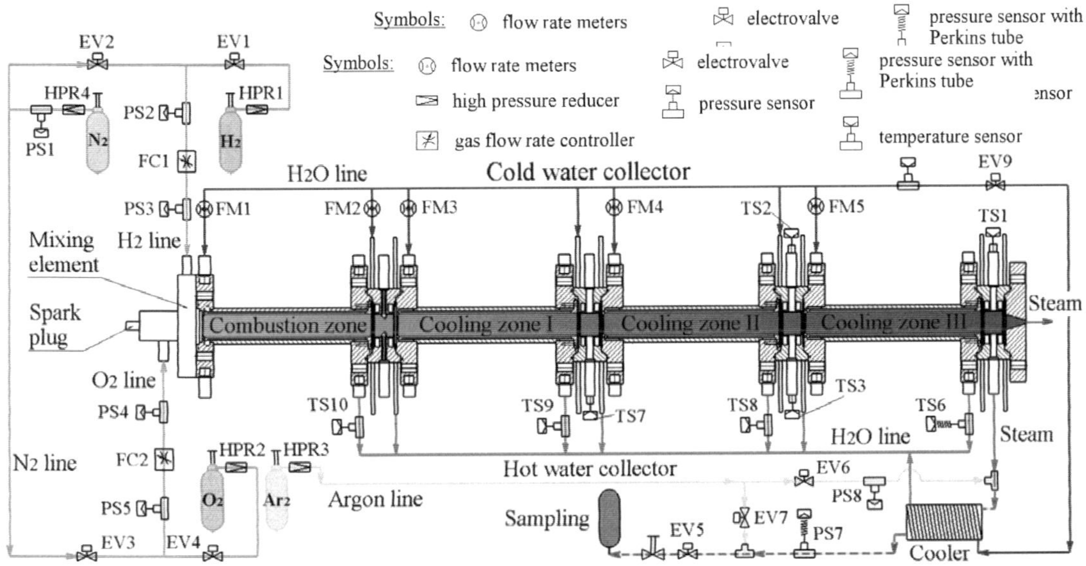

Figure 1. Schematic diagram of the diagnostic and control elements of the test stand. Symbols on the diagram: HPR1...HPR4 – high pressure reducer, EV1...EV9 – electrovalve, FC1, FC2 – gas flow controllers, PS1...PS8 – pressure sensors, TS1...TS10 – temperature sensors, FM1...FM5 – liquid flow meters.

The establishment of the studied mode of operation of the steam generator follows the general sequence of procedures. First, the gases-hydrogen and oxygen from cylinders with a pressure of up to 15 MPa-are fed into the system of the experimental stand. The spark plug is triggered. Next, the hydrogen ignites and the flame spreads in the combustion zone. At the same time, the water flow is supplied to cool the combustion chamber. The stationary mode is established by means of gas flow regulators. All elements of the steam generator housing have independent separate cooling from the common cold water collector, which is pumped through the cooling circuit and then drained into the hot water collector (figure 1). When the nominal operating mode is reached, the generated steam is sampled. After sampling, a stop is made by closing the electric valves that supply gas to the steam generator. Next, all gas channels are purged with nitrogen.

If the effect of quenching is minimized during vaporization in the experiment, then when sampling, according to the measurement method, in order to fix the volume of active particles in the resulting steam, it is necessary that the quenching process proceeds as efficiently as possible.

The sampling method is proposed and described in detail in [6, 8, 12]. Before sampling, argon is purged through the channels of the sampling system (figure 2). Then, after the pressure disturbances have calmed down, sampling begins. In this case, the extracted steam is diluted with argon and enters the capacitor (Pos. 5 figure 2). The remaining non-condensed components (hydrogen and oxygen in the argon atmosphere) are fed into the control sampling tank (Pos. 7 figure 2), which is pre-evacuated. Then the control sample is subjected to chromatographic analysis.

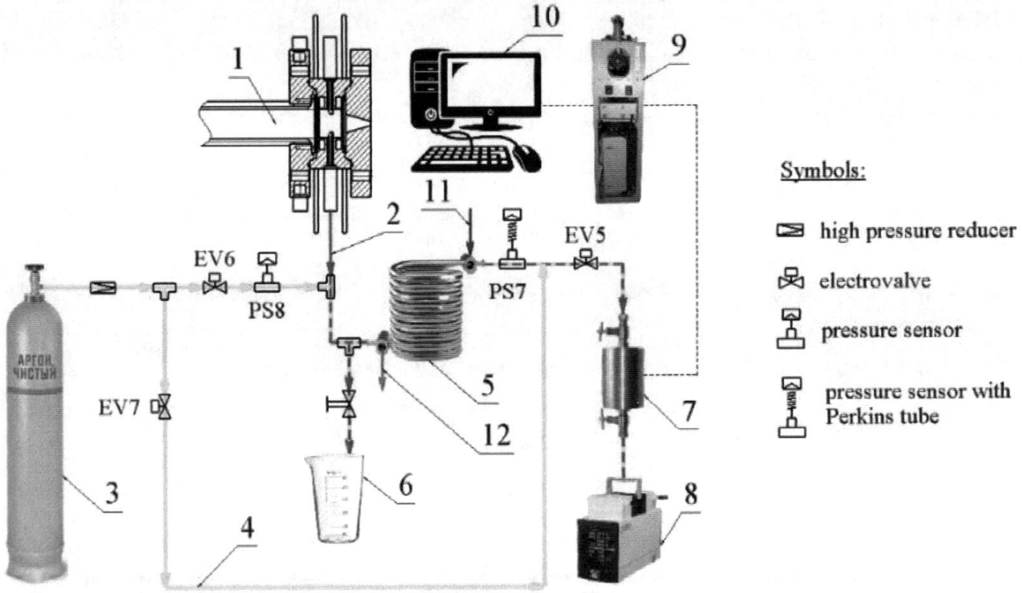

Figure 2. Sampling scheme. Symbols on the scheme: 1 – steam generator, 2 – steam selection, 3 – argon tank, 4 – purge line, 5 – cooler, 6 – condensate tank, 7 – sampler with pressure gauge, 8 – vacuum pump, 9 – chromatograph, 10 – personal computer, 11 – cold water, 12 – hot water.

All procedures (start - up, exit to the nominal mode, sampling, stop) are carried out according to the specified cyclogram of the experiment. A variant of the cyclogram for one of the modes is shown in figure 3. As can be seen from the example of the cyclogram, the time from start to stop is 35 seconds with a quick change of states of the control mechanisms of the stand. Therefore, the development of rapidly changing operations on the cyclogram was implemented using the created automated control system.

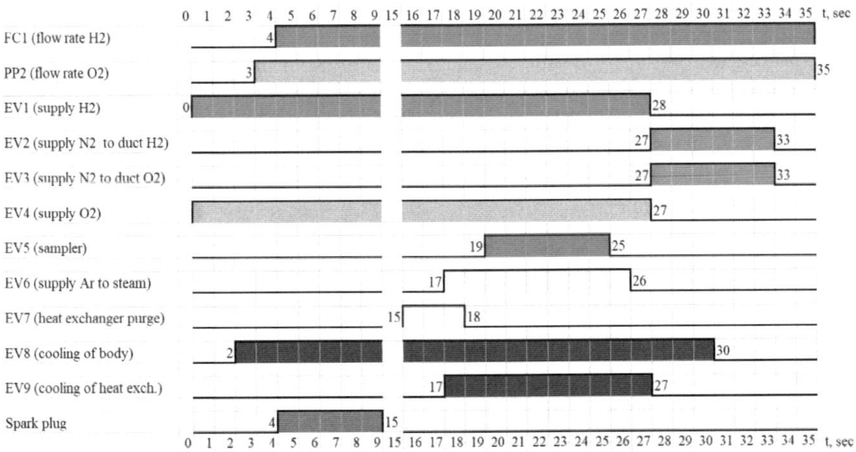

Figure 3. Example of an experiment's cyclogram.

3. Automation and control infrastructure

The development strategy of the automated system was based on the following principles:

- a use of hardware components analog-to-digital input-output converters, which will be redundant in their characteristics in terms of accuracy and frequency of both measurements and discrete control. This may allow us to find out the permissible uncertainty of the measured parameters for the characteristic time of the process and, in the future, create prerequisites for the general industrial performance of a HOSG automation in the hardware, which has higher performance indicators and is available at a reasonable price
- the use of software tools, in which the above-mentioned infrastructure will allow you to study any selected mode of operation of the steam generator, by having the ability to implement a variety of flexible control algorithms, which, among other things, can be based on the processing and analysis of the measured parameters of the experiment in real time

In accordance with the chosen strategy, the LabVIEW graphical engineering software environment with National Instruments hardware was used to complete the task. A flowchart based on the selected hardware is shown in Figure 4.

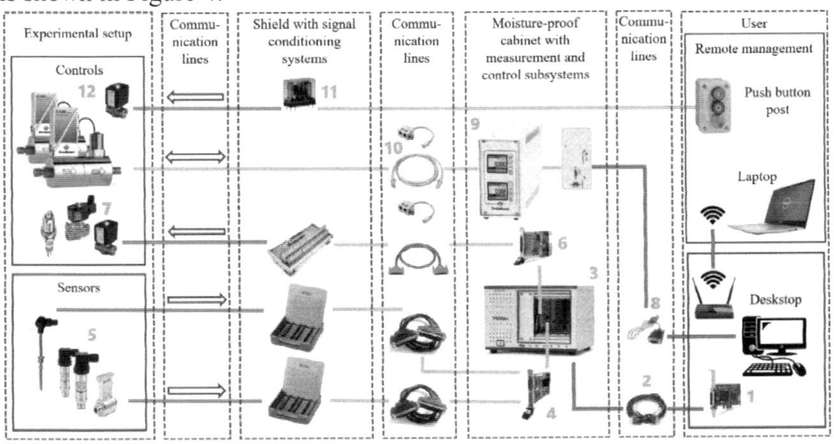

Figure 4. Structural diagram: 1 – electronic board PCIe-8361; 2 – cable MXI-Express; 3 – NI PXI-1078; 4 – analog I / O module NI PXI-6289; 5 - pressure sensors, thermocouples, turbine-type liquid flowmeters; 6 – relay module NI PXI-2568; 7 - gas electric valves, liquid electric valves, car spark plug, gas flow regulators Bronkhorst; 8 – cable RS232; 9 - display, control and power system E-8000; 10 – cable Ethernet; 11, 12 – push-button post relay and electric valve.

Journal of Physics: Conference Series **2039** (2021) 012023 doi:10.1088/1742-6596/2039/1/012023

To implement the cyclograms of experiments, a variant of one of them is shown in figure 5, a classical finite state machine was used, which allows:

- perform automaton steps (different states of control mechanisms) in non-overlapping time periods
- synchronize the switching of control mechanisms (electric valve, spark plug, gas flow regulators) between the steps of a state machine

Figure 5. Implementation of the start cycle diagram on the GUI.

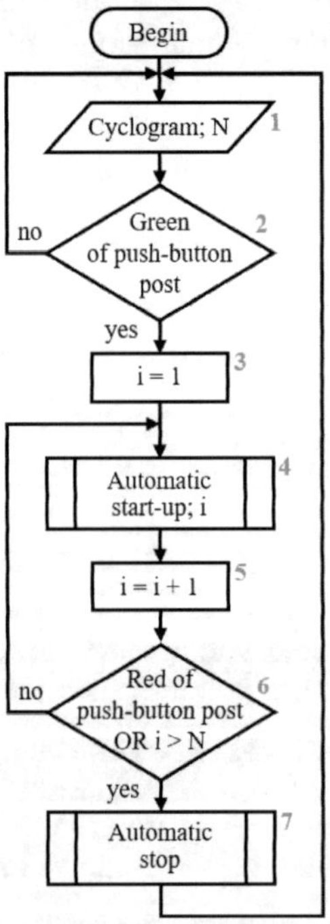

Designations:

1 – input of the experiment cyclogram, input of the number of steps of the state machine N

2 – a condition under which the green light on the remote button post is lit

3 – step number of the trigger state machine

4 – execution of the i-th step of the automatic start-up state machine

5 – moving to the sequence number of the next step of the state machine

6 – a condition in which the red button on the remote push button post is pressed or the steps of the machine start state of the installation have ended

7 – state machine stop

Figure 6. Simplified algorithm of the program.

To enter a cyclogram, a table with explicit transitions is set on the graphical user interface (GUI) (figure 5). When the state machine is turned on, the column states are executed sequentially from left to right, changing the position of the control mechanisms in accordance with the assigned duration of the individual state.

When the program is executed, seven cycles are executed in parallel, which exchange data streams with each other using queues. The data display cycle on the GUI and the data logging cycle are continuously performed at a high frequency and, accordingly, display and record the states of the controls and the readings of the diagnostic elements. A simplified view of the program's algorithm is shown in figure 6.

4. Results and discussion

Figure 7 and table 1 show comparative tests for stoichiometric combustion at two different hydrogen flow rates and at an excess of oxygen. In all experiments, a stable mode of operation of the HOSG is observed. With the same flow rate of hydrogen at stoichiometry, the curve goes higher, because the excess oxygen slightly lowers the temperature of the steam. In PXI systems, the input modules provide data sequentially, so the frequency of polling the input channels for temperature, pressure, and cooling water flow was lower than expected. This did not affect the validity of the automated process control system created for the study of the completeness of hydrogen combustion at this stand, more detailed results of which are given in [7, 8]. The accuracy of the measurement method is described in [6].

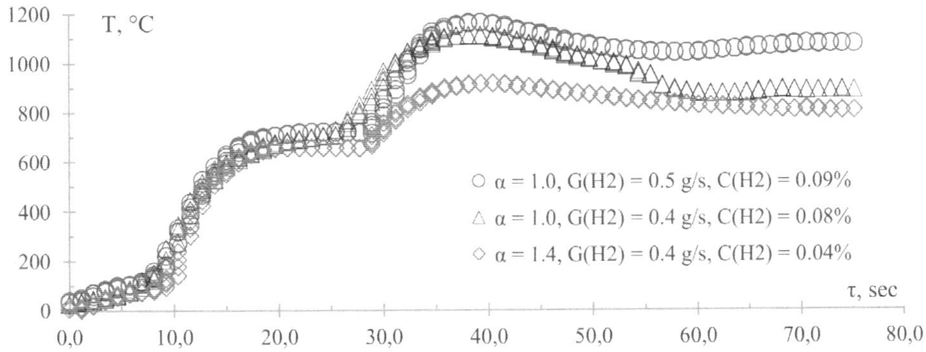

Figure 7. Temperature in the core of a generated steam at the outlet, α – stoichiometry coefficient, C(H2) concentration of a remaining hydrogen gas.

Table 1. Comparative tests of HOSG operating modes.

Average core temperature at the outlet, ° C	Oxidizer excess ratio	Average pressure, atm	Approximate power, kW	The hydrogen content of the generated steam, vol.%	
				Combustion zone	At the outlet of the cooling zone III
945	1.0	7.56	100	5.70	0.09
778	1.0	4.98	80	5.00	0.08
749	1.4	2.99	80	3.50	0.04

5. Conclusion

In this work, the technological development of the automated control system of thermal processes for a new sample of the HOSG is performed. The developed automated process control system in terms of software and hardware was installed and implemented on an experimental stand, on which a new sample of the HOSG was installed.

The software and logic control of the created automated process control system allows you to flexibly implement and test any control algorithms for promising HOSG that will be integrated into various power plant operation schemes. It is obvious that for general industrial applications, the created control and measurement system is redundant in its component base. However, it showed a high efficiency of obtaining experimental data due to the capabilities of the software package used, namely, the presence of SCADA and a flexible graphical programming environment.

During the study of the completeness of hydrogen combustion in an oxygen atmosphere at the stoichiometric ratio, for the first time the underburning of hydrogen at the exit from the HOSG lies in the range of 0.04 – 0.09 vol.%, which is an order of magnitude lower than at the previous stage of the study (0.26 – 1.21 vol.%) [8]. But the experimental values of the unreacted hydrogen content turned out to be slightly higher than the calculated values, which were estimated to be up to 0.006% of unreacted hydrogen [6]. Hence, we can conclude that it is possible to further improve the design of the steam generator.

The considered automated process control system under experimental conditions shows the reproducibility of the operating modes of the HOSG, the required efficiency of obtaining experimental data and increased safety during experimental work.

References

[1] Schastlivtsev A I and *et al* 2020 *High Temperature* vol 58 № 5 pp 733-743
[2] Schastlivtsev A I and Borzenko V I 2017 *Journal of Physics: Conf. Series* vol 891 № 1 p 012213
[3] Malyshenko S P and *et al* 2012 *High Temperature* vol 50 № 6 pp 765-773
[4] Aminov R Z and *et al* 2020 *Int. journal of hydrogen energy* vol 45 № 29 pp 14614-14624
[5] Malyshenko S and Schastlivtsev A 2015 *High Temperature* vol 53 № 4
[6] Aminov R Z and *et al* 2018 *Combining nuclear power plants with multifunctional power plants* (Moscow: Nauka) chapter 2 pp 60–62
[7] Schastlivtsev A I and *et al* 2019 *Journal of Physics: Conf. Series* vol 1370 № 1 p 012010
[8] Borzenko V I and Schastlivtsev A I 2018 *High Temperature* vol 56 №6 pp 927-932
[9] Schastlivtsev A I and *et al* 2020 *Journal of Physics: Conf. Series* vol 1652 № 1 p 012040
[10] Shapiro V I and *et al* 2011 *Thermal engineering* vol 58 № 9 pp 741-747
[11] Aminov R Z and *et al* 2020 *High Temperature* vol 58 № 3 pp 410-416
[12] Lédé J and *et al* 1983 Production of Hydrogen by Direct Thermal Decomposition of Water *(Int. J. Hydrogen Energy)* vol 8 № 9. p 675

Effect of high-temperature oxidation on the surface properties as applied to quenching of high-temperature bodies

I A Molotova[1], A R Zabirov[1,2], V V Yagov[1], E V Terentyev[1], N N Antonov[2], I M Molotov[2], A V Tumarkin[3] and M M Kharkov[3]

[1] National Research University "MPEI",
Russia, 111250 Moscow, Krasnokazarmennaya, 14
[2] Joint Institute of High Temperature,
Russia, 125412 Moscow, Izhorskaya, 13, Bd. 2
[3]National Research Nuclear University MEPhI
(Moscow Engineering Physics Institute),
Kashirskoe sh. 31, Moscow, 115409 Russia

Irina_molotova1705@mail.ru

Abstract. This paper contains the results of studying the surface properties before and after high-temperature oxidation. For this, the plate zirconium samples with chromium, gold and silver coatings were prepared. Cut profiles of the samples were obtained to study the structure of coatings and the thickness of the oxide layer. The measurements of contact angles were carried out. The results showed that a porous heterogeneous oxide layer was formed on the samples after high-temperature oxidation. At the same time, the wettability of the samples was improved. The thickness of the oxide layer on the chrome-coated zirconium sample was the smallest. Using of electroplated silver coating for experiments involving heating to high temperatures seems inappropriate because it was damaged after the oxidation tests. It is assumed that the main factor which influence on the rise of the transition temperature to the intensive cooling regime during quenching is the appearance of the oxide layer, rather than the improved wettability and wickability. High-temperature oxidation leads to the simultaneous formation of an oxide layer with a low thermal effusivity and to an improvement in wettability, therefore the contribution of each of these two effects on quenching can be confused or overestimated.

1. Introduction

Investigation of the cooling process of high-temperature bodies in liquids is of interest for many industries. These include thermal power engineering, metallurgy and nuclear power engineering. For nuclear energy and safety, after the accident at Fukushima in 2011 research works have begun in many scientific centers to find accident-resistant fuel for nuclear power plants (NPP) and to study its characteristics. New types of the fuel and cladding materials of fuel rods are named "accident-tolerant fuel" (ATF). Its concepts are aimed to delay the start of destructive process in the case of accidents by reducing the oxidation rate at high temperatures, as well as by potentially delaying the explosion. It is also necessary to investigate properties of the new materials in the event of a loss-of-coolant accident (LOCA) at NPP. When the core of a nuclear reactor is re-flooded, a film boiling regime takes place. It is characterized by low heat transfer coefficients and cooling rates. For fast cooling of the fuel rods and prevention of a catastrophe, it is necessary to achieve a transition from film boiling to a more intensive heat transfer regime. It is known that during quenching process, a special regime of heat transfer with a very high intensity may occur [1, 2]. This regime observes under certain conditions even at temperatures of the cooled surface significantly higher than the critical temperature for the cooling liquid.

The properties of the cooled surface significantly influence on the cooling process of high-temperature bodies in liquids, that confirmed by different researches. The surface temperature at which intensive heat transfer regime starts (further – transition temperature T_{tr}) can be significantly increased by changing in the surface structure, for example, by mechanical or physicochemical modifications. The fact that the rate of cooling processes is influenced by the thermal effusivity of the cooled body is confirmed by cooling experiments of high-temperature samples made of different metals. For example, the copper sphere cooled slower than the nickel and stainless steel [3]. In [4] the process of cooling high-temperature cylindrical samples (which simulate fuel rods) were studied. This experiments also confirm the influence of the cooled body properties on the cooling intensity. In another work [5] experiments were also carried out on the cooling of cylindrical samples in relation to the tolerant fuel concept. Zircaloy sample was cooled faster than stainless steel, copper and niobium due to its relatively low thermal conductivity.

Many authors apparently believe that an increase of wettability of the cooled surface leads to increasing in the T_{tr}. In some works, also used the concept of wickability and determined the effect of this parameter on the T_{tr} [6]. However, the etched porous surfaces after chemical etching, in addition to superhydrophilicity, also were rough. In [7] it was concluded that the T_{tr} is significantly influenced by nanoporosity, not only high surface wettability. In other works, the role of superhydrophilicity of the surface during cooling in subcooled water of a cylinder [8] and a sphere [9] was studied. Cooled zirconium samples were fabricated by anodic oxidation. It is reported that the superhydrophilicity had a strong effect at low subcooled temperatures, and its effect decreased at high temperatures.

There is a difficulty in analyzing the influence of the surface structure on heat transfer. Perhaps there is a misleading in the assessment of the wettability effect and the effect of a low thermal conductivity coating (oxide layer) on the rise of the T_{tr} during cooling of high-temperature bodies. Surface oxidation leads not only to the appearance of an oxide layer with a lower thermal effusivity ($\sqrt{\rho c \lambda}$), but also to an increase in its roughness, porosity, wettability. In addition, the results of different quenching studies point to the influence of surface oxidation on heat transfer during cooling of high-temperature bodies. In [10] oxidized zirconium and SiC samples had a higher transition temperature. In another study [11] the T_{tr} increased due to the formation of the oxide layer during repeated experiments on cooling a high-temperature zirconium sample. It in turn increased the wettability of the sample. The same effect is observed on an oxidized steel sample when quenched in saturated and subcooled water [12]. Based on the large array of our experimental data on the cooling of metal samples in liquids [2-4], an approximate physical model was presented in the paper [2]. In it the influence of the thermal effusivity of a cooled body on the transition temperature was taken into account. The model has confirmed its efficiency for cooling of high-temperature samples made of various metals.

Direct contact of liquid and solid surface is impossible during film boiling, so wettability can hardly affect cooling at cooled surface temperatures above critical for the cooling liquid. It is possible only short local liquid-solid contact. Therefore, wettability can be a secondary factor, which can increase by the formation of the oxide layer on the surface. In [13] oxidized samples also had a higher T_{tr} in comparison with unoxidized samples. Capillary wicking in unoxidized Zircaloy-4 was higher than in the Zircaloy with chromium coating. According to the authors' assumption, it is due to the oxidation of Zircaloy-4 in the environment. In addition, oxidized surfaces have greater roughness than non-oxidized samples, which also leads to a faster transition to intensive heat transfer regime for all samples.

As is known, cryoliquids have good wettability. Thus, during quenching in cryoliquids (usually liquid nitrogen), a change in the T_{tr} can be caused only by the thermophysical or geometrical properties of the cooled body or coatings [14, 15].

On the other hand, coatings to protect against the formation of the oxide layer are of great interest. With regard to cladding materials of fuel rods, the thickness of the oxide layer limits the service life of a nuclear reactor. Based on this, zirconium samples (since zirconium is a material for fuel cladding) with coatings with different thermophysical properties were selected for the experiments. Tests were carried out on high-temperature oxidation of these samples, followed by cooling in water. These experiments were conducted in order to study changes in wettability after high-temperature oxidation of the samples, as well as to study the structure of coatings before and after tests on high-temperature oxidation.

2. Test samples and methods

For the experiments, flat round samples of zirconium E110 with a diameter of 12.5 mm and a thickness of 1 mm were prepared. Three types of coatings were applied to the samples: chromium with a thickness of about 1 μm, gold and silver with a thickness of about 3 μm. Before applying the coatings, the samples were polished. Chromium was deposited by magnetron sputtering, details on the sputtering method are presented in [16, 17]. The dense coating was obtained as a result, an arithmetic average profile deviation (Ra) was 1 μm (Figure 1a). Gold and silver were applied by electroplating (Figure 1b, c), Ra was 0.5 and 1.7 μm for gold and silver coatings, respectively. Figure 1 shows the results of optical microscopy of the surface of zirconium samples with these coatings.

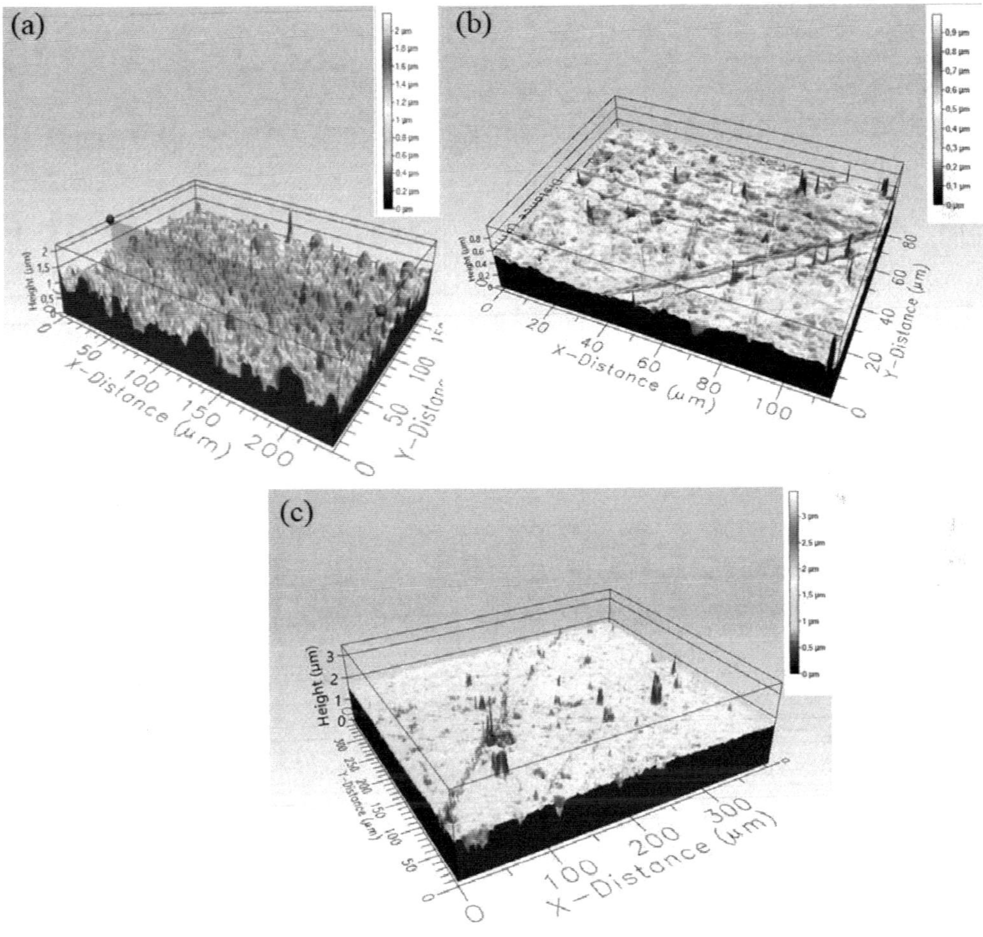

Figure 1. Optical microscopy of the zirconium samples with coatings of (a) chromium; (b) gold; (c) silver.

Before the experiments, the contact angles were measured for the samples with different coatings. The oxidation of the samples was carried out in a crucible melting furnace, which was preheated to 700 °C. The samples were fixed on a stainless steel wire and immersed in the crucible for 30 minutes. The heated samples were cooled in water at room temperature (24 °C). Thereafter, the high-temperature oxidation test was repeated. Between and after oxidation tests, contact angles were also measured. Also, part of the samples was cut for studying the structure and thickness of the coatings and the thickness of the formed oxide layer. It was done before and after oxidation tests and cooling in water.

3. Results and discussion

It should be noted that zirconium with chromium coating is considered as a tolerant material for fuel rods in nuclear reactors, so the oxidation of this sample is of particular interest for nuclear power. Cut profile of the zirconium sample with chromium coating showed that the thickness of the chromium layer is 1 μm (Figure 2a). The coating was enough uniform and non-porous. After oxidation tests, a layer of porous and heterogeneous oxide formed on the sample surface with a thickness of 4-7 μm (Figure 2b).

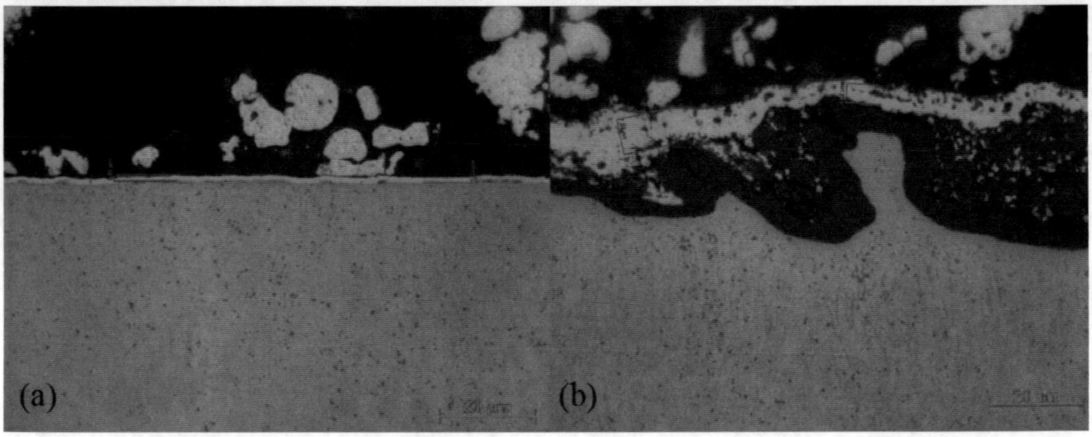

Figure 2. Cut profiles of the zirconium sample with chromium coating before (a) and after (b) high-temperature oxidation tests.

The thickness of gold electroplating coating on zirconium was about 3 μm (Figure 3a). After oxidation tests, oxide layer also had pores. The thickness of the oxide layer was greater than that of the chromium-coated zirconium and was about 10 μm (Figure 3b).

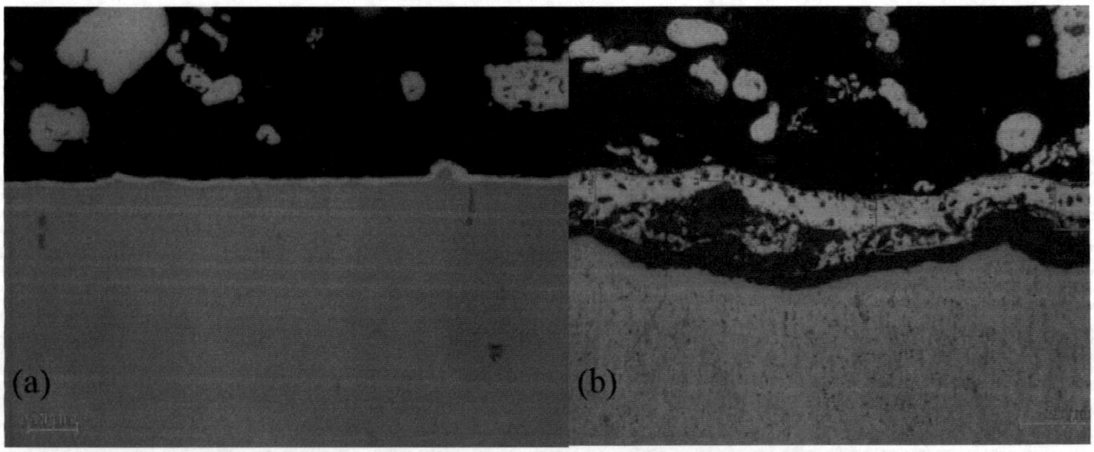

Figure 3. Cut profiles of the zirconium sample with gold coating before (a) and after (b) high-temperature oxidation tests.

The thickness of silver electroplating coating on the zirconium was also about 3 μm (Figure 4a). For better adhesion of silver to zirconium, a copper sublayer with a thickness of 1-3 μm was applied, which is visible in Figure 4a. The silver coating did not pass the high-temperature oxidation tests, and cracks were observed on the coating with a thick oxide layer (thicker than 10 μm) (Figure 4b). Apparently, the coating burned out during experiments on high-temperature oxidation. So, silver electroplating coating is not suitable for protecting of the samples that will be heated to high temperatures of about 700 °C.

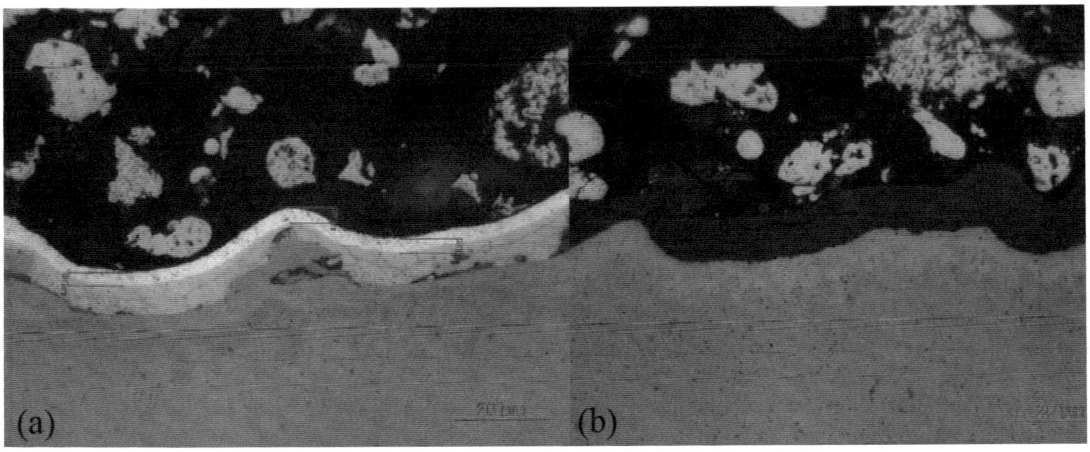

Figure 4. Cut profiles of the zirconium sample with silver coating before (a) and after (b) high-temperature oxidation tests.

Wettability of the samples was measured before the high-temperature oxidation tests (Figure 5a), between oxidation, and also after experiments (Figure 5b). In Figure 5 it is clearly seen the improved of the wettability on the oxidized samples. A drop of water spreads very rapidly when it is deposited on the sample after high-temperature oxidation tests. The average results of the contact angles were summarized in Table 1. As evidenced by the results, the contact angles for all types of coatings on the flat zirconium samples decrease several times after the oxidation of the samples.

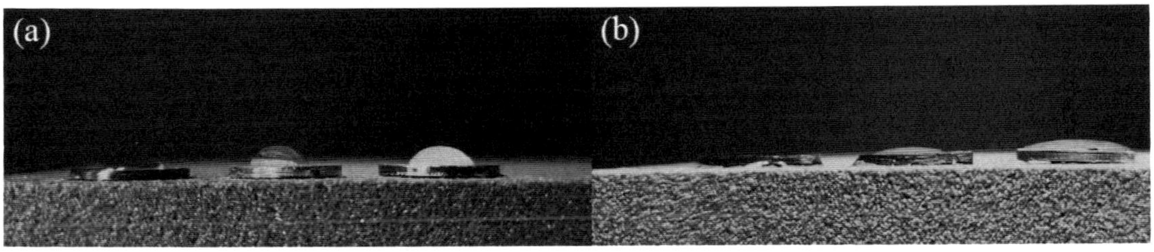

Figure 5. Wettability of the samples before (a) and after (b) two high-temperature oxidation tests, from left to right: chromium, gold, silver coatings.

Table 1. Contact angles of zirconium samples with coatings before and after high-temperature oxidation tests.

Zr with chromium coating	Zr with gold coating	Zr with silver coating
	Before oxidation	
58	59	47
	Between oxidation tests	
35	30	38
	After oxidation	
9	17	17

Thus, the oxide layer with low thermal effusivity was formed on all tested samples. In this case, the thinnest oxide layer was on the zirconium sample with chromium coating. High-temperature oxidation, in turn, has led to improved wettability for all types of coatings (especially for chrome coating). It supports the idea that the formation of the oxide layer usually leads to improved wettability.

4. Conclusion

For the experiments on high-temperature oxidation, the flat round zirconium samples with different coatings were prepared: chromium, gold and silver. After two cycles of high-temperature oxidation in a crucible furnace, porous and heterogeneous oxide layers formed on all samples. The silver coating did not pass these tests. It was suggested that the increasing of the transition temperature to intensive cooling regime during quenching is influenced by the growth of the oxide layer, more precisely, its thermal effusivity. In this case, the wettability is a secondary factor, which growth is caused by the formation of this oxide layer. This hypothesis was confirmed by the conducted experiments. In the experiments presented in this paper the formation of the oxide layer on samples with different coatings led to an improvement in their wettability (the contact angles decreased from 50-60° to 10-20°). Further research will focus on experiments on cooling high-temperature cylindrical zirconium samples without and with different coatings in relation to the tolerant fuel concept (including chromium-coated zirconium). These studies will make it possible to experimentally investigate the transition temperatures to the intensive cooling regime and to verify the developed approximate model of the occurrence of the intensive heat transfer regime during cooling of high-temperature bodies in liquids.

Acknowledgments

This work was supported by the Russian Foundation for Basic Research, Grant No. 19-38-60057.

References

[1] Aziz S, Hewitt G F and Kenning D I 1986 *Int. Heat Transf. Conf. Dig. Libr.* **5** 2149-2154
[2] Yagov V V, Zabirov A R, Kanin P K 2018 *Int. J. Heat Mass Transf.* **126** 823–830
[3] Yagov V V, Minko K B and Zabirov A R 2021 *Int. J. Heat Mass Transf.* **167** 120838
[4] Molotova I A, Zabirov A R, Yagov V V, Vinogradov M M and Belyaev I A 2020 *IOP Conf. Series: J. Phys. Conf. Ser.* **1683** 032043
[5] Lee C Y and Kim S 2017 *Nucl. Eng. Des.* **313** 118-128
[6] Li Q, Mou L W, Zhang J Y, Zhang Y H and Fan L W 2018 *Int. J. Heat Mass Transf.* **125** 494-505
[7] Kim H, Truong B, Buongiorno J and Hu L-W 2011 *Appl. Phys. Lett.* **98** 083121
[8] Kang J Y, Lee G C, Kim M H, Moriyama K and Park H S 2018 *Int. J. Heat Mass Transf.* **117** 538-547
[9] Kang J Y, Kim T K, Lee G C, Park H S and Kim M H 2018 *Int. J. Heat Mass Transf.* **120** 399-410
[10] Kang J-Y, Kim T K, Lee G C, Kim M H and Park H S 2018 *Ann. Nucl. Energy* **112** 794–807
[11] Lee K G, In W K and Kim H G 2019 *Nucl. Eng. Des.* **347** 10-19
[12] Zabirov A, Kanin P, Vinogradov M, Sliva A, Dedov A, Fedorovich S and Yagov V 2018 *Mat. Tod.: Proc.* **5 (12)** 26171-26178.
[13] Seshadri A and Shirvan K 2018 *Nucl. Eng. Des.* **338** 5-15
[14] Zabirov A R, Kanin P K, Vinogradov M M and Sharafutdinov A M 2018 *IOP Conf. Series: J. Phys. Conf. Ser.* **1128 (1)** 012015
[15] Pavlenko A N, Tsoi A N, Surtaev A S, Kuznetsov D V and Serdyukov V S 2016 *High Temp.* **54(3)** 370-376
[16] Tumarkin A V, Kaziev A V, Kolodko D V, Pisarev A A, Kharkov M M and Khodachenko G V 2015 *Phys. Atom. Nucl.* **78 (14)** 1674-1676
[17] Kaziev A V, Tumarkin A V, Leonova K A, Kolodko D V, Kharkov M M and Ageychenkov D G 2018 *Vacuum* **156** 48-54

Investigation of the influence of narrowing annular channel and Reynolds number on formation of Tayler- Görtler vortexes

E V Pankratov

Northern (Arctic) Federal University named after M.V. Lomonosov, Russia, 163002 Arkhangelsk, Severnaya Dvina Emb., 17

e.pankratov@narfu.ru

Abstract. Abstract. The article investigates the Taylor-Görtler vortices arising in a swirling flow when the annular channel narrows. Several options for the geometry of the narrowed annular channel are researched. In the first case, the outer cylinder with a constant diameter and the inner cone with a variable cone angle are considered. In the second case, on the contrary an inner cylinder with a constant diameter and an outer cone with a variable cone angle are considered. All geometries were tested at different Reynolds numbers Re. = $8.3 \cdot 10^3 \ldots 21 \cdot 10^3$. As a result, the analysis of the propagation of secondary vortices along the length of the annular channel is presented.

1. Introduction

Hydrodynamically unstable flows in a coaxial annular channel were first experimentally investigated by Taylor in 1923 [1]. He proved the existence of toroidal vortices in which fluid moves downwash periodically and then in which fluid moves upwash. Further development of research was continued by Goertler in 1955 [2]. He studied the stationary vortices formation that arises due to the imbalance of pressure and centrifugal forces in the boundary layer along concave walls. To describe the effect of instability on the Taylor vortex mechanism, we consider a coordinate system rotating together with the flow core. Then the second Taylor number can be written as:

$$\mathrm{Ta}_2 = \frac{w_\varphi^2 r_{\mathrm{in}} \left(r_{out} - r_{\mathrm{in}}\right)^3}{\upsilon_{eff}^2}, \tag{1}$$

where w_φ – characteristic tangential velocity; r_{in} – internal radius of the annular channel; r_{out} – external radius of the annular channel; υ_{eff} – effective kinematic viscosity.

The critical number for the formation of secondary currents is $\mathrm{Ta}_2 = 1700$. This value is close to the Grashof number. Since the value of the critical Grashof number for the Rayleigh-Benard instability is also close to 1700.

In most of the studies that were considered earlier, vortex flows were formed as a result of the surface rotation. An attempt to investigate secondary vortices in a swirling flow with fixed surfaces was made by Kim and Choi [6]. It was a continuation of Euteneuer's research [7]. In the study hydrodynamic instability in a swirling flow was caused by the impulsively decelerating of a rotating cylinder. It was shown that a swirling flow is formed when there is the surface impulsively decelerating. Taylor-Görtler vortices develop in this swirling flow. However, the characteristics of secondary flows in stationary annular channels at constant flow swirling have not been studied fully.

2. Methods

In this article, we investigate the narrowing annular channels with a swirling flow. Geometry options for narrowing annular channels are shown in Figure 1.

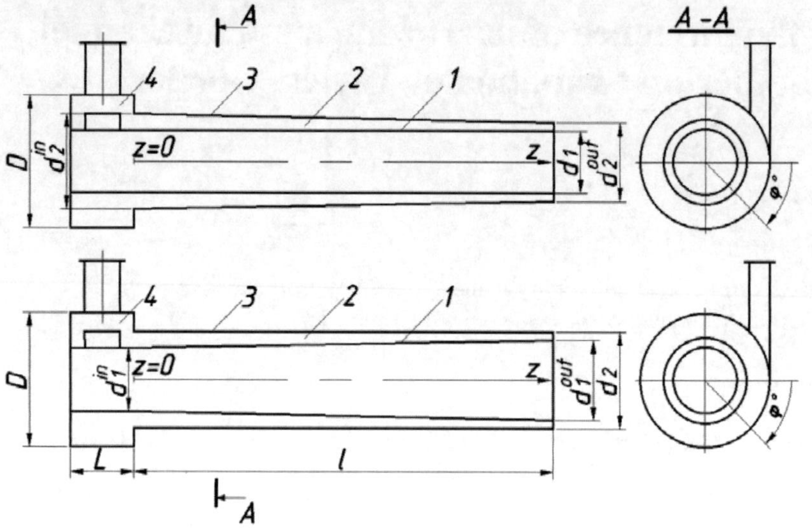

Figure 1. Schematic drawing of the investigate geometry.

A swirling flow is formed in a swirl generator with diameter $D = 259$ mm and length $L = 126.5$ mm. In this study we use a swirl generator with one tangential channel 35 × 70 mm. The annular channel narrows due to changing of the ratio of the areas at the exit from the annular channel f_{ch}^{out} at $z = l = 840$ mm, and at the entrance to it f_{ch}^{in} at $z = 0$. The investigated geometric and operating characteristics of the narrowing annular channels at various ratios f_{ch}^{out}/f_{ch}^{in} re are given in Table 1. The studies were carried out with changing of Reynolds number Re from $8.3 \cdot 10^3$ to $21 \cdot 10^3$.

Table 1. Plan of experiment

	d_1	d_2^{out}	Cone angle	f_{ch}^{out}/f_{ch}^{in}	Re		d_1^{out}	d_2	Cone angle	f_{ch}^{out}/f_{ch}^{in}	Re, 10^3
1	152	184	0	1	8300	**16**	152	184	0	1	8300
3					21700	**18**					21700
4		178.1	0.47	0.80	8300	**19**	158.9		0.41	0.80	8300
6					21700	**21**					21700
7		171.9	0.92	0.60	8300	**22**	165.5		0.82	0.60	8300
9					21700	**24**					21700
10		165.5	1.36	0.40	8300	**25**	171.9		1.26	0.40	8300
12					21700	**27**					21700
13		158.9	1.78	0.20	8300	**28**	178.1		1.71	0.20	8300
15					21700	**30**					21700

Numerical simulation is carried out in three-dimensional formulation with using of the software package ANSYS Fluent 15.0. The solution is made on a structured hex mesh. A mesh-independent solution was found. As a result, a grid with 9 million cells is used, where the minimum size is $2.5 \cdot 10^{-6}$ m (at the surfaces $y^+ \approx 1$) and the maximum dimension is $3.5 \cdot 10^{-3}$ m (in the swirling generator). The numerical simulation was verified on the basis of the physical experiment. The relative errors of the experimentally determined values did not exceed 2.7% for the air flow rate and 0.4% for the velocity. Three turbulence models were compared: standard k-ε with curvature corrections; SST k-ω with curvature corrections and Reynolds stress model. Figure 2 shows a comparison of the experimental and numerical distributions of dimensionless tangential $w_\varphi = w_\varphi/V_{in}$ and axial $\bar{w}_z = w_z / V_{in}$ (V_{in} - is the average velocity in the input channel) components of the velocity vector in the cross section. The measurements were carried out in the sections of the annular channel at z = 290 mm, z = 538 mm, and z = 808 mm (z is the longitudinal coordinate measured from the swirl generator along the annular channel axis).

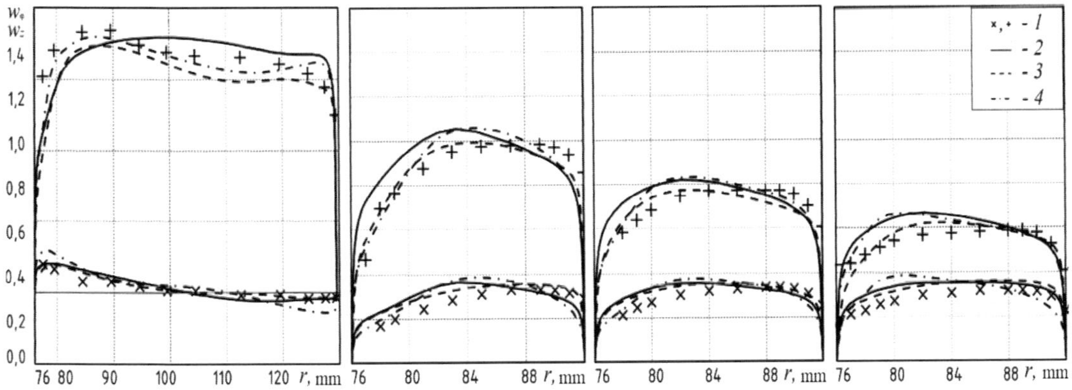

a *b* *c* *d*

a - in the middle section of the swirl generator at φ = 180°; *b* - in section *z* = 290 mm at φ = 0°; *c* - in section *z* = 538 mm at φ = 0°; *d* - in section *z* = 808 mm at φ = 0°; 1 - experimental data; 2 - standard *k*-ε model; 3 - SST *k*-ω model; 4– model RSM

Figure 2. Full velocity components in the cross section of the swirl generator.

The velocity distributions most accurately describe the results of calculations obtained using Menter's two-parameter model of shear stresses SST *k*-ω with curvature corrections, and the solution with the standard *k*-ε with curve correction model gives the lowest accuracy.

3. Results

The study was carried out for dry air at an inlet temperature of 20 °C. when entering the Reynolds number $Re_{in} = 8.3 \cdot 10^3 \ldots 21 \cdot 10^3$ and Prandtl number $Pr_{in} = 0.7$. One-sided local input into the swirling generator determines azimuthal non-uniformity of total velocity distribution and its components. It leads to asymmetric airflow into the annular channel [1]. Figure 3 shows the distribution of the total velocity components (tangential w_φ, axial w_z and radial w_r) in the cross section of the flow swirl generator.

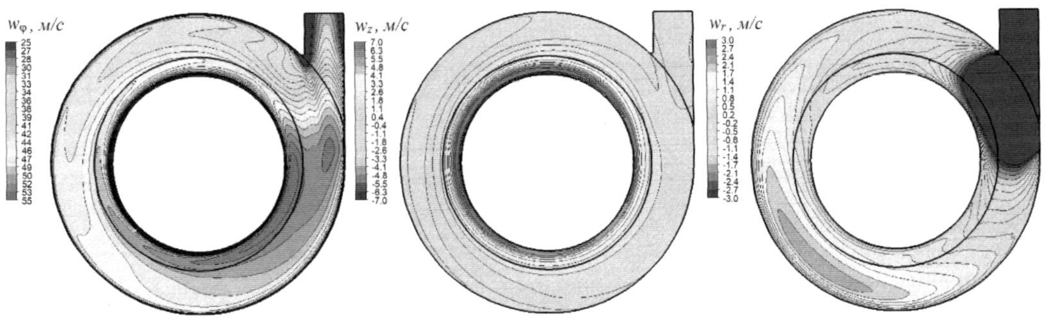

Figure 3. Full velocity components in the cross section of the swirl generator

Based on the figure, it can be confirmed that the tangential component of the velocity is prevalent among all the components of the total velocity. The axial and radial components of the velocity have

lesser significant influence. It should be noted that the swirl generator was smooth in all geometry modifications. Consequently, the aerodynamics in each of the considered cases was identical.
It should also be noted that periodic large-scale unsteady vortices are formed from the leading edge of the outer cylinder of the swirl generator. These vortex structures are characterized by Strouhal numbers are equal to 0.22. In turn it leads to the formation of structures like Taylor - Görtler vortices near the outer wall. These secondary streams are distributed over the entire space of the annular gap. In this case the axes of the secondary vortex structures coincide with the direction of motion of the main swirling flow.

In Figure 4 (a), where a physical experiment is visualizing, confirmation of the presence of secondary flows in annular channels during flow swirling was obtained. This secondary flow has a complex, three-dimensional and unsteady separation flow. The manifestation of instability is explained by the action of centrifugal forces in the boundary layer when flowing around concave surfaces. Figure 4 (b) shows a numerical calculation which qualitatively repeats the experimentally obtained data.

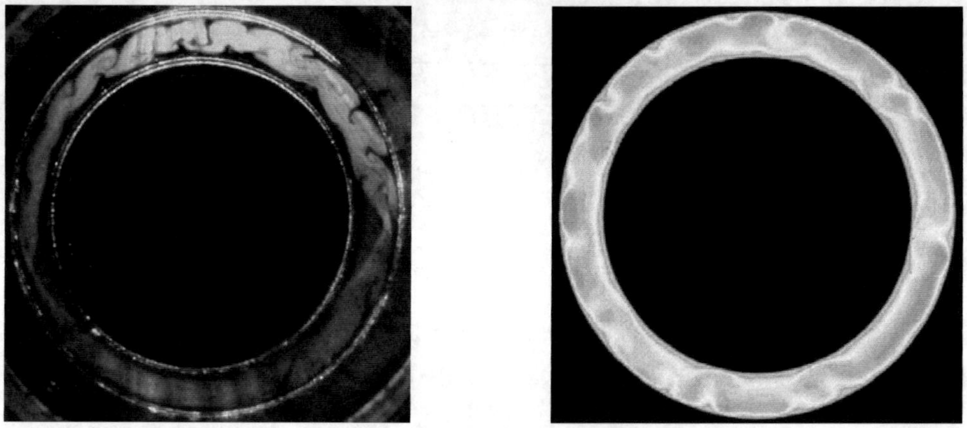

(a) (b)

Figure 4. Secondary flows similar to Taylor-Görtler vortices

First, we consider the case when the diameter of the outer surface d_2 doesn't change. In this case the diameter of the inner surface at the outlet d_1^{out} changes in accordance with the data in Table 1. Figure 5 shows the distribution of the radial velocity along the length of the annular channel for various ratios f_{ch}^{out}/f_{ch}^{in} and at Re. = $21\cdot10^3$.

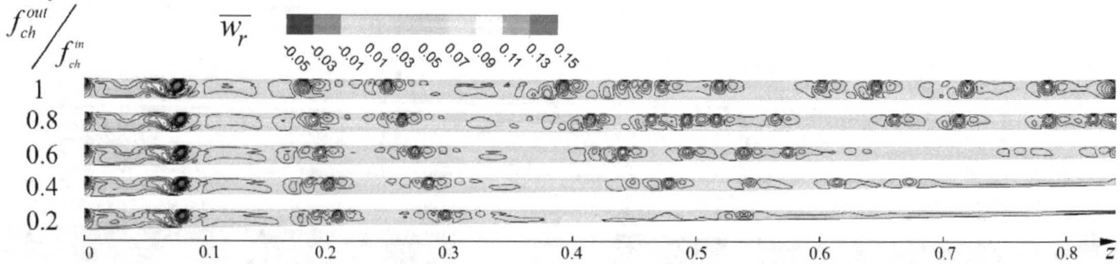

Figure 5. Distribution of the radial velocity along the length of the annular channel when the diameter d_1^{out} changes with different ratios f_{ch}^{out}/f_{ch}^{in} and at Re. = $21\cdot10^3$.

When d_2 is unchanged, the formation of secondary Taylor-Görtler vortices for all f_{ch}^{out}/f_{ch}^{in} begins at $z = 0.08$. It is due to the insignificant effect of the narrowing in the initial sections of the annular channel. At ratios f_{ch}^{out}/f_{ch}^{in} from 1 to 0.8 secondary vortices propagate along the entire length of the annular channel. It should be noted that the number of vortices increases when there is moving along the annular channel. At $f_{ch}^{out}/f_{ch}^{in} \geq 0.6$ the secondary vortices begin to decay at $z \geq 0.5$. It is due to the decrease in the intensity of the flow swirl. Along the length of the annular channel, the axial component of the velocity increases, while the tangential and radial components decrease.

Figure 6 shows the distribution of the radial velocity along the length of the annular channel at various ratios f_{ch}^{out}/f_{ch}^{in} and at Re. = $8.3 \cdot 10^3$.

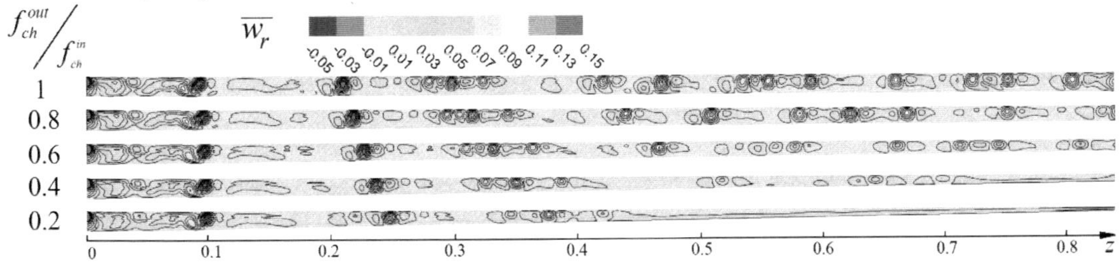

Figure 6. Distribution of the radial velocity along the length of the annular channel when the diameter d_1^{out} changes with different ratios f_{ch}^{out}/f_{ch}^{in} and at Re = $8.3 \cdot 10^3$.

A decrease in the value of Reynolds number leads to the later formation of secondary Taylor-Görtler vortices at $z \geq 0.1$. Due to the decrease in the input speed, more intense decrease in the flow swirl occurs. However, when there is decrease of Reynolds number, the decay of secondary vortex structures occurs at $f_{ch}^{out}/f_{ch}^{in} \geq 0.6$. The decay of secondary vortices also occurs at the higher values of the length z.

Next, we consider the case when the diameter of the inner surface d_1 doesn't change. In this case, the diameter of the outer surface at the outlet d_2^{out} will change in accordance with the data in Table 1. Figure 7 shows the distribution of the radial velocity along the length of the annular channel for various ratios f_{ch}^{out}/f_{ch}^{in} and at Re. = $21 \cdot 10^3$.

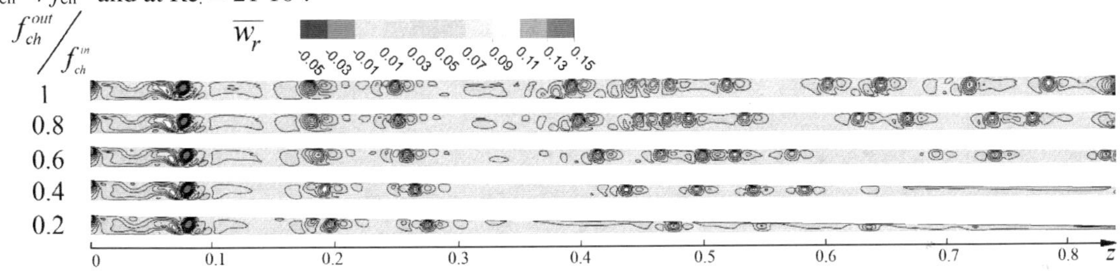

Figure 7. Distribution of the radial velocity along the length of the annular channel when the diameter d_2^{out} changes with different ratios f_{ch}^{out}/f_{ch}^{in} and at Re = $21 \cdot 10^3$.

When d_1 is unchanged, the formation of secondary Taylor-Görtler vortices for all f_{ch}^{out}/f_{ch}^{in} begins at $z \geq 0.08$. Disintegration occurs when the secondary flow $f_{ch}^{out}/f_{ch}^{in} \leq 0.4$. In this case the length of the decay of vortex structures is $z \geq 0.6$. Figure 8 shows the distribution of the radial velocity along the length of the annular channel at various ratios f_{ch}^{out}/f_{ch}^{in} and at Re. = $8.3 \cdot 10^3$.

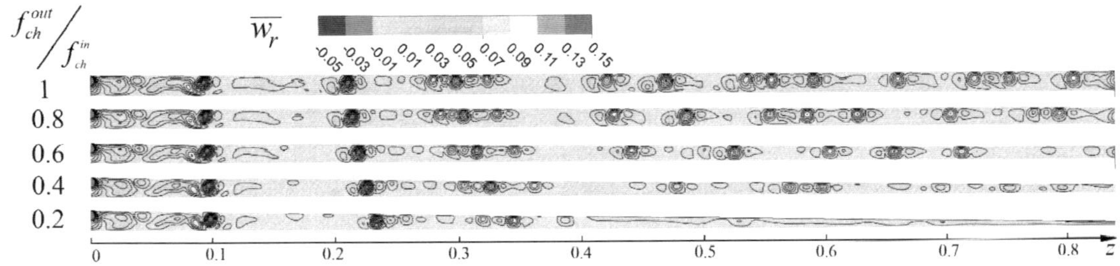

Figure 8. Distribution of the radial velocity along the length of the annular channel when the diameter d_2^{out} changes with different ratios f_{ch}^{out}/f_{ch}^{in} and at Re = $8.3 \cdot 10^3$.

When Reynolds number Re decreases, decay of secondary vortices occurs later. Thus, secondary vortex structures are retained at $f_{ch}^{out}/f_{ch}^{in} \geq 0.4$. At $f_{ch}^{out}/f_{ch}^{in} = 0.2$ the decay of vortices begins with a length $z \geq 0.55$.

In all the cases, that were considered, the secondary Taylor-Görtler vortices propagate along the entire length of the annular channel and disintegrate into smaller structures. These structures are destroyed downstream due to secondary instabilities and the development of small-scale turbulence. However, there is a region in which secondary vortices will dominate the flow before they will decay.

4. Conclusion

The results obtained for conical annular channels indicate a noticeable effect of the channel width on the state of secondary vortices. There is a decrease in the flow swirl with a decrease in the width of the beneficial section area f_{ch}^{out}. This follows from the stability criterion of the second Taylor number, in which the value of the width of the annular gap is raised to a cube. The restructuring of the flow, that is related to its acceleration and derotation leads to the disintegration and gradual degeneration of secondary flows such as Taylor - Görtler vortex flows. Secondary vortices exist for the longest time in straight annular channels with lower Reynolds numbers Re. However, when there is narrowing of the annular channel, the more uniform distribution of secondary vortices occurs in the cross section.

Acknowledgments

This work was supported by a grant from the Young scientists of Pomorye № 13-2021a.

5. References

[1] Taylor G I 1923 Stability of a viscous liquid contained between two rotating cylinders *Phil. Trans. R. Soc. Lond. A* **223** 289–343

[2] Görtler H 1955 Dreidimensionales zur Stabilitätstheorie laminarer Grenzschichten *Z. Angew. Math. Mech.*

[3] Mitrofanova O V 2010 *Hydrodynamics and heat transfer of swirling flows in channels of nuclear power plants [in Russian]* (Moscow: FIZMATLIT)

[4] Shchukin V K 1980 *Heat Exchange and Hydrodynamics of Internal Flows in Fields of Body Forces [in Russian]* (Moscow)

[5] Wimmer M 1995 An experimental investigation of Taylor vortex flow between conical cylinders *Journal of Fluid Mechanics* **292** 205–27

[6] Kim M C and Choi C K 2004 The onset of Taylor-Görtler vortices in impulsively decelerating swirl flow *Korean J. Chem. Eng.* **21** 767–72

[7] Euteneuer G-A 1972 Die Entwicklung von Längswirbeln in zeitlich anwachsenden Grenzschichten an konkaven Wänden *Acta Mechanica* **13** 215–23

Investigation of the influence of Taylor-Görtler vortexes on heat transfer of annular channel

E V Pankratov

Northern (Arctic) Federal University named after M.V. Lomonosov,
Russia, 163002 Arkhangelsk, Severnaya Dvina Emb., 17

e.pankratov@narfu.ru

Abstract. This article is about the influence of Taylor-Görtler vortices on heat transfer in concentric annular channels with turbulent decaying swirling flows. The study shows that the occurrence and transformation of secondary vortex structures has a significant effect on the distribution of heat flux over the annular channel surface. An explicit is relationship between the radial velocity fluctuations and the heat flux density distribution. The highest intensity of heat transfer on the outer surface is observed in the areas of positive radial velocity values, while on the inner surface it is observed in the areas of negative radial velocity values.

1. Introduction

Secondary flows often arise in annular channels under different conditions. The presence of additional vortex structures allows for more efficient dissipation of thermal energy. Secondary flows are divided into several types. The circulating motion in the shape of paired vortices is indicated by a secondary flow of the first kind. [1,2]. This type of flow is formed when the centrifugal force F at any point satisfies the condition $\mathrm{rot}F \neq 0$. Vortex structures with alternation of right and left rotation along the swirling flow at $\mathrm{rot}F = 0$ are indicated by a secondary flows of the second kind [1,2].

M. Couette was one of the first to experimentally study unstable flows limited by two dimensions in 1890 [3]. He studied the flow in an annular channel formed by two concentric cylinders, while the inner cylinder was fixed, and the outer one was rotating. As a result, when the velocity was less than a certain value it was found that the friction loss exerted by the fluid on the inner cylinder was proportional to the velocity of the outer cylinder. When this velocity was exceeded, the friction loss increased. The author explains this change by the transition of the flow from laminar to turbulent motion.

A. Mallock obtained in his study results similar to those of Couette in 1896 [4]. In addition, he conducted a series of experiments covering cases where the inner cylinder was rotating and the outer one was at rest. In this configuration, flow instability was found for all rotation velocity values of the inner cylinder. Also, a large difference was noted in terms of stability between the cases where the inner and outer cylinders were fixed. It shows that in the case when the inner cylinder rotates while the outer cylinder is stationary, the secondary flows have a complex three-dimensional character.

One of the most important studies related to secondary vortices was carried out by D. Taylor in 1923 [5]. In this research he studied the stability of a viscous fluid located between two rotating cylinders. He showed that fluid flow becomes unstable when the angular velocity of the inner cylinder exceeds a certain value. This instability is represented by axisymmetric toroidal vortices, which have alternating left and right rotation.

G. Görtler investigated a similar problem on the rotation of two coaxial cylinders in 1955 [6]. He was the first who showed that the action of centrifugal forces in the boundary layer on concave surfaces leads to flow instability. D. Taylor observed similar secondary vortices in his work. In all previous works, cases were investigated when secondary flows are formed during the motion of an annular channel surfaces.

However, secondary flows can form in a swirling flow when the surfaces of the annular channel are stationary [1,7]. In the author's previous work [8], it was shown that secondary flows of the Taylor-Görtler type can be formed when the flow swirls with fixed surfaces of the annular channel. However, the effect of Taylor-Görtler vortices on the intensification of heat transfer in an annular channel during flow swirling has not been studied enough.

2. Methods

The considered in this work geometry is shown in figure 1. The annular channel (2) is defined by two coaxial cylinders. The inner surface (1) with a diameter of d_1 = 152 mm, and the outer (3) with a diameter of d_2 = 184 mm. Channel length l is 840 mm. Air is supplied through a tangential channel (3) with dimensions of 35 × 70 mm to the swirl generator (4) with a diameter of D = 259 mm and a length of L = 126.5 mm. The flow movement is measured by the angle φ. This angle is measured from the mating point of the tangential channel from the inner surface.

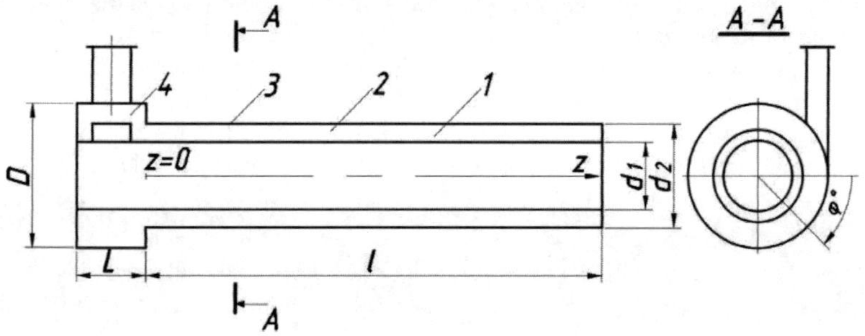

Figure 1. The geometry of the investigated annular channel.

Aerodynamics was investigated using a Laser Doppler Anemometer (LDA). For this, an optically transparent experimental model of the annular channel in full-size was made. The velocity measuring section is equipped with a corrective lens. The lens eliminates unwanted aberrations of the laser beams. This improves the measurement accuracy [9]. The error in measuring the LDA velocity is less than 0.1 %. by the method.

In the experimental bench for measuring aerodynamics, air inoculated with propylene glycol tracer particles with a size of 1 to 1.5 μm was used. The volumetric flow is controlled via a fans with frequency converter. Air flow is measured using the differential pressure method. A flow preparation device is installed on the sensing line, which allows to reduce turbulence in front of the diaphragm. To determine the physical quantities in the experimental bench, the following are used: a thermal resistance sensor, gauge pressure sensors, and a differential pressure sensor. All signals from the sensors are fed into the data collection system, which allows you to process and analyze the signals received from the sensors in real time. The flow measurement error is 2.7%.

Heat transfer was investigated using gradient heat flux sensors. A calorimeter was made to install the sensors. It is a hollow tube to which superheated steam is supplied. The heat flux profiles in the cross section were measured by rotating the calorimeter every 10° around its axis. The length measurement was carried out by moving the calorimeter in the horizontal plane. On the surface of the calorimeter there are four heat flow sensors with a step of 90°. The error of the heat flux gradient sensor is no more than 1%.

ANSYS Fluent was selected as the CFD simulation software. This program is based on the finite volume method. The system of Navier-Stokes's equations is considered to solve the problem. It consists of the equations of continuity, motion and energy. The SST k - ω model with a curvature correction was chosen as a turbulence model [10].

For numerical simulation, the following boundary conditions are set. A velocity profile is set at the entrance to the swirl generator. The velocity profile was obtained as a result of additional numerical modeling of the entrance section. For this, a straight entrance section with a length of 2 meters and an overall dimension of 35 × 70 mm was built. At the entrance to it, the air flow rate is set, similar to the physical experiment, and at the exit, the barometric pressure. The resulting velocity profile at the exit from the stabilization inlet section is exported to the inlet to the swirl generator. The boundary conditions on the heated wall were set equal to 100 °C, which corresponded to the wall temperature in the physical experiment. The unheated wall was considered adiabatic. The boundary conditions at the outlet of the annular channel are specified as atmospheric pressure.

The hexahedral mesh model was implemented in ICEM CFD. To model the boundary layer, special attention should be paid to the mesh model near the walls. To carry out a numerical calculation near the wall, near-wall functions are used, which describe the inner region (viscous and transitional sublayer). Semi-empirical turbulence models resolve the entire internal flow region, including the viscous sublayer. At the same time, the quality of the obtained result increases in the entire computational domain. To implement this approach, it is necessary to provide a mesh resolution of the boundary layer $y^+ \leq 1$. Based on this, a mesh-independent solution was obtained. For each mesh model, a test problem was solved in order to determine the optimal dimension in terms of computational costs and the accuracy of the data obtained. The characteristics of all investigated mesh models are shown in Table 1.

Table 1. Characteristics of the used mesh.

	Dimension, count cells, pcs	Average mesh size, mm	y^+
1	1 819 264	7.5x7.5	1.1
2	6 071 917	5x5	1
3	14 184 394	2.5x2.5	0.9
4	24 841 784	1x1	0.85
5	38 184 612	0.5x0.5	0.85

The verification of the results of numerical simulation was carried out on the basis of experimental data at the input Reynolds number $Re_{in} = 21 \cdot 10^3$. This data was obtained by a physical model of a straight annular channel with close geometric dimensions. figure 2 (a) shows the distribution of the dimensionless tangential velocity profile $\overline{w_\varphi} = w_\varphi / V_{in}$ (where w_φ is the current tangential velocity and V_{in} is the average input velocity) in the swirl generator for various sizes of the mesh model.

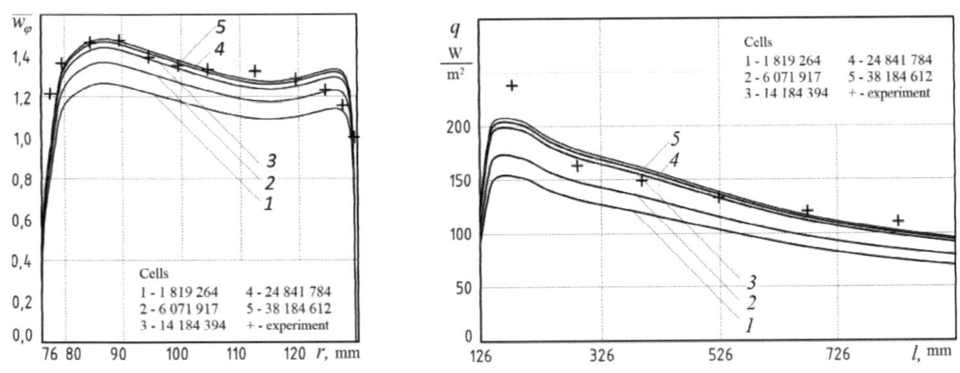

(a) (b)

Figure 2. Profiles of tangential velocity (a) and heat flux density (b) for different dimensions of the mesh model.

Figure 2 (b) shows the distribution of the average heat flux profile along the length of the annular channel for different sizes of the mesh model. The tangential velocity and heat flux density practically don't change when the dimension of the mesh model is more than 14 184 394 cells. Therefore, a grid model was used with a cell size of 2.5 x 2.5 and a y + value of 0.9 and an increase in the wall layer by 20 percent.

3. Results

The study was carried out for dry air at an inlet temperature of 20 °C. when entering the Reynolds number $Re_{in} = 21 \cdot 10^3$ and Prandtl number $Pr_{in} = 0.7$. Aerodynamics in the annular channel depends on the geometry of swirling flow generator. The analysis of velocity for the considered swirl generator geometry shows that the one-way air supply has a significant non-uniformity of the flow throughout the volume. This is due to fact that the air stream enters the swirl generator through the tangential channel and expands in the radial and axial directions. Vortex structures are formed at the lateral boundaries of the jet. The interaction of these vortex structures with the existing air flow at the walls leads to the formation of secondary paired vortices, as it is shown in figure 3 (a).

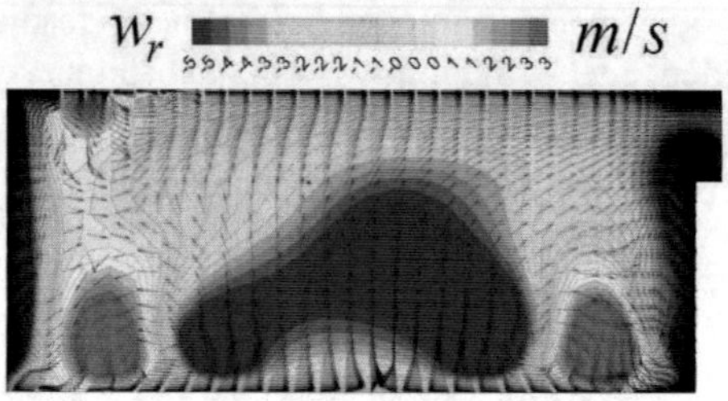

(a) (b)

Figure 3. Radial velocity distribution in the flow swirl generator at φ = 0° and φ = 90°.

Further, the flow propagates along the concave surface of the swirl generator. Part of air from the jet center enters to annular channel. It leads to the loss of flow stability and the appearance of one more pair vortex, as shown in figure 3 (b). The formed vortices with opposite directions of rotation are separated from the surface and enter the annular channel. At the same time, it breaks down into smaller structures.

The swirling turbulent flow is fed into the annular channel asymmetrically and unevenly. This process is periodic and characterized by Strouhal numbers of the order of 0.22. At the entrance to the annular channel, flow separations are observed, generating large-scale unsteady turbulent vortex structures. These vortex structures drift downstream as it is shown in figure 4.

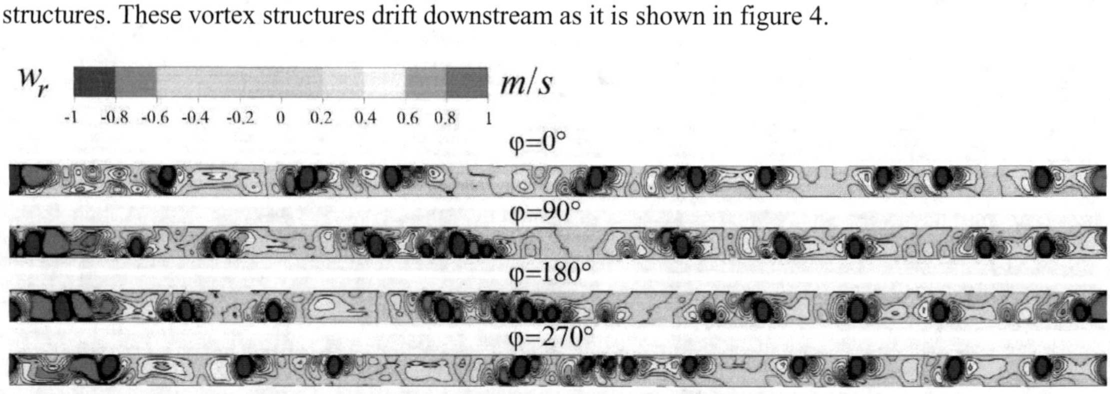

Figure 4. Distribution of radial velocity in a longitudinal section in an annular channel.

The vortices curvature and their subsequent alignment along the flow leads to the formation of structures like Taylor - Görtler vortices. These secondary flows gradually occupy the entire space of the annular channel. The propagation of secondary vortices coincides with the swirl angle of the main flow. In this case, the number of vortices increases in the direction of the outlet section. Thus, for a straight annular channel, with increasing z, the number of well-formed vortices increases from 2 to 14, and the process of forming new ones continues.

Changes of the dimensionless radial velocity $\overline{w_r} = w_r / V_{in}$ (where w_r is the current radial velocity) in the middle section along the length of the annular channel $\bar{z} = z/d_h$ (where d_h = d_2-d_1 is hydraulic diameter) is the current radial velocity) are shown in figure 5. Along the entire length of the channel there is an alternation of maximum and minimum velocity values. This is due to the transfer of vortex flows from the core of the flow to the wall and vice versa. The limiting angles of swirling of the flow on the outer and inner surfaces are different. Therefore, the vortex structures are stretched in the azimuthal direction, and then the large ones are divided into smaller ones.

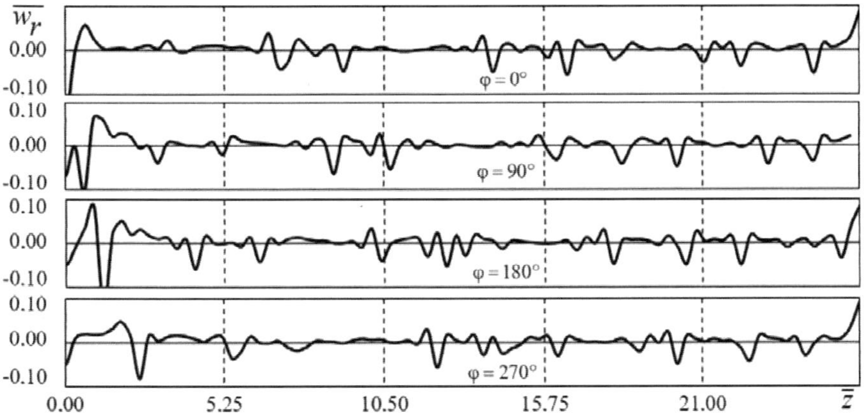

Figure 5. Change of the dimensionless radial velocity in the middle section along the length of the annular channel.

As in channels with non-swirling axial air flow, the heat transfer coefficient decreases as the flow moves towards the outlet section. Moreover, its largest values in the cross section are observed on the trajectory of the swirling jet moving from the swirl generator. Performed using numerical methods calculations make it possible to obtain a qualitatively similar averaged pattern of heat flux density distributions as it is shown in figure 6.

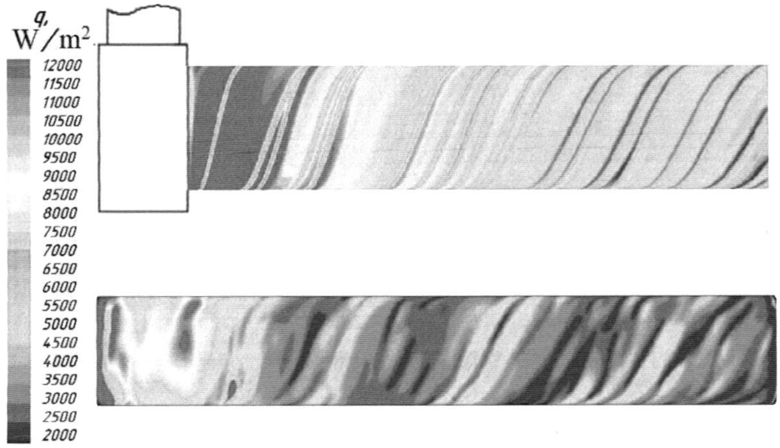

Figure 6. Distribution of heat flux density on the outer and inner surfaces of the annular channel.

The emergence and transformation of vortex structures have a significant impact on the distribution of heat flux over surfaces. Spiral lines of distributions along the length of the heat flux densities coincide with the limiting angles of flow swirling on them. There is a clear relationship between the total velocity fluctuations and the heat flux density distribution, as it is shown in figure 7. It should be noted that the highest intensity of heat transfer on the outer surface is observed in the regions of positive values of the radial velocity, and on the inner surface – in negative ones.

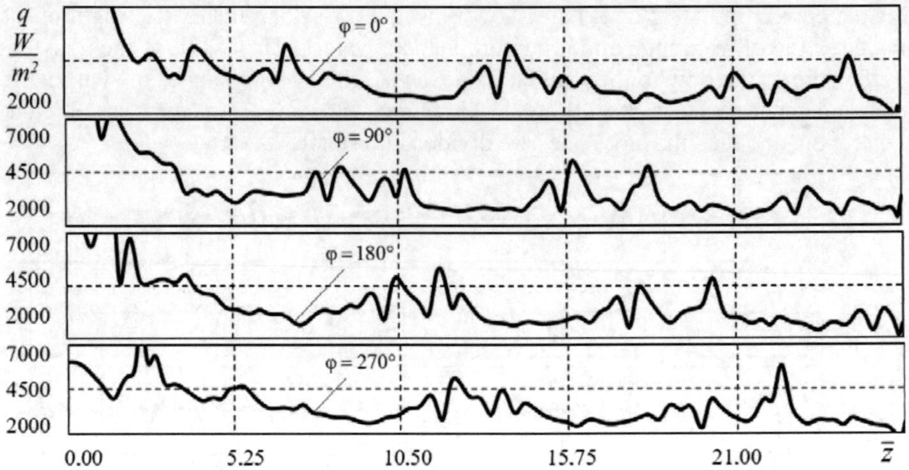

Figure 7. Change in the heat flux density along the length of the annular channel.

Due to the swirling flow decay, there is a significant decrease in the average heat transfer coefficient along the length of the outer surface. On the inner wall the drop in heat transfer is less intense. It is explained by the fact that the transfer of a more tribalized flow from the outer surface to the inner surface by secondary vortices reduces the conservative effect of centrifugal forces on heat transfer.

4. Conclusion

It was found that nonstationary Taylor - Görtler vortices gradually occupy the entire space of the annular gap. The emergence and transformation of these secondary flows affect the distribution of the heat flow over the working surfaces. A clear connection is observed between the fluctuations of the total velocity and the heat flux density on them. The highest intensity of heat transfer on the outer surface is observed in the regions of positive values of the radial velocity, and on the inner surface - in negative ones.

Acknowledgments

This work was supported by a grant from the Young scientists of Pomorye № 13-2021a.

5. References

[1] Mitrofanova O V 2010 *Hydrodynamics and heat transfer of swirling flows in channels of nuclear power plants [in Russian]* (Moscow: FIZMATLIT)

[2] Shchukin V K 1980 *Heat Exchange and Hydrodynamics of Internal Flows in Fields of Body Forces [in Russian]* (Moscow)

[3] Couette M M 1890 Études sur le frottement des liquides *Ann. Chim. Phys.* **21** 433–510

[4] Mallock A 1896 Experiments on Fluid Viscosity *Phil. Trans. R. Soc. Lond. A* **187** 41–56

[5] Taylor G I 1923 Stability of a viscous liquid contained between two rotating cylinders *Phil. Trans. R. Soc. Lond. A* **223** 289–343

[6] Görtler H 1955 Dreidimensionales zur Stabilitätstheorie laminarer Grenzschichten *Z. Angew. Math. Mech.*
[7] Shchukin V K and Khalatov A A 1982 *Heat Exchange, Mass Exchange, and Hydrodynamics of Internal Flows in Fields of Body Forces [in Russian]* (Moscow)
[8] Leukhin Y L, Pankratov E V and Karpov S V 2017 Investigation into aerodynamic and heat transfer of annular channel with inner and outer surface of the shape truncated cone and swirling fluid flow *J. Phys.: Conf. Ser.* **891**
[9] Zhang Z 2010 *LDA Application Methods* (Berlin, Heidelberg: Springer Berlin Heidelberg)

Influence of microadditive of carbon dioxide on the kinetics of boiling-up of superheated *n*-pentane

S A Perminov, E V Lipnyagov and M A Parshakova

Institute of Thermal Physics, Ural Branch of the Russian Academy of Sciences, Russia, 620016 Yekaterinburg, Amundsena St., 107a

parmari@yandex.ru

Abstract. The effect of a low-boiling impurity (CO_2 gas <1.5% mol) on the kinetics of boiling-up of superheated n-pentane in a vertical glass tube have been studied by high-speed video (2050 fps). The method of continuous pressure decrease from 2.00 to 0.10 MPa (in the temperature range of 100.2-145.1 °C), as well as the method of measuring the lifetimes of superheated liquids at 0.10 MPa (90.2-134.1 °C) have been used. The inner surface of the tube has two visible defects, one of which defines the boundary of the attainable superheat. After degassing the system, the defects of tube cease to play an appreciable role, the active centers are redistributed. The temperature of the attainable superheat increases from the initial value by 20 °C in tests with gas and by 10 °C in subsequent tests without it. The result obtained may be related to physical gas adsorption on the glass surface during the process of evacuation of the system.

1. Introduction

Boiling-up and subsequent boiling of liquid is a physical phenomenon related to phase transitions of the first kind [1, 2]. The presence of undissolved gas and suspended particulate in the fluid, poorly wetted areas of vessel walls facilitates this process reducing the lifetime of metastable state. From experimental works [3-6] it follows that gas saturation of liquids leads to a decrease in their superheat. On the other hand, it was shown in [7] that the addition of some impurities (up to 1.5 mol.%) including carbon dioxide to distilled water suppress the cavitation centers of the liquid during its impulsive stretching, although in such small amounts they either increase the equilibrium vapor pressure p_s and reduce the surface tension σ (CO_2, C_3H_6O), or practically do not change them (HCl, CH_2O_2). In the opinion of the author of work [7], one of the reasons may be the binding of free hydrogen ions in water by the indicated impurities. In this work, we investigated the effect of small addition of low-boiling impurity (CO_2 gas up to 1.5 mol%) on the kinetics of spontaneous boiling-up of superheated *n*-pentane in a glass tube.

2. Materials and methods

An automated installation for studying the kinetics of boiling-up of a superheated liquid with simultaneous high-speed video of this process (2050 fps) is described in detail in [8, 9]. In the current work, the installation has been modified. A system of continuous pressure control with high-precision pressure-vacuum sensors was introduced. As in [9], the moment of boiling-up of superheated liquid was determined by an abrupt change in optical density of the image of the measuring cell.

Figure 1 shows a block diagram of the experimental installation. The investigated liquid (*n*-pentane) is in a glass molybdenum tube with an inner diameter of 5.6 mm. The tube is oriented vertically, its top is sealed. The upper (working) part of the tube with a length of 114.6 mm and a volume of $2.8 \cdot 10^{-6}$ m^3 is thermostated in a cell with a coolant from a Huber CC-208B circulation thermostat, but its lower part is kept at room temperature. The temperature difference of between the coolant in the thermostat and the cell is measured with a precision temperature meter MIT-8 using three differential thermocouples. The temperature T in the thermostated cell is maintained with an accuracy of 0.1 °C. The superheated state of the liquid under investigation is realized by reducing the pressure p from 2.00 MPa to 0.10 MPa at a fixed temperature T. The pressure in the installation is created by compressed gas and transferred to the liquid through a metal bellows. The pressure maintenance accuracy is 0.01 MPa.

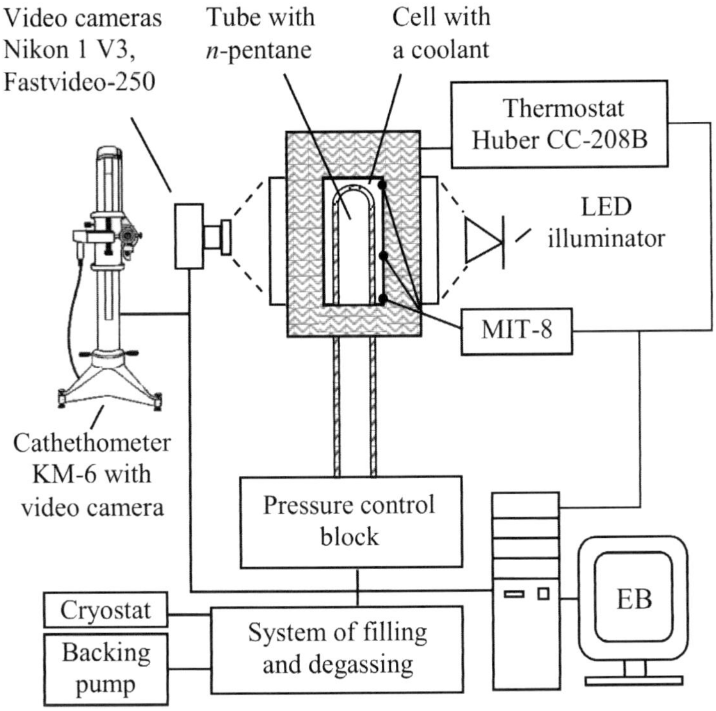

Figure 1. Block diagram of the experimental installation.

The boiling-up process of a superheated liquid is random, so the experiments are carried out many times. At temperatures below the temperature of the attainable superheating, boiling-up occurs predominantly at a steady-state pressure $p = 0.10$ MPa. Then the temperature T and the lifetime of the metastable liquid τ are measured. At higher temperatures, the expected event occurs during the depressurization process. Then the random variable is the pressure at which the liquid boils-up. The first method of studying the kinetics of boiling-up of a superheated liquid is called the method of measuring the lifetimes of the superheated state, the second method is known as the method of continuous pressure decrease [1, 2]. The experiment includes preliminary ''running-in'' of the surface of the tube with the liquid under investigation. The holding time of the system under the pressure of 2.00 MPa (between tests) is 90 s. The measurement cycle is automated.

Gas saturation of n-pentane is as follows. First, the system under study being kept at room temperature and evacuated down to -0.095 MPa (relative to the atmosphere) is filled with carbon dioxide up to -0.087 MPa. Then the vapors of n-pentane are added to the system until the mixture pressure is 0.10 MPa. After filling the measuring cell, the resulting vapor-gas mixture is liquefied at

p = 2.00 MPa. In a mixture of ideal gases, the mole fraction can be expressed as the ratio of the partial pressure to the total pressure of the mixture [10]. Thus, for the gaseous mixture-filled tube the molar fraction of CO_2 in n-pentane is estimated as ~ 8.4 mol%. This gas concentration is selected empirically. With an increase in the CO_2 content in the closed system under study, a poorly soluble vapor-gas bubble is formed at the upper sealed end of the glass tube. The bubble significantly reduces the superheat of the liquid. The experiments were carried out in the temperature range of 90.2 - 145.1 °C. From the experimental data on the phase equilibrium of the CO_2 + *n*-pentane system [11] and the equation of state of the mixture [12], it follows that the molar fraction of CO_2 in *n*-pentane at a pressure of ~ 2.00 MPa does not exceed 1.5 mol% in this temperature range. Note that the solubility of a gas in a liquid increases with increasing pressure.

3. Results and discussion

The experiments on the study of the boiling-up kinetics of the superheated gas-saturated liquid were carried out by the method of measuring the lifetimes of a superheated liquid, as well as by the method of continuous pressure decrease. Statistical processing of data such as lifetimes is described extensively in [13]. The temperature dependence of the mean lifetime of superheated *n*-pentane (with and without CO_2 addition) at p = 0.10 MPa is plotted in figure 2a. It is conventional to explain the typical “plateaus” of the dependence by the exposure of background radiation (α-, β- and γ-particles) [14, 15]. Their positions is practically independent of the pressure and temperature of the liquid. The boundary of the attainable superheat 4 is calculated according to the theory of homogeneous nucleation for pure *n*-pentane. It follows from figure 2a that the temperature of experimentally attainable superheat in series 2 (*n*-pentane with gas) increases by 20 °C compared to series 1 (*n*-pentane without gas). At a temperature of 120 ° C and below, the probability of boiling-up in the n-pentane + CO_2 system does not depend the value of T. In this case, the mean lifetime of the superheated liquid on the "plateau" turns out to be shorter than in experiments with pure *n*-pentane and amounts to ~ 20 s. After degassing and refilling the measuring cell with pure *n*-pentane, the data of series 1 are reproduced (see series 3), but the temperature of the attainable superheat increases by 10 °C.

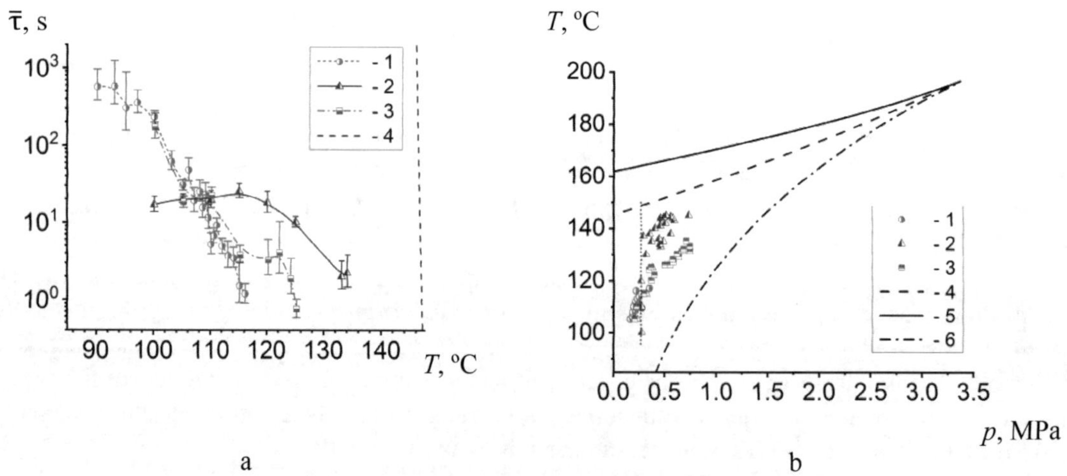

Figure 2. The dependence of the mean lifetime of superheated *n*-pentane on temperature at p = 0.10 MPa (a) and the phase diagram of *n*-pentane (b): points – experimental data: 1, 3 – *n*-pentane; 2 – *n*-pentane + CO_2; 4 – the boundary of the attainable superheat; 5 – spinodal; 6 – binodal.

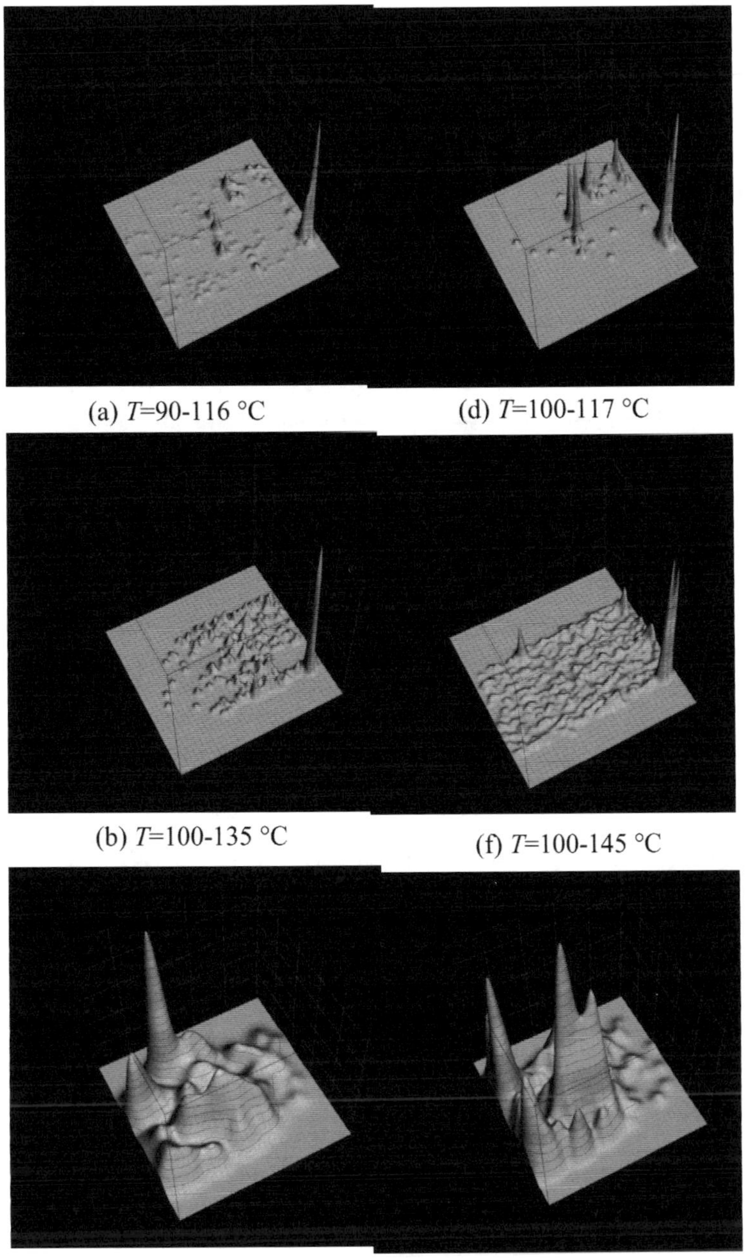

(a) T=90-116 °C (d) T=100-117 °C

(b) T=100-135 °C (f) T=100-145 °C

(c) T=100-126 °C (e) T=100-135 °C

Figure 3. The probability density of boiling-up site distribution on the image plane of the observed region in all range of temperature T: (a - c) – boiling-up at 0.10 MPa, (d - e) – at continuous pressure decrease; (a, d) - *n*-pentane, series 1 (see figure 2); (b, f) - *n*-pentane + CO_2, series 2; (c, e) - *n*-pentane, series 3. The coordinates are reduced to dimensionless form and are given as a percentage of the size of the observed inner region of the tube.

In figure 2b shows the superheat temperature corresponding to the average pressure at which the investigated liquid boils-up without reaching 0.10 MPa. At some temperatures, part of the events occurs during pressure decrease and part of them occurs at a steady-state value of p=0.10 MPa. In this range, the *n*-pentane + CO_2 system has a special feature. The average pressure at which the gas-saturated liquid boils-up is practically independent of temperature and is ~ 0.27 MPa. In other cases, active vaporization centers limit the depth of entry into the metastable region (p_s-p) to approximately the same value at different temperatures, therewith its value for series 2 (*n*-pentane with gas) is greater than ones for series 1 and 3 ("pure" *n*-pentane).

The coordinates of the sites of appearance of the first vapor bubbles in the superheated liquid were found from the video frames of the onset of boiling-up. To visualize them, we used the Parzen – Rosenblatt nuclear density estimation method [9, 16]. Figure 3 shows the probability density of boiling-up site distribution over the area of the glass tube image for each series of experiments. Activation centers of boiling-up at steady-state pressure and non-activation centers of boiling-up during the drop of pressure were separately processed. Coordinates x and h are reduced to dimensionless form and are given as a percentage of the diameter and height of the observed region of the measuring cell, respectively. Visual analysis of figure 3 shows that initially the total number of active vaporization centers is limited, and they are distributed nonuniformly. The glass tube has two visible defects on the inner surface, one of which defines the boundary of the attainable superheat. The probability distribution of the boiling-up site on the inner surface of the tube outside the defects becomes uniform with the addition of CO_2, i.e. the gas suppresses the other most active centers of vaporization. After re-filling the tube with "pure" *n*-pentane, defects cease to play a noticeable role, active centers are redistributed. In this case, the boundary of the attainable superheat, as elsewhere, is determined by one of the most active centers.

4. Conclusion

The kinetics of boiling-up of superheated *n*-pentane has been studied by the method of continuous pressure decrease from 2.00 to 0.10 MPa and by the method of measuring the lifetimes at 0.10 MPa by high-speed video. The experiments were carried out in the temperature range of 90.2-145.1 °C with both gas-saturated (CO_2 up to <1.5% mol) and pure *n*-pentane (before and after experiments with a volatile impurity). It follows from the video data that initially boiling-up occurs with a high probability on two visible glass defects, one of which determines the boundary of the attainable superheat. The probability distribution of the boiling-up site on the inner surface of the tube outside the defects becomes uniform with the addition of CO_2, i.e. the gas suppresses the other most active centers of vaporization. The temperature of the attainable superheat in tests with gas increases by 20 °C as compared with the initial value for pure *n*-pentane. After degassing and refilling the glass tube with pure *n*-pentane, the data without gas are reproduced but the temperature of the attainable superheat increases by 10 °C. In this case, two visible defects cease to play a notable role, active centers are redistributed. The result obtained may be related to physical (reversible) gas adsorption on the glass surface during the process of evacuation of the system.

Acknowledgments

This study was financially supported by the Russian Foundation for Basic Research (project no. 20-08-00270).

References

[1] Skripov V P, Sinitsyn E N, Pavlov P A, Ermakov G V, Muratov G N , Bulanov N V, Baidakov V G 1988 *Thermophysical Properties of Liquids in the Metastable (Superheated) State* (London: Gordon and Breach Science Publishers)

[2] Baidakov V G 2007 *Explosive Boiling of Superheated Cryogenic Liquids* (Weinheim: Wiley-VCH)
[3] Mori Y, Hijikata K, Nagatani T 1976 *Int. J. Heat Mass Transfer.* **19** 1153-1159
[4] Forest T W, Ward C A 1977 *J. Chem. Phys.* **66** (6) 2322-2330
[5] Pavlov P A, Skripov P V 1985 *High Temperature* **23** (1) 67–73
[6] Baidakov V G, Kaverin A M 2020 *Int. J. Heat Mass Transfer.* **163** 120498
[7] Vinogradov V E 2009 *Tech. Phys. Lett.* **35** (2) 54–56
[8] Lipnyagov E V, Parshakova M A, Perminov S A, and Ermakov G V 2013 *Int. J. Heat Mass Transfer.* **60** 612–615.
[9] Lipnyagov E V, Parshakova M A, and Perminov S A 2017 *Int. J. Heat Mass Transfer.* **104** 1353-1361
[10] Landsberg G D 1988 *Elementary Textbook on Physics* vol. 1 Mechanics. Heat. Molecular physics (Moscow: Mir Publishers)
[11] Cheng H, Pozo de Fernindez M E, Zollweg J A, and Streett W B 1989 *J. Chem. Eng. Data* **34** 319-323
[12] Peng Ding-Yu and Robinson D B 1976 *Ind. Eng. Chem. Fundam.* **15** (1) 319-323
[13] Lipnyagov E V, Parshakova M A, and Perminov S A 2017 *Int. J. Heat Mass Transfer.* **104** 1362-1371
[14] Glaser D A 1953 *Phys. Rev.* **91** (3) 762-763
[15] Sinitsyn E N, Uskov V S 1986 *Thermodynamical investigations of metastable liquids* (Sverdlovsk: Urals Scientific Center of the Academy of Sciences of the USSR) 59-62 (In Russian)
[16] Silverman B W 1986 *Density Estimation for Statistics and Data Analysis* (New York: Chapman & Hall / CRC)

Experimental investigation of heat and mass transfer of an annular impinging jet

M V Philippov, I A Chokhar, V V Terekhov, V I Terekhov and I N Baranov

Kutateladze Institute of Thermophysics SB RAS,
Russia, 630090 Novosibirsk, Ac. Lavrentieva av., 1

mfilippov181096@gmail.com

Abstract. This work presents an experimental study of a turbulent flow and heat transfer of an annular impinging jet for organizing effective surface cooling. Heat and mass transfer of the impinging annular jet was studied at Re = 5500. At that, a distance from the nozzle to the wall was varied. The focus was made on configurations with small nozzle-to-wall distances. It is shown that, depending on the indicated distance, fundamentally different flow regimes with characteristic features of heat transfer distribution are observed.

Introduction

Annular impact jets have become widely used in many modern industrial applications, such as heating, cooling and drying processes due to their significantly higher efficiency in comparison with circular jets [1]. Numerous studies [1–11] of the heat transfer characteristics of an annular impinging jet in various configurations have been carried out over many years. These studies revealed the main characteristics of the flow and heat transfer of annular jets. Heat transfer of an impinging annular jet depends on the Reynolds number, nozzle geometry, and distance between the nozzle and the surface. It should be noted that, in some regimes, the phenomena of bistability and hysteresis, when heat transfer of the annular impinging jet can differ for the same parameters [10], are observed. In addition, the flow field of an annular impinging jet at low Reynolds numbers is asymmetric [8-9]. In this regard, the study of an annular impinging jet is relevant and important also for the construction of a fundamental theory. This work is aimed at studying the aerodynamic and thermal processes in annular impinging jets, taking into account the influence of a large number of factors affecting the structure of turbulence and near-wall heat and mass transfer.

Experimental setup

The experimental setup for studying the aerodynamics and heat transfer (figure 1a) is a two-component laser-Doppler anemometer with adaptive time-gated channels, velocity measurement range V = 0-120 m/s, and a 3-component KPD (coordinate positioning device) with a moving step of 0.00125 mm. The accuracy of velocity measurement was 2 percent and the accuracy of velocity pulsation measurement was 3 percent. An aerosol formed during recondensation of glycerin vapors using a generator created by the authors was used as a source of light-scattering particles. The compressor injected an air flow through the aerosol generator to the experimental section. Heat transfer was investigated with a testo thermal imager for thermal imaging with a resolution of 1280x800. An annular jet cooled a substrate made of titanium foil of 260x200x0.1 mm, heated uniformly using a power supply unit (heating power of 36 W). A distance from the jet to the substrate was controlled using a one-component coordinate device with a moving step of 0.2 mm. The experimental section (figure 1b) was an annular impinging jet with a ring width b = 3 mm and a jet width D = 22 mm. The distance to the z/D plane was varied from 0.5 to 1.5. The Reynolds number was $Re_D = 5000$, $Re_b = 682$. The Nu_D number was calculated as in [12].

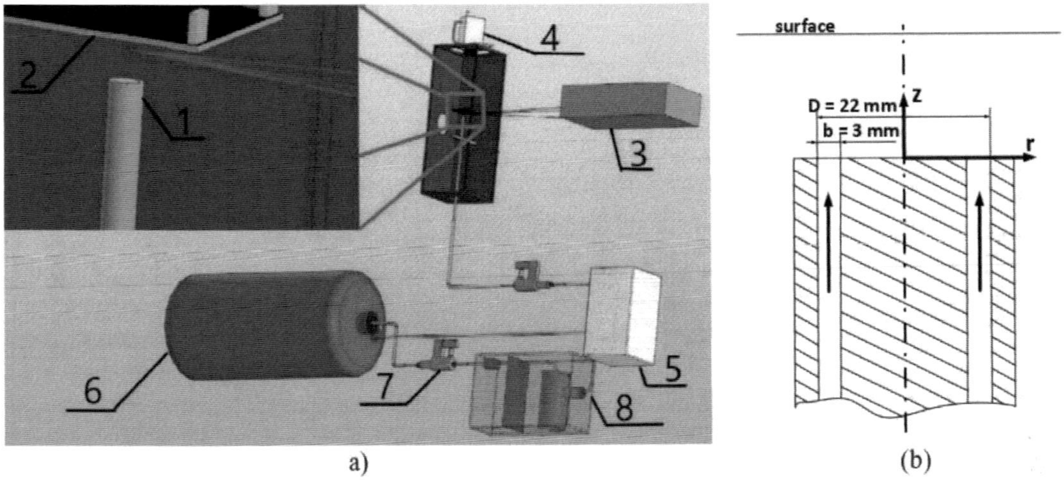

Figure 1. (a) experimental setup: 1 – nozzle; 2 – heated surface; 3 – LDA; 4 – thermal imager; 5 –receiver; 6 – compressor; 7 – flow meter; 8 – aerosol generator; (b) experimental section.

Experimental results

1. Aerodynamics of an annular impinging jet

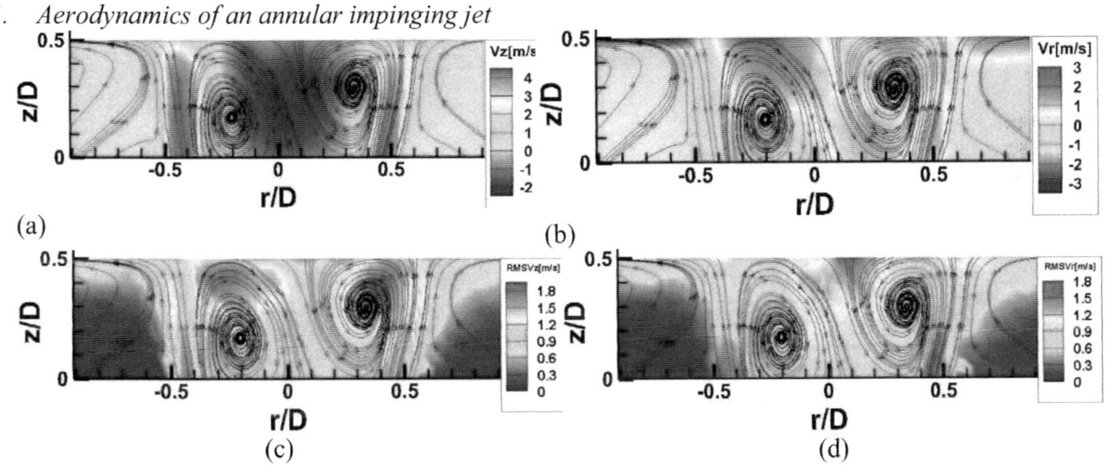

Figure 2. z/D = 0.5: (a) streamlines and velocity V_z (b) streamlines and velocity V_r (c) streamlines and pulsations of velocity V_z (d) streamlines and pulsations of velocity V_r.

In this paper, we consider the aerodynamics of an annular impinging jet for three distances between the nozzle and the wall z/D = 0.5, 1, and 1.5. Area ratio was $\frac{(D-2b)^2}{D^2} \approx 0.5$. With this area ratio, no hysteresis was found. Despite this, two regimes resulting from the changes in z/D were experimentally investigated. The first regime is shown in figures 2 and 3. It is characterized by adhesion of vortex structures to the wall. At z/D = 0.5, the highest ejection and the highest negative velocity are observed in the heart of the annular jet (figures 2-4a). Moreover, in the near-wall zone, the radial velocity and

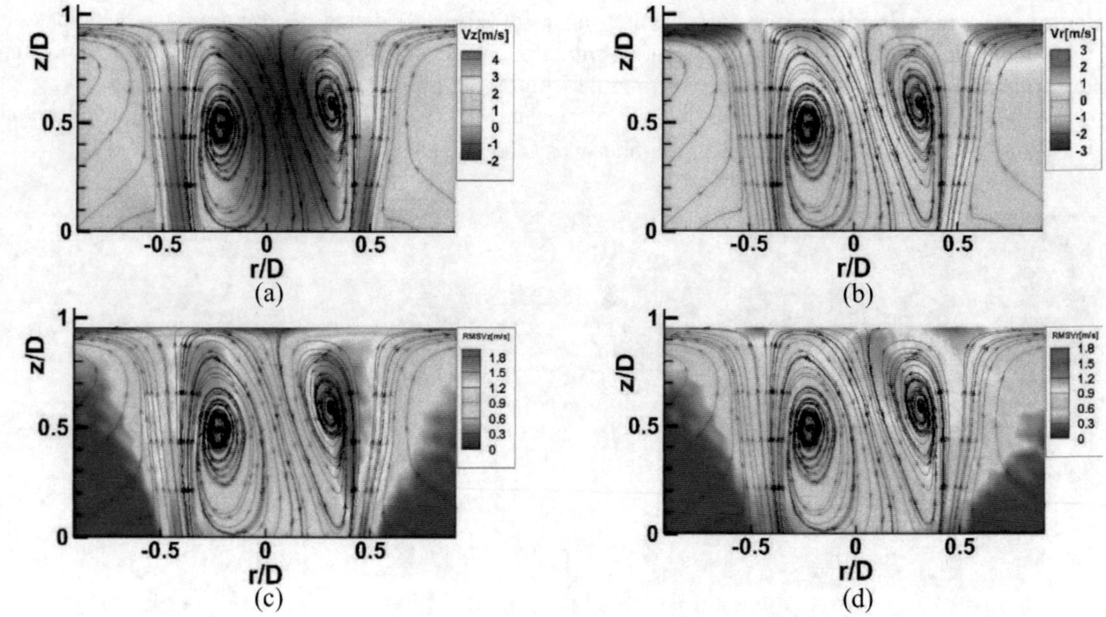

Figure 3. z/D = 1: (a) streamlines and velocity V_z (b) streamlines and velocity V_r (c) streamlines and pulsations of velocity V_z (d) streamlines and pulsations of velocity V_r.

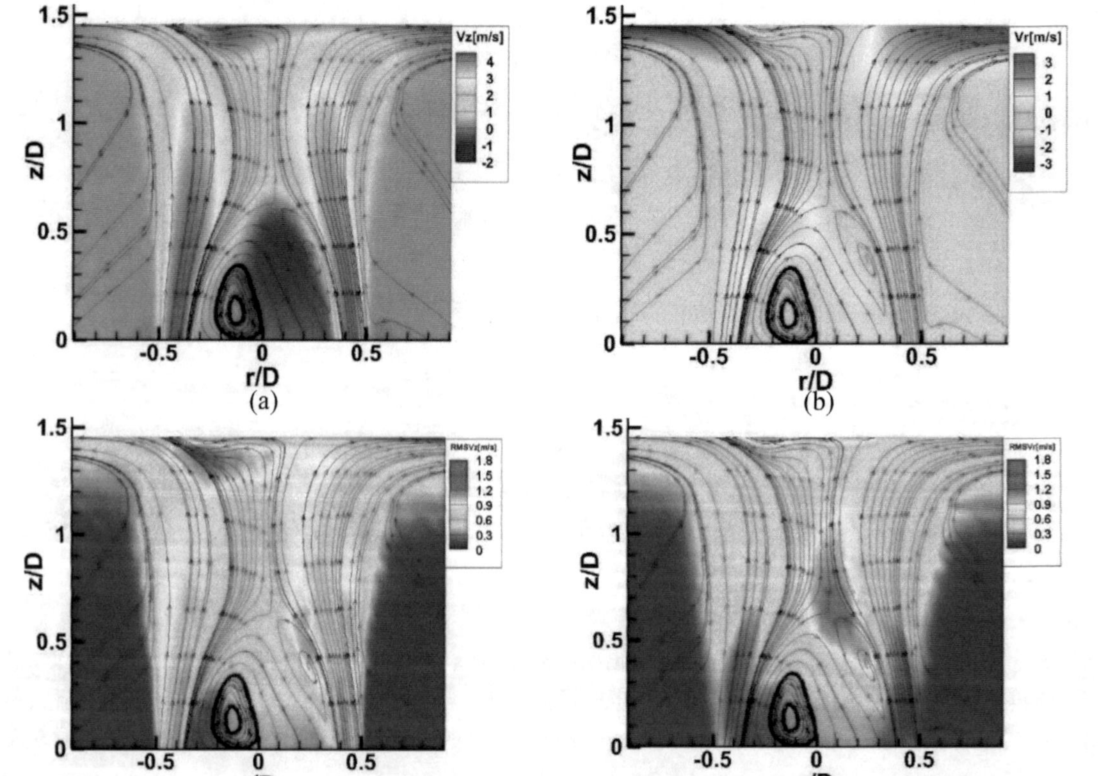

Figure 4. z/D = 1.5: (a) streamlines and velocity V_z (b) streamlines and velocity V_r (c) л streamlines and pulsations of velocity V_z (d) streamlines and pulsations of velocity V_r.

pulsations of the axial and radial velocities demonstrate the highest values in comparison with z/D = 1, 1.5 (figures 2-4bcd). In addition, with symmetric injection, an asymmetry of velocity and pulsations is observed, which is associated with bifurcations at low Re numbers [8-9]. As it can be seen in figure 3, at z/D = 1, the regime of vortex adhesion to the wall is implemented. The vortices become more elongated, and the magnitude of inverse velocity decreases.

Regime switching occurs at z/D = 1.36 both with increasing and decreasing z/D. Let us consider the second flow regime. It is characterized by the closure of the axial jet; moreover, the axial velocity field is symmetric, except for the region of the toroidal vortex. The volume of a toroidal vortex is approximately equal to its volume for an annular free jet [11]. According to this, it can be assumed that hysteresis does not form for such a configuration due to the large slot width. The momentum of the outgoing jet is rather large to ensure vortex deformation and faster closure of the annular jet.

2. Heat transfer of an annular impinging jet

Let us consider heat transfer for an annular impinging jet (figure 5). The asymmetry in velocity and pulsations affects distribution of the heat transfer coefficient Nu_D But for z/D = 1.5, distribution of the heat transfer coefficient is symmetrical. In general, the heat transfer coefficient for an annular impinging jet Nu_D corresponds approximately to the heat transfer coefficient for a circular impinging jet (except for the central region) [12]. The distribution of the Nu_D number is characterized by a cavity in the central zone for the first regime, whose area is equal to the area of vortex adhesion. According to the values of the integral heat transfer coefficient $\overline{Nu_d}$, calculated as in [10] (figure 5d), we can observe 2 zones: the near zone, where the second regime efficiency is prevailing, and the far zone, where the values of the wall velocity and pulsations of the integral heat transfer coefficient $\overline{Nu_d}$ are the best for z/D = 0.5 due to higher ejection.

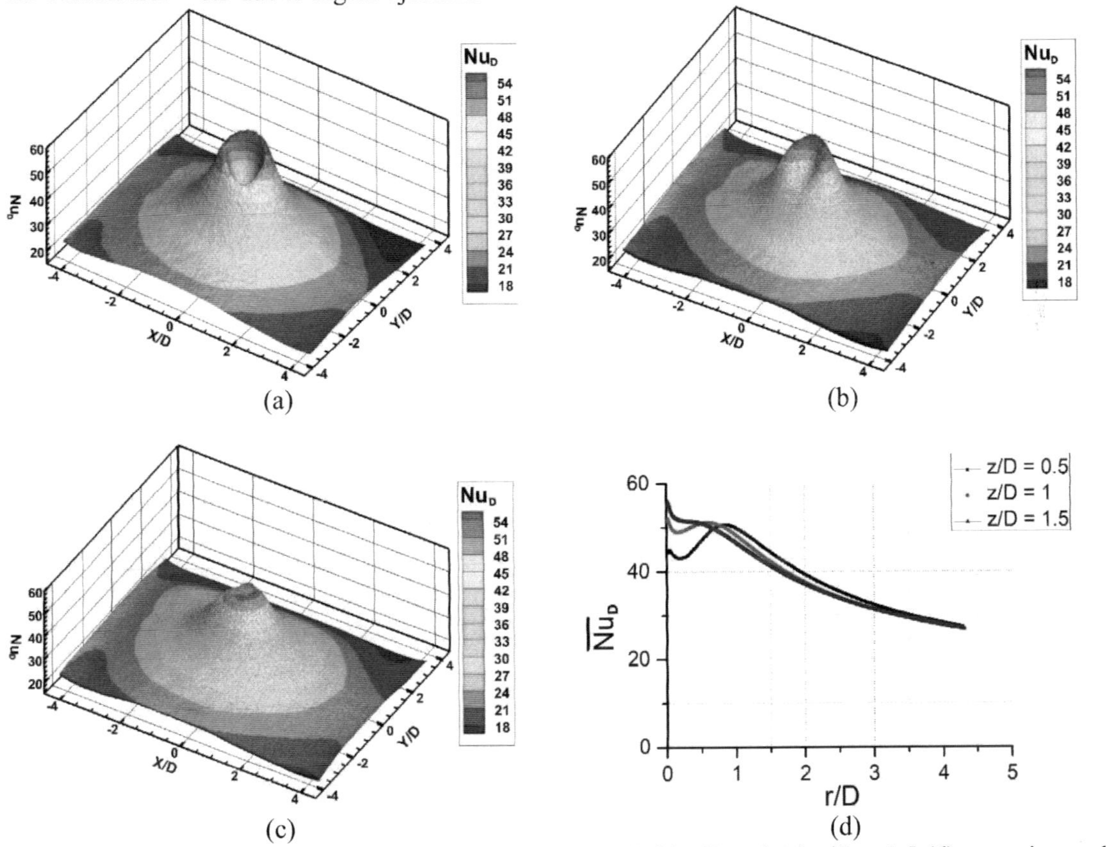

Figure 5. The heat transfer coefficient Nu_D: (a) z/D = 0.5 (b) z/D = 1 (c) z/D = 1.5 (d) mean integral Nusselt number $\overline{Nu_d}$ (black color - z/D = 0.5, red color - z/D = 1, blue color - z/D = 1.5).

4. Conclusion

The paper demonstrates an experimental study of aerodynamics and heat transfer for the annular impinging jets with varying nozzle-to-wall distance. It is shown that when the ratio of the ring area to the nozzle area is ~ 0.5 and Re = 5000, no hysteresis and bistability are observed. The transition between the characteristic flow regimes occurs at H/D = 1.36. The first of the regimes is characterized by spreading of an annular jet along the surface, which leads to a decrease in heat transfer near the axis of symmetry. In the second regime, the closure of the annular jet into a round one is observed until it reaches the surface. Asymmetry of the fields of the mean and pulsation velocities, noted earlier in [8-9], was observed for all considered distances z/D.

Acknowledgements

The study was financially supported by the RFBR (Project No. 20-58-26003).

References

[1] Terekhov V Kalinina S Sharov K 2016 *ICJWSF2015* pp 19–27
[2] Yang H Q Kim T Lu T J Ichimiya K 2010 *International Journal of Heat and Mass Transfer* 53(19-20) pp 4092–4100
[3] Afroz F Sharif M A R 2020 *International Journal of Thermal Sciences* 151 p 106257.
[4] Hammad K J 2018 *International Journal of Heat and Mass Transfer* 121 pp 28–36.
[5] Wawrzak K Boguslawski A Tyliszczak A Saczek M 2019 *International Journal of Heat and Fluid Flow* 79 p 108460
[6] Vanierschot M Van Dyck K Sas P Van den Bulck E 2014 *Physics of Fluids* 26(10) p 105110
[7] Vanierschot M Ogus G 2019 *Experimental Thermal and Fluid Science* 106 pp 148–158
[8] Vanierschot M Percin M van Oudheusden B *Experiments in Fluids* 2021 p 62:77
[9] Del Taglia C Moser A Blum 2009 *ASME. J. Fluids Eng.* 131(8) p 081202
[10] Trávníček Z Tesař V 2013 *Experimental Thermal and Fluid Science* 44 pp 565–570.
[11] Philippov M V Chokhar I A Terekhov V V Terekhov V I 2020 *Journal of Physics: Conference Series* 1565(1) p 012070
[12] Gulati P Katti V Prabhu S V 2009 *International Journal of Thermal Sciences* 48(3) pp 602–617

Experimental study of the adiabatic wall temperature of a cylinder in a supersonic cross flow

S S Popovich, N A Kiselev, A G Zditovets and Y A Vinogradov

Institute of Mechanics, Lomonosov Moscow State University, Russia, 119192 Moscow, Michurinsky Prospect, 1

pss@imec.msu.ru

Abstract. The results of an experimental study of the adiabatic wall temperature for a supersonic air flow across the cylinder are presented. The temperature was measured contactlessly using an InfraTEC ImageIR 8855 thermal imager through a ZnSe infrared illuminator. The freestream Mach number was 3.0, input flow total temperature was 295 K, and the total pressure 615 kPa. The Reynolds number calculated from the cylinder diameter (30 mm) was about 10^6. It is shown that it is possible in principle to determine the high-speed flow total temperature by defining the maximum temperature of a cylindrical probe at the front critical point. Thermograms of the wall temperature distribution along the profile of the cylinder were obtained. The research was performed at the experimental facilities of the Institute of Mechanics of Lomonosov Moscow State University.

1. Introduction

Determination of the adiabatic wall temperature - the temperature of the near-wall gas layers on a heat-insulated wall - is one of the main problems in the study of heat transfer in supersonic flows, since its direct measurement in the experiment is problematic [1, 2]. The value of the adiabatic wall temperature determines the magnitude and direction of the heat flux. In a dimensionless form, the adiabatic wall temperature is determined through the temperature recovery factor r (1), which shows the fraction of the dynamic temperature component $u^2/(2C_p)$ recovered in the boundary layer [3].

$$r = \frac{T^*_{aw} - T}{T^*_0 - T} = \frac{2C_p}{u^2}\left(T^*_{aw} - T\right). \qquad (1)$$

In the case of a flow around a plate, the difference between the adiabatic wall temperature and the flow total temperature is the greater, the greater the flow velocity and the more the Prandtl number differs from unity. For example, for an air flow (Pr=0.7) with a Mach number M=3, the adiabatic wall temperature is 0.93 of the undisturbed flow total temperature (r=0.89). At the same time, when an incompressible (low-speed) gas flow is flowing around the plate, these temperatures are practically the same (r=1). External influences (pressure gradient, wall permeability, surface shape and relief, shock waves, separated flows, condensation and evaporation of the working fluid) can lead to both an increase in aerodynamic heating in a localized area and a wall cooling [4-8].

An interesting effect of decreasing the temperature recovery factor was obtained in the work of Eckert and Weise, later described by Eckert in a review article on energy separation [9]. The authors experimentally investigated the transverse subsonic flow of a hollow cylinder with thermocouples pressed into its surface. At the front of the cylinder with a laminar boundary layer, the temperature recovery factor was as expected from theory. However, at a flow rotation angle of 180°, *r* decreases to a value of -0.1, which means a decrease in the cylinder wall temperature below the thermodynamic temperature in the input flow. This phenomenon was called "aerodynamic cooling" by analogy with the effect of aerodynamic heating.

The results of the experiments of Eckert and Weise were at first questioned by L. Prandtl [9], but were soon confirmed by Ryan [10] and Thomann [11]. Ryan [10] also demonstrated that the effect of energy separation extends further on in the wake behind the cylinder. The problem of energy separation in a transverse flow across a cylinder was again investigated by Goldstein and He [12], Sanitaja and Goldstein [13]. The cylinder diameter was 28.55 mm, and the maximum flow velocity reached 100 m/s. The local wall temperature was measured with copper-constantan thermocouples along the circumference of the cylinder. The problem was also investigated by direct numerical simulation in the work of Aleksyuk and Osiptsov [14]. The range of the main considered parameters included: Reynolds number $Re<10^3$, Prandtl number $0.1<Pr<10$ and Mach number $M<0.6$, which corresponds to the periodic vortex shedding regime. In the region of supersonic velocities, Eber [15] investigated the flow around cones and cylinders in short-time tubes at Mach numbers 2.87 and 4.25. The paper [16] investigated the flow around a cylinder 1 inch (25.4 mm) in diameter with a fixed Mach number $M_0=3.9$ in the range of Reynolds numbers $Re_d=2.1\times10^3\div6.7\times10^3$. Figure 1 shows a comparison of experimental results for subsonic and supersonic flow across a cylinder.

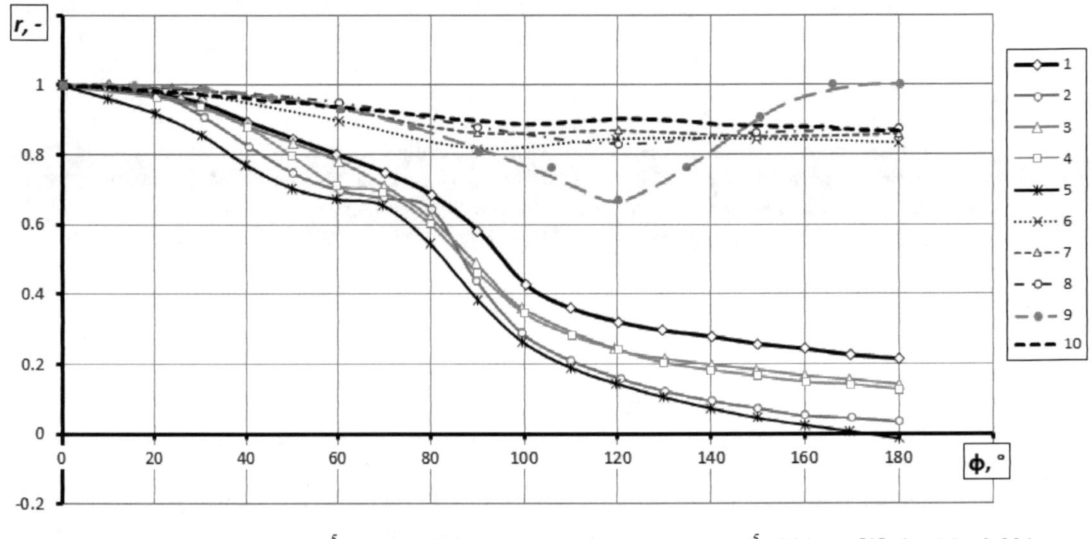

1 – $M_o=0.882$, $Re_d=2.57\times10^5$, Ø20 мм [9]; 2 – $M_o=0.526$, $Re_d=1.81\times10^5$, Ø20 мм [9]; 3 – $M_o=0.831$, $Re_d=1.21\times10^5$, Ø10 мм [9]; 4 – $M_o=0.50$, $Re_d=8.55\times10^4$, Ø10 мм [9]; 5 – $M_o=0.575$, $Re_d=9.3\times10^4$, Ø10 мм [9]; 6 – $M_o=1.56$, $Re_d=1.88\times10^5$, Ø5 мм [15]; 7 – $M_o=1.86$, $Re_d=1.49\times10^5$ [15]; 8 – $M_o=2.50$, $Re_d=8.68\times10^5$ [15]; 9 – $M_o=3.9$, $Re_d=6.7\times10^3$ [16]; 10 – present experiment: $M_o=3.00$, $Re_d=10^6$

Figure 1. Comparison of the results of measuring the temperature recovery factor of a cylinder in a subsonic (1-5) [9] or a supersonic cross flow (6-9) [15, 16].

As seen from figure 1, the adiabatic wall temperature is equal to the flow total temperature at the front critical point (r=1) for both subsonic flow velocities and supersonic ones. According to the works of Eber [15], Walter and Lange [17], the temperature recovery factor does not decrease so much around the cylinder profile at supersonic velocity, in contrast to subsonic one. In this case, the value at the rear critical point increases from 0.1 to 0.87 when the Mach number changes from 0.5 to 2.5. In [16], on the contrary, the recovery factor varied from 1 at the front critical point on the cylinder to 0.67 at a bypass angle of 120° and then again tends to 1 at the rear critical point.

This work is aimed at investigating the potential for reducing the temperature recovery factor in a supersonic transverse flow across a cylinder. The data of previous studies are contradictory, while the potential for temperature reduction in a supersonic flow is much greater than in a subsonic one, since the dynamic temperature component is much larger. Flow effect resulting in low values of temperature recovery factor can make it possible to propose new ways of energy separation [18-21] and reduction of aerodynamic heating in compressible gas flows.

2. Methodology

The experimental study is carried out on the basis of the AR-2 supersonic wind tunnel [22, 23]. The working part of the installation has a rectangular cross-section with dimensions of 70 × 98 mm (Fig. 2). The freestream Mach number is M_0=3.0. Total flow temperature was 295 K, total pressure – 615 kPa. The Reynolds number, calculated from the cylinder diameter, was about 10^6.

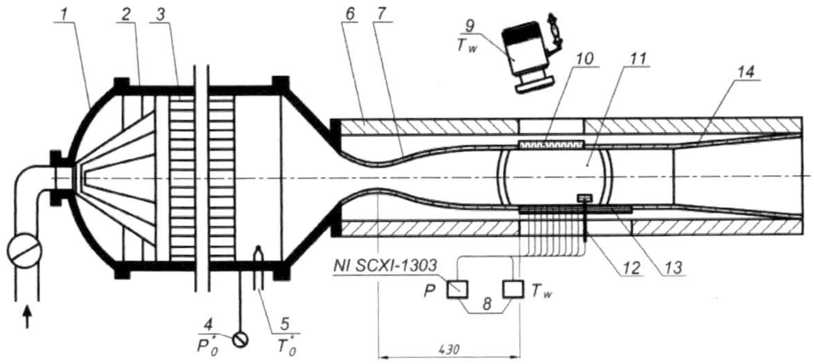

a

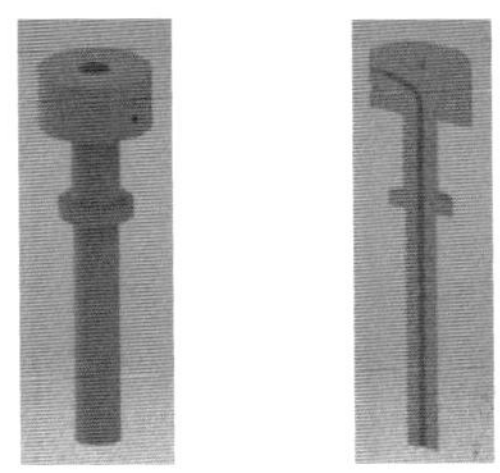

b

Figure 2. a) Scheme of the experimental facility: 1 – prechamber; 2 – assembly of cones; 3 – honeycomb; 4 – sensor for measuring total pressure; 5 – thermocouple for measuring total temperature; 6 – working channel; 7 – flat adjustable supersonic nozzle; 8 – static pressure and temperature sensors; 9 – infrared thermal imager; 10 – ZnSe illuminator; 11 – optical glass illuminator, 12 – model cylinder, 13 – plexiglass bottom wall; 14 – diffuser; b) 3D-design of the model cylinder.

The capacity of the compressor feeding the installation (up to 10 kg/s) makes it possible to maintain constant parameters of the main flow in the working section for about an hour or more and to carry out experiments practically in steady-state regime. The side walls of the channel are formed by a set of flat plates, among which there are plates with round windows with optical protective glasses. Rearrangement of the plates makes it possible to position the plates with illuminators in the required places of the working section and to visualize and photograph the flow pattern. Also, instead of the side wall of the test section of the wind tunnel, an infrared illuminator was installed, made of a ZnSe polycrystal, a material transparent to the infrared region of the spectrum. Thus, it was possible to carry out a contactless survey with the InfraTEC ImageIR 8855 thermal imager of the model cylinder.

The investigated cylinder was made of photopolymer resin on a UNIZ 3D printer, which made it possible at the stage of 3D solid modeling to provide channels for sampling static pressure and supplying thermocouples to the investigated bypass of the cylinder profile. The cylinder diameter is 30 mm. To exclude the influence of the growing boundary layer on the bottom wall, the cylinder was removed on a narrow foot from the flow to a height of 30 mm.

3. Results and discussion

As a result of the investigations carried out, a number of thermograms of the temperature distribution along the profile of the cylinder were obtained (Fig. 3). In this case, the study was carried out in a steady-state regime, i.e. readings were taken after thermal equilibrium was established. The supersonic flow was directed from left to right. The contour of the profile in the middle of the cylinder height was considered in order to exclude the influence of the upper and lower edges. The cold spot in the upper part of the thermogram is a reflection of the thermal imager matrix.

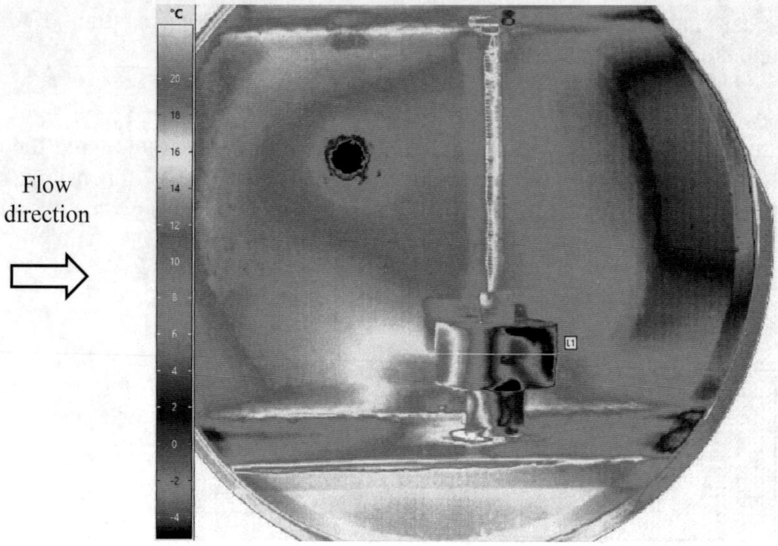

Figure 3. Thermogram of a supersonic flow around a cylinder.

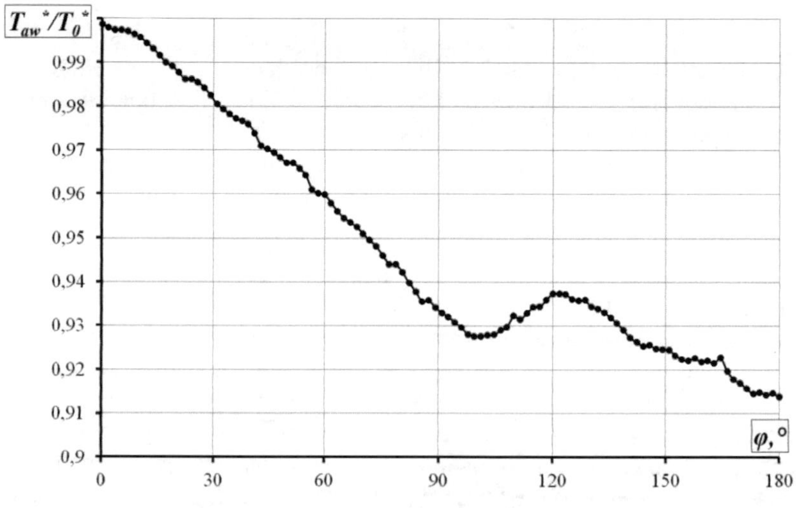

Figure 4. Distribution of the ratio of the adiabatic wall temperature to the input supersonic flow total temperature along the profile of the cylinder.

Fig. 4 shows the distribution of the temperature recovery factor calculated from the measured local wall temperature and the parameters of the input flow. As in the previously mentioned works [15-17], it can be seen that the temperature value at the front critical point of the cylinder is practically equal to the flow total temperature measured in the prechamber of a supersonic wind tunnel (as is known, the flow total temperature does not change during expansion in the Laval nozzle). Based on this result, it is possible to propose a method for measuring the temperature of a high-speed flow by measuring the maximum temperature (at the front critical point) of cylindrical probes. Such measurements can be carried out both by pressing a thermocouple in the front of the cylinder and non-contacting using a thermal imager through an illuminator. However, this method of measuring the total temperature in a supersonic flow does not allow measuring the total temperature profile in the boundary layer, since the cylinder in this case becomes a source of disturbances in the near-wall region.

In figure 1, Graph 10 shows the distribution of the temperature recovery factor obtained in the experiment according to the data in figure 4. The wall temperature along the cylinder profile decreases from the total temperature (r=1) - at the front critical point of the cylinder to the level of the adiabatic wall temperature specific for a turbulent flow around a smooth wall (r=0.89) - in the preseparation region. In the boundary layer separation region, a local maximum temperature is observed and a further decrease to a level slightly lower (r=0.87) - at the rear critical point of the cylinder. Thus, a significant effect of a wall cooling, by analogy with subsonic flow velocities, has not been revealed at the moment for a supersonic flow velocity.

Acknowledgements
This work is supported by the Russian Science Foundation (Grant 19-79-10213).

References

[1] Hayes J R and Neumann R D 1992 *AIAA PrAA* **142**
[2] Kutateladze S S and Leontiev A I 1990 *Heat Transfer, Mass Transfer, and Friction in Turbulent Boundary Layers* (Taylor and Francis)
[3] Shapiro Ascher H 1953 *The Dynamics and Thermodynamics of Compressible Fluid Flow,* Ronald Press. Vol. 2. New York, 553 p.
[4] Leontiev A I, Popovich S S, Vinogradov U A and Strongin M M 2018 *JPCS* **1129**
[5] Leontiev A, Popovich S, Strongin M and Vinogradov Y 2017 *EPJ Web Conf* **159**
[6] Leontiev A I, Popovich S S, Vinogradov U A and Strongin M M 2019 *JPCS* **1369** 012040
[7] Vinogradov Y A, Zditovets A G, Kiselev N A, Medvetskaya N V and Popovich S S 2020 *Fluid Dynamics* **55** 5
[8] Popovich S S, Zditovets A G, Kiselev N A and Vinogradov Y A 2020 *JPCS* **1683** 022064
[9] Eckert E R G 1986 *Int Commun Heat Mass Transf* **13**
[10] Ryan L F 1951 *Experiments on aerodynamic cooling* (Ph. D. thesis, Eidgen. Tech. Hockschule, Zurich)
[11] Thomann H 1959 *The Aeronautical Research Institute of Sweden FFA Report* **84**
[12] Goldstein R J and Boyong He 2001 *Transactions of the ASME* **123**
[13] Sanitjia S and Goldstein R J 2004 *Int J Heat Mass Transf* **47**
[14] Aleksyuk A I and Osiptsov A N 2018 *Int J Heat Mass Transf* **119**
[15] Eber G 1941 *German Archive Report* **66/57**
[16] Goodwin G, Creager M O and Winkler E L 1956 *NACA RM* **A55H31**
[17] Walter L W and Lange A H 1953 NAVORD Rep. 2854 (U.S. Naval Ordnance Lab)
[18] Burtsev S A and Leontiev A I 2014 *High Temp* 52 pp 297–307
[19] Leontiev A I, Zditovets A G, Kiselev N A, Vinogradov Y A and Strongin M M 2019 *Exp Therm Fluid Sci* **105**
[20] Makarov M and Makarova S 2017 *MATEC Web of Conferences* 33rd Siberian Thermophysical Seminar **09001**
[21] Aleksyuk A I 2021 *J Fluid Mech* **915**
[22] Vinogradov Y A, Zditovets A G, Leontiev A I, Popovich S S and Strongin M M 2017 *JPCS* **891**
[23] Leontiev A I, Popovich S S, Vinogradov U A and Strongin M M 2018 *Proc 16th Int Heat Transf Conf* **212244** (Beijing)

Enhancement of heat transfer for boiling in nanofluid

S Z Sapozhnikov, V Yu Mityakov, A V Pavlov, P G Bobylev, Yu V Andreev, V M Proskurin and M D Vinogradov

Peter the Great St. Petersburg polytechnic University,
Russia, 195251 St. Petersburg, Politekhnicheskaya, 29

pavlovav196@mail.ru

Abstract. The paper considers heat transfer during boiling of subcooled water with suspended nanoparticles Al_2O_3 using a suspension from 0.32% to 4%. On a spherical model, the local heat flux per unit area was measured by the method of gradient heatmetry for model temperature of 464 °C and water temperature of 64 °C. The results are compared with the data obtained at the same temperature conditions for pure water. Enchancement of heat transfer was revealed in the entire concentration range - with a maximum at a particle concentration close to 1%.

1. Introduction

Application of Al_2O_3 nanoparticles as a heat exchange intensifiers at boiling controversial, and there is still no consensus on the impact of suspensions on heat transfer [2,3]. Considered temperature conditions (model temperature of t_w = 464 °C, fluid temperature of t_f = 64 °C) implemented on pure water demonstrates classic regimes [1]. Figure 1 clearly shows film (I), transient (II), and nucleate (III) boiling regimes. Boiling patterns are confirmed by high-speed imaging. We used these results as a basis for comparison.

Gradient heatmetery is the advanced method for direct measurement of heat flux per unit area (HFPUA). The method is based on the use of heterogeneous gradient heat flux sensors (HGHFS) (figure 2), the action of which is based on the transverse Seebeck effect [4]. When a heat flux of q passes through the sensor, it appears proportional to transverse thermoEMF

$$E_0 = S_0 \cdot A \cdot q \quad (1)$$

where E_0 – thermoEMF, mV; S_0 – volt-watt sensitivity of HGHFS, mV/W; A – area of HGHFS, m^2 ; q – heat flux per unit area, W/ m^2 .

HGHFS were successfully used for study of boiling and they made it possible to obtain not only the distribution of HFPUA at the surface of the model, but also instantaneous values of its quantity [5].

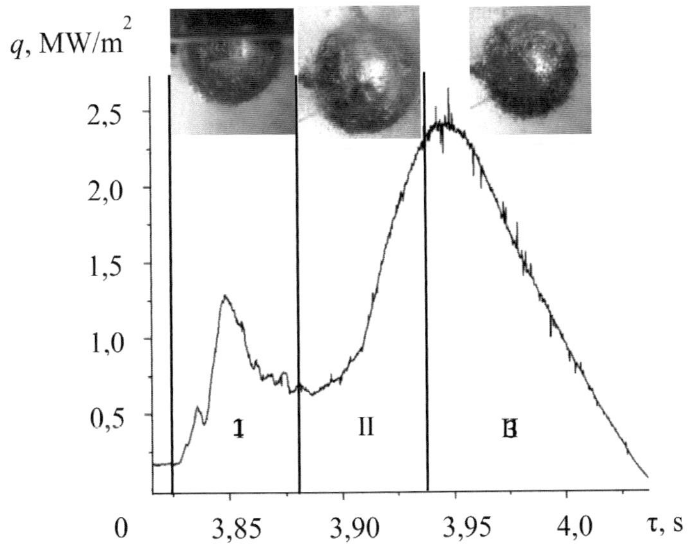

Figure 1. Dependence of heat flux per unit area on time (T_w = 464 °C; T_f = 64 °C).

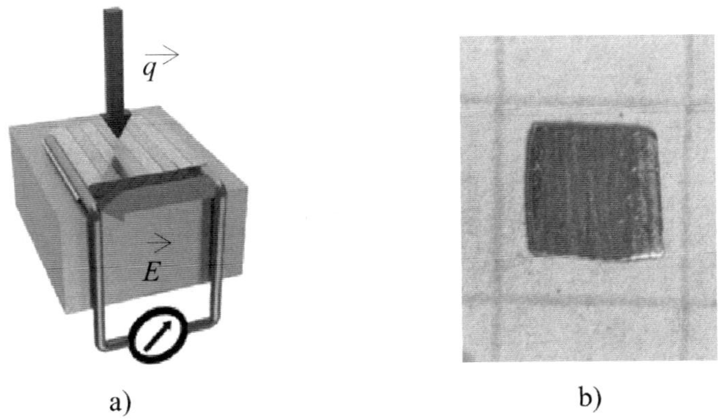

Figure 2. Heterogeneous gradient heat flux sensor (HGHFS): a –action scheme; b – photo (in 3×3 mm square).

2. Experimental setup

The installation scheme of our setup is shown in figure 3. Experimental model – sphere 1 – made of VT22 titanium with a dia of d = 34 mm with a HGHFS with dimensions of 3 × 3 × 0.3 mm installed at its surface. Using the holder 3, sphere 1 is fixed in the channel furnace 2 and is heated to the required temperature. Uniform heating is controlled by two copper-constantan thermocouples, mounted on the surface and in the core of the sphere, respectively. Upon reaching the desired temperature, the holder 3 releases the model 1, and it is lowered into the water tank 4. All data from the HGHFS and thermocouples are recorded by the measuring computer complex 8 of the *NIPXI-1050* type, and the water temperature is monitored by a 5 *Fluke 289* multimeter with its own thermocouple. In experiments with pure water was we have used high-speed camera 6 model *Evercam 1000-4-M* with an illuminant 7.

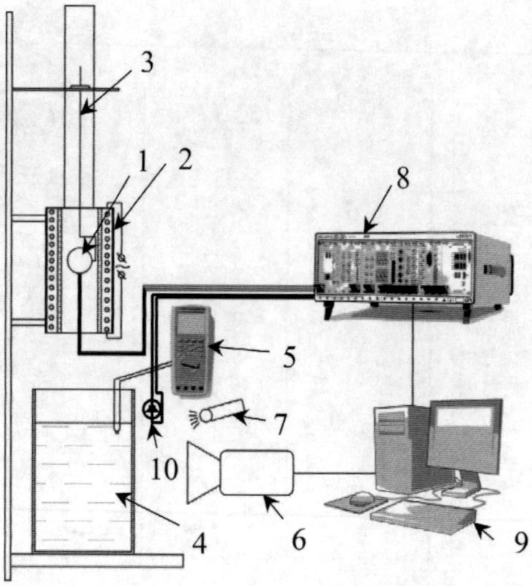

Figure 3. Experimental setup: 1 –model; 2 – furnace; 3 –holder; 4 – water tank; 5 – *Fluke 289* with thermocouple; 6 – high speed *Evercam 1000-4-M* camera; 7 – illuminant; 8 – *MCC NIPXI-1050*; 9 – computer; 10 – light emitting diode.

3. Results

The mass concentration of particles of Al_2O_3 were in the range of 0.32% – 4% [6,7], while the particle size did not exceed 1 μm. At mass concentrations of Al_2O_3 equal to 0.32% (figure 4 a) and 1% (figure 4 b), HFPUA significantly increases. Absence of stable film boiling regime reduces the cooling time of the model. Visualization in these modes is impossible due to the optical impermeability of the medium, but the curves show that there is no pronounced nucleate boiling. This qualitatively coincides with the boiling regime in pure subcooled water [1]. τ = 0 – the moment of time when the sensor touches the liquid surface.

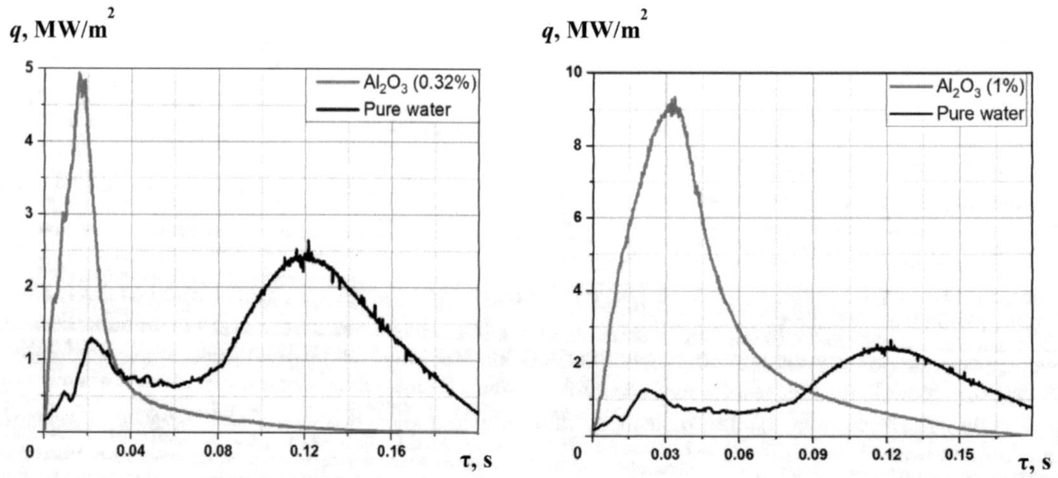

Figure 4. Dependence of heat flux per unit area on time: a – for pure water vs Al_2O_3 mass fraction of 0.32 %; b – for pure water vs Al_2O_3 mass fraction of 1 %.

According to thermometry data, the section of intensive cooling of the model coincides with the development of nucleate boiling, measured by gradient heatmeatry (figure 5). As can be seen from figure 5, with the addition of nanoparticles, an increase in the heat flux is observed, and as a consequence, the cooling rate of the model increases, which can be observed from the slope of the temperature curves.

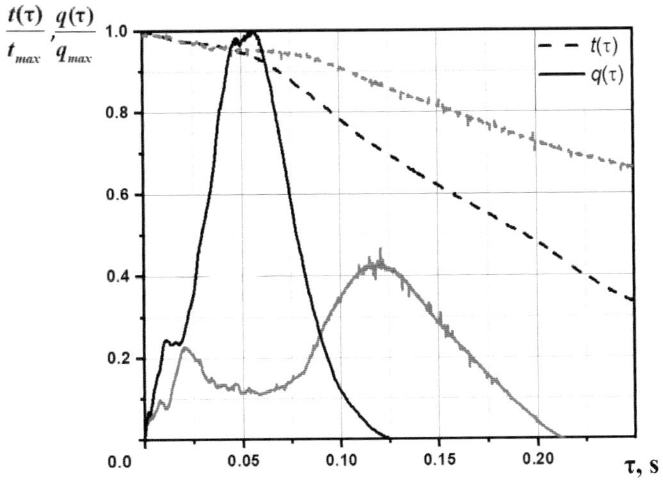

Figure 5. Time dependence of dimensionless temperature and heat flux per unit area in pure water and in Al_2O_3 with a mass concentration of 2%.

At concentration of Al_2O_3 particles equal to 4% (figure 6), HFPUA also increases. In this mode, the film boiling regime is qualitatively recorded, which corresponds to low values of heat flux at the beginning of the process. (figure 6). The lifetime of the film and transient regimes is much shorter than in pure water. At such a concentration of particles, developed nucleate boiling is observed, with fluctuations near the maximum value of HFPUA.

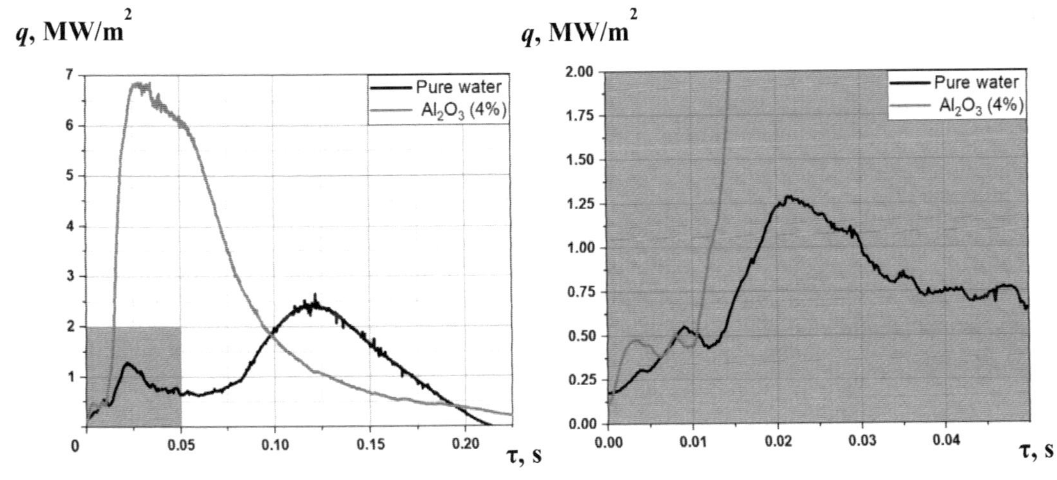

Figure 6. Dependence of heat flux per unit area on time in pure water and Al_2O_3 mass fraction of 4 %.

Figure 7 shows the dependence of HFPUA on concentration of Al_2O_3 particles and a comparison with values of heat flux during boiling in pure water. There is a pronounced maximum near 1% concentration.

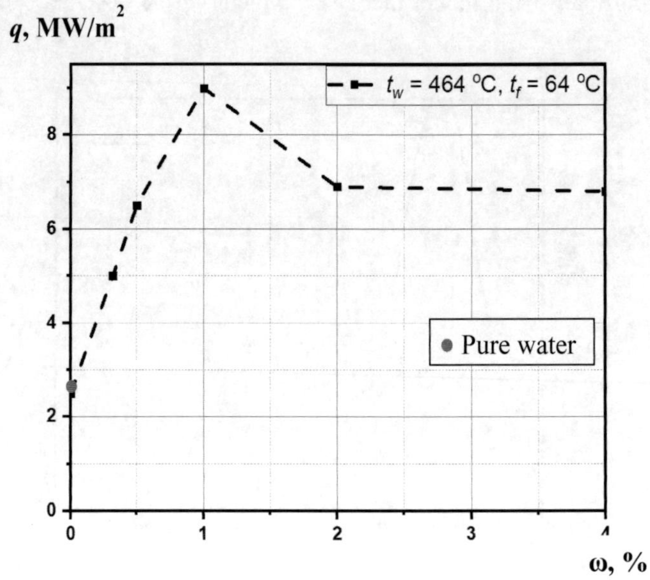

Figure 7. Critical heat flux vs Al_2O_3 mass fraction.

4. Conclusion

It was found that addition of nanoparticles of Al_2O_3 to subcooled water intensifies heat transfer in entire range of particle concentrations (0.32 - 4%). The optimal result was achieved at a concentration of 1% when the critical value of HFPUA reaches 9 MВт/м^2, which is 3.6 times more than for pure water. It is planned to continue research for other temperature conditions.

References

[1] Sapozhnikov S Z, Mityakov V Yu, Pavlov A V and Bobylev P G 2020 *Gradient heatmetry for boiling of underheated water on spherical surface* (Journal of Physics: Conference Series) vol 1683

[2] Dedov A V 2019 *A review of modern methods for enhancing nucleate boiling heat transfer* (Teploenergetika) vol 12 pp 18-54

[3] Terekhov V I, Kalinina S V and Lemanov V V 2010 *The mechanism of heat transfer in nanofluids: state of the art (review). Part 2. Convective heat transfer* (Thermophysics and Aeromechanics) vol 17

[4] Sapozhnikov S Z, Mityakov V Yu and Mityakov A V 2020 *Heatmetry The Science and Practice of Heat Flux Measurement* (St.-Petersburg, Springer International Publishing) 209 p

[5] Sapozhnikov S Z, Mityakov V Yu, Pavlov A V, Bobylev P G and Vinogradov M D 2021 *Gradient heatmetry in the study of boiling on spherical surface* (Journal of Physics: Conference Series) vol 1867

[6] Das S K, Putra N and Roetzel W 2003 *Pool boiling characteristics of nano-fluids* (Int. J. of Heat and Mass Transfer) vol. 46 pp 851–862

[7] Wen D and Ding Y 2005 *Experimental investigation into the pool boiling heat transfer of aqueous basedc-alumina nanofluids* (Springer, Journal of Nanoparticle Research) vol. 7 pp 265–274

Heat transfer enhancement during condensation of water steam on inclined pipe

S Z Sapozhnikov, V Yu Mityakov, A V Mityakov, A Yu Babich and E R Zainullina

Peter the Great St. Petersburg Polytechnic University,
Russia, 195251 St. Petersburg, Politekhnicheskaya, 29

zaynullinaelza@gmail.com

Abstract. This paper presents experimental study of heat transfer during film condensation of saturated water steam on the outer surface of the inclined pipe by gradient heatmetry. Heat flux per unit area was measured by gradient heat flux sensors made of a single-crystal bismuth. The experimental results are presented in the graphs of heat flux per unit area dependence on time and azimuthal angle. The highest average heat transfer coefficient during condensation of $\alpha = 6.94\ \text{kW/(m}^2 \cdot \text{K)}$ was observed when the pipe was inclined at the angle of $\psi = 20\ °$. This value exceeds one obtained on a vertical pipe by 14.9 %. Heat transfer enhancement during condensation of saturated water steam on inclined pipe is associated with changes in condensate film flow. Another part of experiments was made by simultaneously using of gradient heatmetry and condensate flow visualization. Experimental results confirmed the applicability and high informative content of proposed comprehensive method. Comprehensive study of heat transfer during condensation confirmed that heat flux per unit area pulsations may be explained by the formation of individual drops, their coalescence, and drainage from the sensor surface.

1. Introduction

The most common methods of heat transfer enhancement during condensation of water steam are associated with surface development. Ribbing or irradiation of the surface complicate the design of heat exchanger and increase its cost. One of the simplest and underestimated ways to increase heat transfer coefficient (HTC) during film condensation of saturated water steam is pipe inclination.

On the inclined pipe the condensate film flow breaks up into two zones [1]: the main one and the bottom, which is developed in the lower sector of the pipe. Change in the condensate film thickness occurs both along the length and the circumference of the pipe (figure 1). Because of condensate flow redistribution over the surface, it is necessary to install sensors around the circumference and along the length of the pipe. But installation of many invasive measuring devices is unacceptable due to the distortion of natural flow of condensate film.

Surface-averaged measurements [2] allow to determine the optimal angle of pipe inclination with minimal distortion during the condensate flow, but it turns out to be undescriptive. The use of non-invasive optical research methods [3] makes it possible to obtain information about condensate flow distribution over the pipe surface, but turns out to be inapplicable for industrial experiments.

T. Garrett [4] applied optimal sensors' mounting method. Thermocouples were installed at the same pipe generatrix, and experimental setup could be inclined and rotated around its own axis.

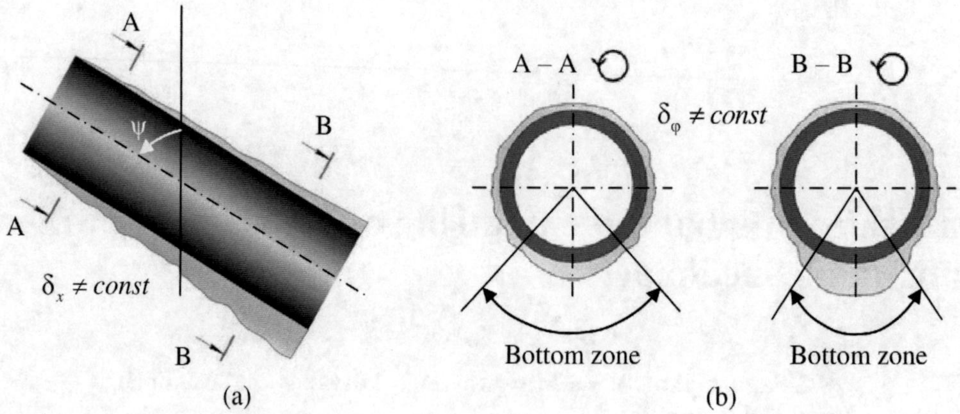

Figure 1. Condensate film flow distribution along the length (a) and around the circumference of two sections (b) of the inclined pipe.

Since 2015, in Science Education Centre «Energy Thermophysics» of Peter the Great St. Petersburg Polytechnic University, heat transfer during saturated water steam condensation has been studied by gradient heatmetry [5]. Gradient heatmetry is innovative technology based on use of gradient heat flux sensors (GHFSs). GHFS action is based on transverse Seebeck effect: when heat flux passes through a plate with anisotropy of thermal and electrophysical properties, thermoemf occurs with vector of intensity normal to heat flux one [5]. Signal of GHFS E is associated with heat flux per unit area q passing through the section of the sensor, as

$$E = q \cdot F \cdot S_0,$$

where F is GHFS' area, and S_0 is its volt-watt sensitivity.

Gradient heatmetry allows to measure local heat flux per unit area, to calculate local HTCs and to estimate the condensate film flow distribution over heat exchange surface. The main purpose of this study is to assess heat transfer enhancement during saturated water steam condensation on inclined pipe by gradient heatmetry.

2. Gradient heatmetry in study of heat transfer during condensation on inclined pipes

The previous works [6, 7] have confirmed applicability of gradient heatmetry to study of heat transfer during condensation. Capabilities of heatmetry and thermometry were successfully combined to calculate local HTCs.

Experimental setup consists of two coaxial pipes (figure 2, a). The inner one made of stainless steel (d = 20 mm, δ = 2 mm, l = 1000 mm) is fixed with rubber plugs in the outer (casing) one which is made of reinforced rubber sleeve (d = 0.065 m, δ = 5 mm). During the experiments, saturated water steam was fed into annular space from above at atmospheric pressure; its flow rate was of 2.8 g / s. Cooling water with temperature of t = 20 ° C and flow rate of 200 ml / s entered into the inner pipe. Condensate was discharged into the condensate trap.

5 GHFSs made of single-crystal bismuth with dimensions of 10.5 × 2.5 × 0.3 mm were installed at the same generatrix of the pipe, flush with the outer surface of it (figure 2, b). Surface temperature was controlled by 5 semi-artificial thermocouples made of a stainless steel + copper composition. The measuring section was equipped with devices for inclination in the range of ψ = 0 ... 90 ° with a step of 10 ° and rotation in the range of an azimuthal angle of φ = 0 ... 180 ° with a step of 15 °. This design of the setup allows to minimize distortions in condensate film flow.

Figure 3, a, as an example, shows the time heatgram – the dependence of heat flux per unit area on time – when the measuring section is inclined by the angle of ψ = 20 ° and rotated by an azimuthal angle of φ = 180 ° (all GHFSs are located on the lower generatrix of the pipe). All sensors register heat flux pulsations, which confirms heat transfer nonstationarity during condensation.

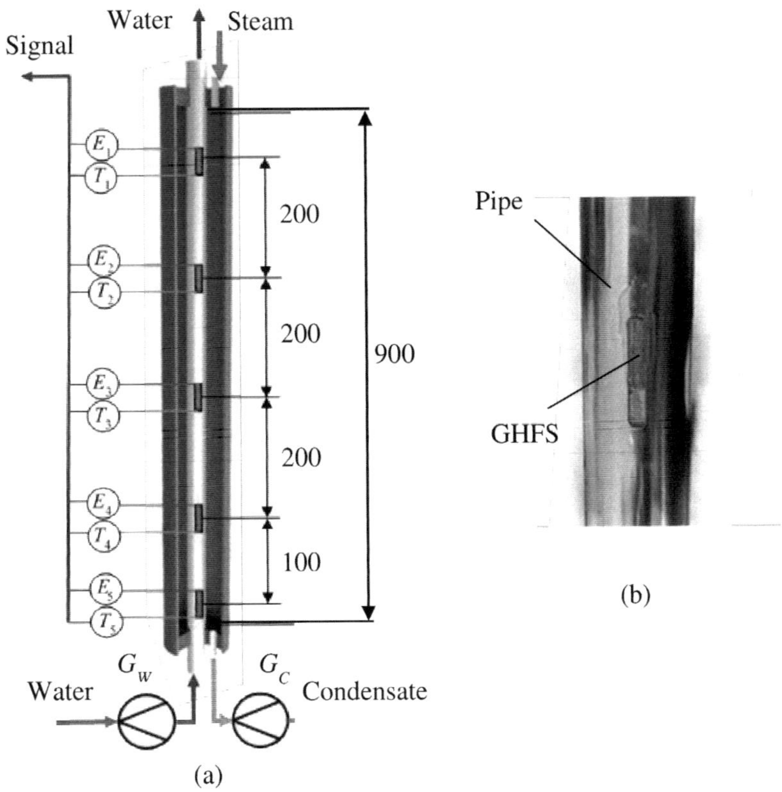

Figure 2. Experimental section: scheme (a) and photo of GHFS installation on the pipe (b).

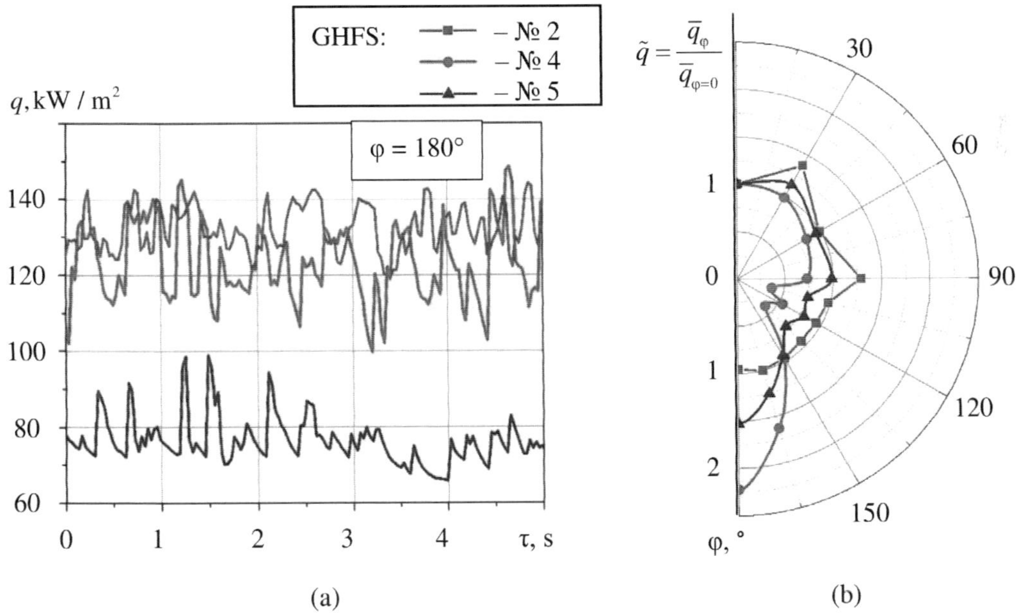

Figure 3. Experimental results on the pipe inclined at the angle of $\psi = 20$ °: time (a) and angular (b) heatgrams.

According to the signal of GHFS №2 at the distance of $x = 300$ mm from the upper cut of the pipe, average heat flux per unit area is of 130 kW / m^2, its pulsations are of 10% of average signal. At the distance of $x = 700$ mm, GHFS № 4 registers decreasing of average heat flux per unit area up to 120 kW / m^2 and increasing of pulsation amplitudes up to 35% of the average level. Signal of GHFS № 5 installed at the distance of $x = 800$ mm corresponds to average heat flux per unit area of 76 kW / m^2 while pulsations are of 22% of the average level. Large amplitude of heat flux pulsations can be associated either with formation of waves on the condensate film, or with runoff of rivulets or separate condensate drops.

Summarizing of experimental results of our study is carried out in polar coordinates. Figure 3, b, shows distribution of dimensionless local heat flux per unit area: its value measured when the pipe is rotated by the azimuth angle of φ, is related to heat flux on the upper generatrix, where $\varphi = 0$ °. The signals from GHFSs №4 and №5 indicate an increasing of heat flux per unit area in lower sector in the range of an azimuthal angle of $\varphi = 150 - 210$ °. Heat transfer enhancement in the lower sector is associated with changes in condensate film flow.

Temperature measurement at the place of sensor installation allows to calculate local HTCs. Figure 4 shows the dependence of average HTCs on the angle of pipe inclination. Average HTC increases when the pipe is inclined by an angle of $\psi = 10 \ldots 40$ °. At $\psi = 20$ °, the average HTC is maximal ($\alpha = 6.94\,\text{kW} / \text{m}^2 \cdot \text{K}$) and exceeds the value obtained on a vertical pipe by 14.9 %.

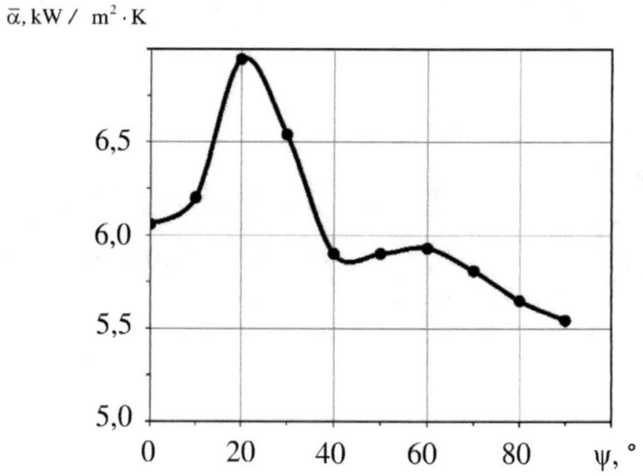

Figure 4. Average HTC for the inclined pipe.

To assess heat transfer enhancement during condensation on inclined pipes, it is necessary to determine the reason for occurrence of heat flux pulsations. Therefore, it is necessary to combine gradient heatmetry with condensate film flow visualization.

3. Simultaneous using of gradient heatmetry and film flow visualization

The first experimental series on combining of gradient heatmetry and film flow visualization was performed on the model of a cooled vertical plate. The model is a box-shaped structure with fittings for supplying and removing of cooling water with temperature of $t = 24$ ° C and flow rate of 40 ml / s. The heterogeneous gradient heat flux sensor (HGHFS) made of a copper + nickel composition with dimensions of 10 × 15 × 0.2 mm was installed on the plate surface (figure 5). The experimental set up was an "open system": steam was supplied from the steam generator from below to a plate installed vertically, therefore, in these experiments flow rates of steam and condensate were not measured.

Figure 6 shows the time heatgram and photographs during saturated water steam condensation on the vertical plate. The photographs confirm that condensate drops are formed on the surface of the HGHFS, as well as on the plate surface. Visualization of the condensation revealed that at low condensate flow rate, a continuous condensate film does not form. In this case, the HGHFS registers pulsations associated with formation and drainage of condensate drops.

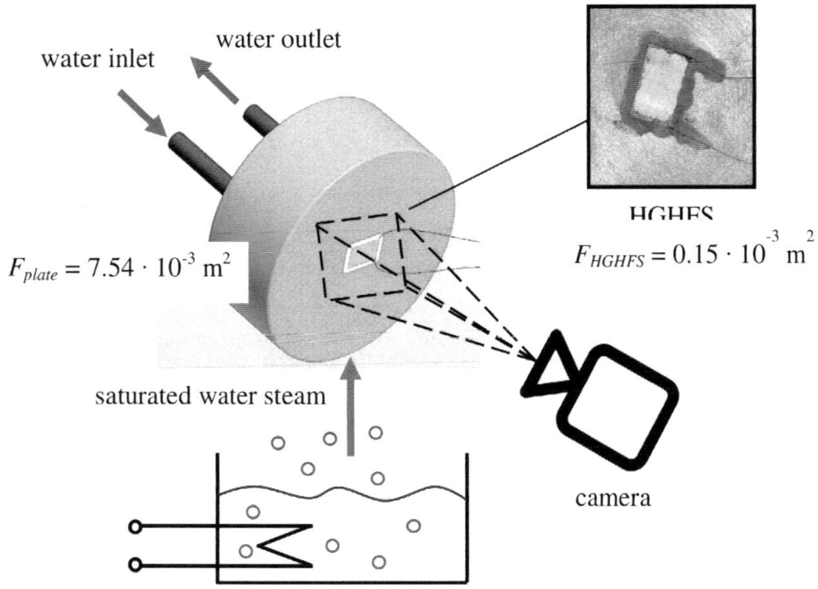

Figure 5. Experimental section scheme for combining gradient heatmetry and flow visualization.

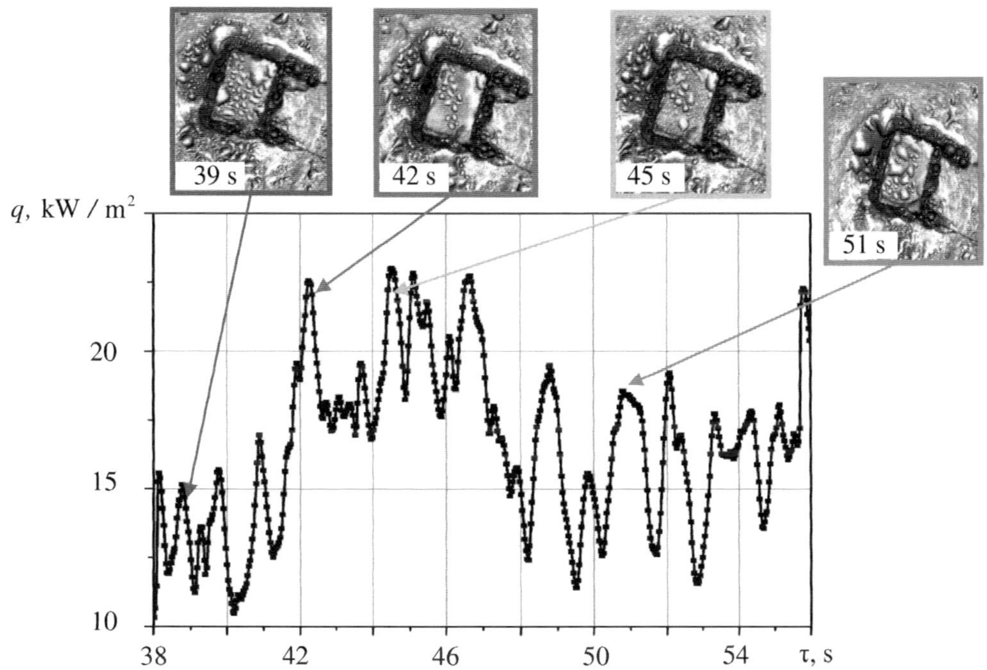

Figure 6. Time heatgram for condensation of saturated water vapor on a vertical plate.

The average heat flux per unit area was of 12.5 kW / m^2 with formation of condensate droplets on the surface of the HGHFS; after the condensate draining from the surface, the value increased up to

18 kW / m^2. After the condensate drains off, the process of condensate formation is repeated, and heat flux per unit area decreases up to 15.5 kW / m^2.

To create the effect of dripping condensate, in addition to supplying steam from the bottom, separate water drops with a temperature of about t = 70 °C are supplied to the HGHFS surface through the injector (figure 7).

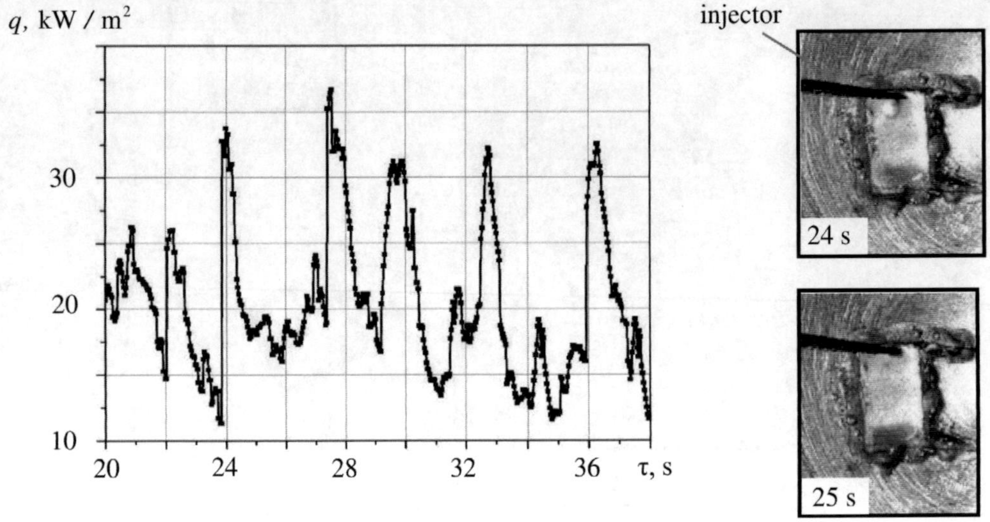

Figure 7. Time heatgram for condensation of saturated water vapor on a vertical plate.

When water is supplied through the injector, the condensate film forms on the surface of the HGHFS, which indicates the film condensation flow at high condensate flow rate. The HGHFS registers increase of heat flux per unit area up to 45 ... 55 kW / m^2 when heated drops are supplied, and a reverse decrease up to 10 … 20 kW / m^2 when the surface is released from them.

4. Conclusion

Experimental study of heat transfer during condensation on the outer surface of inclined pipe have shown that HTC is intensified on pipe inclined at angle of ψ = 10 ... 40 °. At ψ = 20 °, the average HTC has maximum of $\alpha = 6.94\,\text{kW} / \text{m}^2 \cdot \text{K}$ and exceeds its average value for vertical pipe by 14.9 %. Series of experiments was performed to combine gradient heatmetry and visualization of condensate film flow on a vertical plate. It was revealed that HGHFS registers formation of condensate drops on all the surface, their coalescence and runoff. Experiments carried out with an artificial supply of water have confirmed that the HGHFS registers condensate drops flowing down from above.

References

[1] Sheinkman A G, Yanchenko V F and Linetsky V N 1967 *Tr. Ural Polytechnic. Inst.* p 166
[2] Nada S A, Hussein M S 2016 *Int. J. of Thermal Sciences* vol 100 pp 391–400
[3] Lel V V, Al-Sibai F and Renz U 2005 *Experiments in Fluids* vol 39 pp 856–864
[4] Garrett T W and Wighton J L 1964 *Int. J. of Heat and Mass Transfer* vol. 7 pp 1237–1243
[5] Sapozhnikov S, Mityakov V and Mityakov A 2020 *Heatmetry: The Science and Practice of Heat Flux Measurement* (Switzerland: Springer Int. Publ.)
[6] Sapozhnikov S Z, Mityakov V Y, Mityakov A V and etc. 2019 *Technical Physics Letters* vol. 45, pp 321-323
[7] Sapozhnikov S Z, Mityakov V Y, Mityakov A V and etc. 2020 J. of Physics: Conf. Series 1683(2), 022028

Features of thermodynamic and thermal processes in hydrogen combustion units and systems on their basis

A I Schastlivtsev and V I Borzenko

Joint Institute of High Temperature,
Russia, 125412 Moscow, Izhorskaya, 13, Bd. 2

h2lab@mail.ru

Abstract. The main types and designs of hydrogen combustion units (HCU), including hydrogen-oxygen steam generators, superheaters and air heaters of various power levels, are considered. The main problems arising in the development, creation and testing of such installations are determined, including the problems of cooling the most heat-stressed units, mixing of the main components of the fuel and oxidizer, mixing of high-temperature combustion products and ballasting components, problems associated with the completeness of hydrogen combustion and ensuring safety during operation.

1. Introduction

Currently, the world energy industry is undergoing a period of important structural changes, and one of the main trends is the constant growth of the annual commissioning of renewable energy sources (RES) capacities. To ensure the sustainability of power systems and guarantee the quality of power supply, the spread of RES must be supported by energy storage systems. Hydrogen is seen as an efficient and environmentally friendly secondary energy source for energy storage in the energy systems of the future. However, modern hydrogen technologies remain relatively expensive, especially in areas related to the production and use of hydrogen, and require significant government support. Meanwhile, the problem of using hydrogen in the field of high capacities can be solved by hydrogen combustion units (HCU), the technologies for the production of which are well mastered, including in Russia.

Hydrogen as a fuel for power plants has a number of important features: high combustion heat, wide concentration limits of ignition and stable combustion, high flame propagation rates (7 times higher than for methane), low activation energies during combustion in oxygen and air. For power engineering, it is of great importance that when a stoichiometric mixture of hydrogen burns in oxygen, pure superheated steam is produced - the working fluid of modern steam turbine plants. In this case, the transfer of heat from the fuel to the working fluid can be carried out most efficiently - by mixing the combustion products (superheated steam with a high temperature) with the working fluid (water or steam), i.e. with minimal losses of fuel exergy.

The technology of combustion of hydrogen in oxygen in order to obtain high-temperature steam or air, implemented in hydrogen-oxygen steam generators, superheaters and air heaters, makes it possible to create installations with a high specific power and no harmful emissions during operation. The high speed of the processes occurring in them (combustion of hydrogen, heat transfer, mixing of components, etc.) provides a minimum start-up and recovery time, which is especially important in the case of using such technologies as backup energy sources.

2. The main types of hydrogen combustion units

Hydrogen-oxygen plants can be classified by the type of coolant used (ballasting component), by the technologies used to create the combustion chamber, and by the type of mixing of hydrogen and oxygen. By the type of ballasting component, two main types of hydrogen-oxygen plants can be distinguished:

- hydrogen-oxygen steam generators and superheaters [1-3];
- hydrogen-oxygen air heaters [4, 5].

In most cases, the ballasting component is used simultaneously to cool the walls of the combustion chamber. The type of ballasting component used determines the field of application of the HCU. When using water and steam, the final product of the HCU is superheated steam, which can then be used as a working medium for steam turbines [1, 6-8]. When using air at the outlet, air is obtained, heated to a certain temperature, which is then used as a working medium for gas turbines. [5].

By the type of technologies used to create a combustion chamber, 2 main types of HCU can also be distinguished.:

- rocket technologies;
- gas turbine technologies.

Experimental models using rocket technologies were created in Russia [1, 9, 10] and Germany [8], and using gas turbine technologies in Japan [11].

The most promising for HCU are rocket technologies, since they provide a higher completeness of hydrogen combustion, are more compact, and have a shorter start-up and operating time.

3. The main problems and features of the study of thermal processes in the HCU

The use of steam from the HCU in steam turbines necessitates their high efficiency. The main losses in the HCU are associated with incomplete combustion of hydrogen and heat losses to the environment. With cooling water, up to 5...7% of the total released heat can be removed from the heat-stressed parts of the steam generator [12]. Therefore, in order to achieve high efficiency, it is necessary to use all the cooling water for steam production. This, in turn, entails the need to create such a design of the compressor station, which will provide the necessary cooling with a limited flow of water.

Ensuring reliable cooling of the most heat-intensive units of a hydrogen-oxygen steam generator is an important design problem. In particular, in [13] it is shown that usually the heat fluxes in the combustion chamber of the HCU are up to 20 MW/m^2.

To solve this problem, a combined system for cooling the walls of the combustion chamber is mainly used. Firstly, external cooling of the combustion chamber walls is used, which are made of heat resistant bronze, which provides high thermal conductivity, preventing the walls from heating up by more than 150 degrees. Secondly, internal cooling is used by sprinkling the wall layer with a small amount of water, which also lowers the temperature of the combustion chamber walls and ensures a longer service life. By dividing the cooling water flows into internal and external, a decrease in the concentration of steam in the combustion zone is achieved, which leads to an increase in the completeness of fuel combustion.

When steam is supplied to the flow path of steam turbines, an important requirement for its parameters is temperature irregularity, which should be no more than 15...30 degrees along the radius. This is due to the peculiarities of the rotor blades, for which a large nonuniformity of temperatures can lead to additional thermal stresses and their premature destruction. When creating an internal combustion chamber both in Russia and in Germany, it was possible to obtain an uneven temperature at the exhaust of 7...10 degrees at installations of a megawatt power level [9, 12], by mixing a high temperature flow with a large amount of cooling water.

The presence of unburned hydrogen in the generated steam is especially dangerous for the subsequent use of this steam in closed cycles of steam turbine plants. Even a small content of it (less than 1% (vol.)) Can lead to the formation of an explosive mixture in the stagnant zones of the elements of power equipment, in particular in capacitors. It is known from the experience of rocket and gas turbine engineering that in order to ensure the most complete combustion of the fuel, it is necessary to ensure its good mixing with the oxidizer. There are 2 main design options for organizing the mixing of fuel and oxidizer jets:

- Centrifugal nozzles;
- Jet nozzles (cross jet and coaxial jet).

Centrifugal nozzles have a wide and relatively short spray cone, while the spray is thinner than for jet nozzles, therefore they are most widely used when supplying fuel or an oxidizer in the form of a liquid, and since HCU mainly uses gaseous components in them, they do not will apply.

Jet nozzles are smaller and simpler than centrifugal nozzles and this allows more of them to be placed on the mixing element. For example, in the most advanced DLR mixing element [12] placed 58 coaxial nozzles on a diameter of 100 mm, which significantly complicates its design. The flow rate of jet nozzles is 2.5...3 times higher than that of centrifugal nozzles, which makes it possible to ensure their high throughput. Their main disadvantage is a relatively long range and a small spray angle, and therefore, for the best mixing, the intersection of the jets of fuel and oxidizer, or a large difference in the speeds of the components (in the case of coaxial nozzles) is used.

The advantage of such a mixing element is its reliable operation in a wide power range. But at the same time, to ensure high completeness of hydrogen combustion, a significant complication of the design is required.

The use of coaxial jet nozzles is also advisable to use in an internal combustion unit with a combustion chamber that provides additional mixing of the components due to the injection of a ballasting component into the combustion zone. However, in this case, a significant slowdown of chemical reactions in the combustion zone can occur and a decrease in the completeness of hydrogen combustion, as a result.

The completeness of hydrogen combustion in oxygen is also strongly influenced by the rate of cooling of high temperature steam. It depends to the greatest extent on the type of heat carrier used (water or steam), as well as the method of its mixing with combustion products. Too high cooling rate leads to a "quenching effect" of the high temperature steam. A significant influence of this effect on the completeness of hydrogen combustion was noted in the work [12]. Its manifestation takes place in three stages. At the first stage a sharp cooling of the high temperature steam near the relatively cold walls of the combustion chamber is possible, however, due to the small amount of high temperature steam subjected to such cooling (about 5%) and the not very high rate of its cooling, the effect of this stage on the combustion efficiency is not great and is up to 0.006% for hydrogen and up to 0.04% for oxygen. An experimental study of this stage and its contribution to the completeness of combustion is very difficult to carry out, since all the experimental studies carried out were carried out with the addition of a coolant to the high temperature steam, the effect of which is much more significant as will be shown below.

At the second stage a cooling component is supplied, the mass flow rate of which is several times greater than the mass flow rate of high temperature steam. In this case, if the cooling component (steam or water) is supplied too quickly, especially directly into the combustion zone, a sharp decrease in the reaction rate occurs due to a decrease in the concentration of active components in this zone. As a result, some of the components and intermediate products of the combustion reaction do not have time to react.

At the third stage, the high temperature steam is abruptly cooled by a cooling component. In this case, due to the fact that at high temperatures the decomposition of water into its main components occurs, the combustion temperature in the combustion chamber is about 3600 K, the mole fractions of hydrogen and oxygen in such a steam are maximum. Therefore, the influence of the quenching effect on the completeness of hydrogen combustion can be very large. For example, let's take the data presented in the work [14] for steam at a pressure of 0.1 MPa (figure 1). Abrupt cooling of the high-temperature flow can lead to significant underburning of the fuel. The authors explain the reason for the decrease in the completeness of combustion by the fact that a lower water temperature provides a faster heat removal from the combustion zone, which reduces the rate of the reaction.

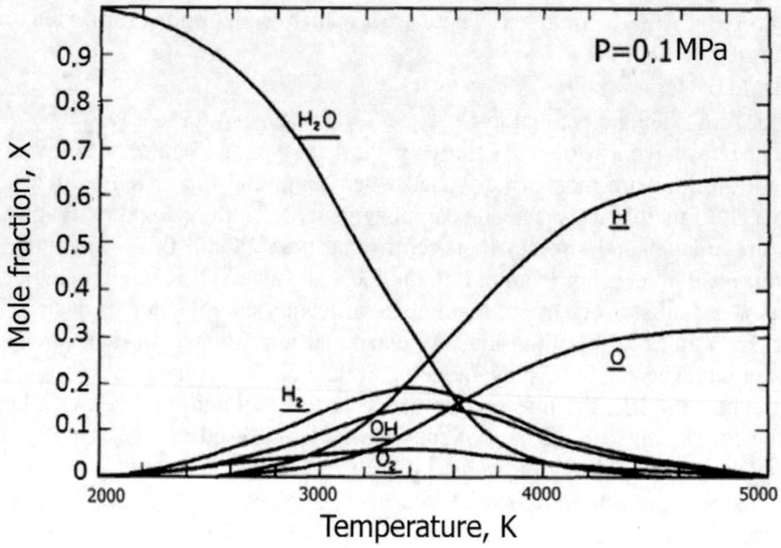

Figure 1. Equilibrium distribution of molar fractions of components in steam at a pressure of 0.1 MPa.

It should be noted that the quenching effect in highly heated steam is used directly in the production of hydrogen by thermal decomposition of water, in which, on the contrary, sharp cooling and dilution of the flow are used to obtain the maximum amount of hydrogen. [14, 15]. Experimental studies carried out in [14] showed that the dependence of the molecular hydrogen content is almost linearly dependent on the cooling rate of the high temperature flow, and the minimum cooling rate should be more than 105 K/s. Thus, to prevent the formation of unburned hydrogen in a HCU, the cooling rate of combustion products should be significantly lower than 105 K/s. It is quite difficult to provide such a cooling rate when used as a water cooler. To do this, it is necessary to ensure its supply in several stages, while the water used to cool the combustion chamber must be preheated in it to a temperature close to the boiling point, in this case, when it is injected into the HCU, it will partially evaporate, which will reduce the rate of cooling of combustion products. A similar cooling scheme was used on all similar devices, as shown above.

A lower rate of cooling of the combustion products of hydrogen and oxygen can be achieved when using water vapor as a cooling component. Significant disadvantages of using water vapor are the lower cooling efficiency of the combustion chamber and the need for the steam itself of the required pressure. As a result, in such structures in the combustion chambers, there is practically no internal cooled insert, which prevents the ingress of cooling steam into the combustion zone. This leads to the fact that the completeness of combustion begins to be strongly influenced by the second stage of the quenching effect, which can significantly reduce the completeness of combustion of hydrogen. The option using an internal insert cooled by low-temperature steam can be applicable in the case of a very large amount of steam, for example, at nuclear power plants with wet steam turbines, but such designs have not yet been developed. Experimental data on the operation of installations with a similar cooling steam supply scheme are also lacking.

4. The main areas of application and thermodynamic features of the HCU

Being universal and compact devices, HCU allow implementing various thermodynamic cycles of energy conversion using modern and promising steam turbine technology. While using them, the restrictions on the temperature of the superheated steam, determined by the materials of the boiler units and steam pipelines, are practically lifted and it is possible to modernize the existing power plants, which makes it possible to fully realize the power reserves of the existing equipment, increase the efficiency of power plants and reduce the specific heat consumption per kilowatt-hour of electricity generated.

Among the main areas of application of the HCU are the following:

- Additional storage superstructures for the production of peak capacities;
- Superheating of steam in front of the turbine;
- Hydrogen energy storage systems.

When using the HCU for turbine generators of turbines K-160-130, K-200-130, K-300-240 and a number of others, it was found that they have a margin of 10% at nominal and maximum power, i.e. modernization of the power plant in order to increase power by 10% by increasing the temperature of steam superheating from 540 °C to 565 °C and its consumption can be carried out with minimal costs by including in the HCU scheme as a superstructure to the turbine, and the turbine itself and the turbine generator can remain unchanged. In this case, the specific fuel consumption is reduced by 1.2...1.5 %. Calculated studies of the thermodynamic efficiency of steam turbine power plants of various types (new high temperature and modernized existing ones) with HCU show that the efficiency of using hydrogen for electricity generation in such plants can exceed 60% [10].

A side effect of using the HCU as a superstructure is the possibility of providing additional superheating steam before the steam turbine. The use of such schemes has recently been considered for nuclear power plants [7]. It should be noted that in the european part of Russia, where the share of nuclear power plants (23.2 million kW) and poorly maneuverable thermal power plants (58 million kW) is relatively large, today the required maneuverability of generating capacities is not provided (there is one Zagorsk pumped storage power plant (PSP) with a capacity 1200 MW).

An important potential area of HCU application is steam superheating before wet steam turbines, at geothermal power plants (GeoPP) [6]. The very name of wet steam turbines indicates that they operate on wet steam, and their efficiency (14...38%), as a rule, is significantly lower than that of turbines operating on superheated steam (up to 52%). At the same time, the use of wet steam turbine plants is most often due to the peculiarities of steam generating plants, for example, at a nuclear power plant, this is a restriction by the parameters of the reactor, and at a GeoPP, it is limited by the parameters in a geothermal well. A partial solution to this problem can be the use of additional superheating of steam to reduce its moisture content in the flow path of the turbine, as well as at the outlet from it. In this case, as a result of a decrease in steam moisture in the flow path of a steam turbine, its internal efficiency increases, which in general makes it possible to increase the overall (taking into account the cost of producing hydrogen) efficiency of the power plant by 3...4% when replacing a wet steam turbine with a turbine using superheated steam. In addition, reducing the moisture content of steam in the flow path of the turbine, and especially at the outlet from it, will avoid premature wear of the turbine blades, and, consequently, will reduce the cost of overhaul, increase reliability and reduce the number of forced shutdowns. It should be noted that the literature sources do not pay sufficient attention to the technical, economic and thermodynamic assessment of such schemes, which is primarily due to the lack of prototypes of hydrogen-oxygen steam superheating systems, which makes it difficult to assess their cost and technical characteristics.

HCU can also be used in hydrogen energy storage systems. One of the latest schemes using hybrid air-hydrogen energy storage is shown in [5]. A combined scheme of an air storage gas turbine power plant is proposed, in which the heat of combustion of hydrogen in oxygen obtained by electrolysis is used to heat the air in front of the turbine. Hydrogen as an energy accumulator has a number of advantages over other storage systems: high density of stored energy (up to 38 kWh/kg (t)), absence of harmful emissions during operation, widespread in nature. However, the main advantage is the very low cost of energy storage, which is about 15 $/kWh, which is unattainable for any other known storage system. The use of hydrogen makes it possible to solve the main problem of compressed air energy storage devices associated with large storage volumes and the use of additional organic fuel for heating by creating a combined hydrogen-air gas turbine storage system. The product of hydrogen combustion in oxygen is high temperature steam, which allows it to be used for heating air from a storage facility by mixing, eliminating the formation of harmful substances, unlike, for example, burning hydrogen directly in air.

5. Conclusion

An analysis of the main features of hydrogen combustion units showed that they must be taken into account in order to create effective and safe structures, especially when used in the field of power engineering. In this case, it is especially important to take into account possible problems arising from the formation of explosive mixtures of hydrogen and oxygen. An especially strong influence on this can be exerted by the underburning of hydrogen, which occurs when the high-temperature steam is not cooled correctly and the mixing process is improperly organized.

Plants based on HCU, superheaters and air heaters, characterized by high power density, environmental friendliness and maneuverability, can be widely used in traditional, distributed and autonomous power generation in the power range from 100 kW to 100 MW. In general, the main critical problems for the successful creation of prototypes of such installations have now been resolved. It was possible to create a wide range of units of various capacities and different types of coolants, combining compactness, efficient cooling of the most heat intensive units and high efficiency of hydrogen combustion. Nevertheless, further additional research is required to improve their effectiveness and expand the field of application.

Due to the fact that the scale of the use of hydrogen technologies in the energy sector is small due to the relative high cost of such basic elements as electrolyzers and storage systems, the field of use of hydrogen-oxygen plants, implying the consumption of large amounts of hydrogen, is limited. It can be assumed that as these elements develop and become cheaper, hydrogen-oxygen plants will also receive additional development.

References

[1] Schastlivtsev A and Borzenko V 2017 *Hydrogen-oxygen steam generator applications for increasing the efficiency, maneuverability and reliability of power production* Journal of Physics: Conference Series **891**(1) 012213

[2] Stathopoulos P, Sleem and C Oliver Paschereit 2017 *Steam generation with stoichiometric combustion of H2/O2 as a way to simultaneously provide primary control reserve and energy storage* Appl. Energy **205** 692-702

[3] Aminov R, Schastlivtsev A and Bairamov A 2017 *On the issue of investigating the kinetics of processes in dissociated water steam* International Journal of Hydrogen Energy **42**(32) 20843-20848

[4] Schastlivtsev A, Dunikov D and Borzenko V 2019 *Experimental study of the processes in hydrogen-oxygen gas generator* International Journal of Hydrogen Energy **44**(18) 9450-9455

[5] Schastlivtsev A and Nazarova O 2016 *Hydrogen–air energy storage gas-turbine system* Thermal Engineering (English translation of Teploenergetika) **63**(2) 107-113

[6] Dunikov D 2018 *Cycle improvement and hydrogen steam superheating at Mutnovsky geothermal power plant* Case Studies in Thermal Engineering **12** 736-741

[7] Aminov R and Egorov A 2019 *Hydrogenoxygen steam generator for a closed hydrogen combustion cycle* International Journal of Hydrogen Energy **44**(21) 11161-11167

[8] Fröhlke K and Haidn O 1997 *Spinning reserve system based on H2/O2 combustion* Energy Conversion and Management **38**(10) 983-993

[9] Malyshenko S, et al. 2012 *Effectiveness of steam generation in oxyhydrogen steam generators of the megawatt power class* High Temperature **50**(6) 765-773

[10] Gryaznov A, et al. 2008 *Hydrogen steam generators for stationary energy applications* 17th World Hydrogen Energy Conference, WHEC 2008 (2008) Brisbane, QLD

[11] Kato S and Nomura N 1997 *Hydrogen gas-turbine characteristics and hydrogen energy system schemes* Energy Conversion and Management **38**(10) 1319-1326
[12] Haidn O, et al. 1998 *Improved combustion efficiency of a H2O2 steam generator for spinning reserve application* International Journal of Hydrogen Energy **23**(6) 491-497
[13] Shternfeld H and Wolfmuller K 1986 *The hydrogen/oxygen spinning reserve system – an approach to the economic generation of electricity from hydrogen* VGB Kraftwerkstechnik **66** 675-683
[14] Lédé J, Lapicque F and Villermaux J 1983 *Production of hydrogen by direct thermal decomposition of water* International Journal of Hydrogen Energy **8**(9) 675-679
[15] Ihara S 1978 *Feasibility of hydrogen production by direct water splitting at high temperature* International Journal of Hydrogen Energy **3**(3) 287-296

A map of regimes of evaporation and boiling in the horizontal liquid layer on the modified surface

D A Shvetsov[1], A N Pavlenko[1], A E Brester[2] and V I Zhukov[1,2]

[1] Kutateladze Institute of Thermophysics,
Russia, 630090 Novosibirsk, Acad. Lavrentiev Ave., 1
[2] Novosibirsk State Technical University,
Russia, 630073 Novosibirsk, K. Marx Ave., 20

shvetsov.kh301@ya.ru

Abstract. The paper presents the results of the study of the evaporation and boiling regimes in thin horizontal layers of liquid on a modified surface in a wide range of changes in the pressure and height of the liquid layer. Depending on the heat flux, pressure, and height of the liquid layer, the formation of various structures was observed. In this paper, maps of the evaporation and boiling regimes are obtained, which show the heat fluxes from the natural convection regime up to the boiling crisis, depending on the height of the liquid layer. The results are compared with the calculation dependencies.

1. Introduction

The evaporation and boiling of thin films of liquid are used in various heat and mass transfer devices (evaporators, distillation columns, heat pumps and pipes, diffusion vacuum pumps, spray cooling systems, air conditioning, etc.). Horizontal thin films of the liquid allow the removal of high heat fluxes at low temperature head. The use of modified surfaces also allows multiplying the heat transfer coefficient.

The aim of the work was an experimental study of heat transfer during evaporation and boiling in a layer of a single-component liquid on a horizontal heating surface with a capillary-porous coating applied in a wide range of changes in the pressure and height of the liquid layer. In the experiments the boiling curves were obtained at different pressure values over the n-dodecane layer. Measurements of the regime parameters were carried out simultaneously with visual observations of the structures formed in the layers. Based on the comparison of the experimental results with the observed structures, maps of the evaporation and boiling regimes are constructed.

2. Experiments

The detailed description of the experimental setup is given in [1]. The working chamber is a cylindrical vessel made of 12Cr18Ni10Ti steel with an internal diameter of 120 mm, a height of 300 mm, and a wall thickness of 1 mm. The spiral heater is located at the bottom of the working surface, allowing to evenly heat the bottom of the working chamber. The working chamber is equipped with viewing windows, through which the video recording of the evaporation and boiling processes taking place in the working chamber was carried out.

The capillary-porous coating was made by using the 3D laser printer [2] (figure 1(a)). The coating consists of 15Ni5Cu4Nb powder sintered by SLS (selective laser sintering). The porosity of the coating is $\varepsilon = 0.44$, the particle size – (20–40) μm. The 2-D profile of the capillary-porous coating is a sinusoid. Distance between fins – 3.5 mm, height of fins – 0.5 mm, maximum thickness of the porous coating used – 0.55 mm. A photo of the surface is shown in figure 1(a). A more detailed description of the modified surface is presented in [3]. N-dodecane was used as the working fluid. The pressure range in the experiments is $P = (33 – 20000)$ Pa or reduced pressure values of $P/P_{cr} = 1.84\cdot 10^{-5}$–$1.1\cdot 10^{-2}$, where P_{cr} is critical pressure, Pa. The pressure in experiments was chosen in such a way that the experimental data obtained complied to the pressure range corresponding to the boundaries of intersection of dependences Yagov [4] and Landau [5]. Eq. [4] can be considered for a comparative assessment of the increase in critical heat flux (CHF) on the coating at low reduced pressures. In Landau's theory, a layer of infinite depth is considered. CHF is independent of the conditions on the heating surface. Eq. [5] should be considered as some limiting case to which CHF can increase. The calculated point, where dependences [4] and [5] cross, corresponds to the n-dodecane saturation temperature of 38.7 C° and pressure of 52 Pa, $P/P_{cr} = 2.9\cdot 10^{-5}$. Experiments with water as a working fluid in the studied pressure range are impossible, since at a pressure of $P/P_{cr} = 2.736\cdot 10^{-5}$ water is in the triple point state and intersection of dependences Yagov and Landau is located far beyond the triple point (figure 1(b)).

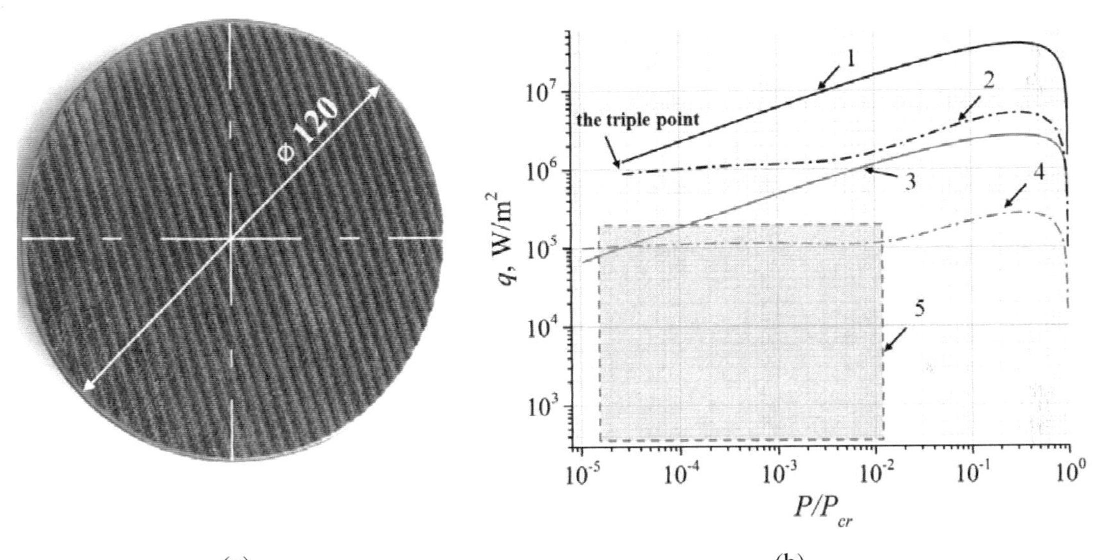

(a) (b)

Figure 1. (a) – photograph of a capillary-porous coating on a heat transfer surface; (b) – the CHF according to the Yagov [4] and Landau [5] dependences: 1 – calculation of [5] for water; 2 – calculation of [4] for water; 3 – [5] for n-dodecane; 4 – [4] for n-dodecane; 5 – the range of pressures and heat fluxes studied in this paper.

The experiments were carried out within the layer height range of $h = 1.7 – 10$ mm or dimensionless values of $(h/l_\sigma) = 0.99 – 5.56$, where h is the layer height, m; (h/l_σ) is a dimensionless layer height; $l_\sigma = \sqrt{\sigma / g(\rho_L - \rho_V)}$ – is Laplace constant, m; σ – is the surface tension, N/m; ρ_L, ρ_V – are liquid and vapour densities, respectively, kg/m³; g – is the gravitational acceleration, m/s².

3. Results and discussion

When changing the height of the liquid layer, the pressure in the volume, and the heat flux, the following structures were formed: "funnels", "craters" and bubble systems. "Funnels" were observed in the 1.7

mm layer in the range P_v/P_{cr} = 1.822·10^{-5}–6.623·10^{-4} from the start of heating to the boiling crisis. “Funnels” look like cavities with a hemispherical bottom on a surface of the liquid layer [1]. It differs in size, and its location is determined by the microstructure of the coating on the heating surface. The number of “funnels” increased with increasing heat flux. “Craters” were observed in the (2.5–4.0) mm layers in the range P_v/P_{cr} = 1.822·10^{-5}–7.358·10^{-5}. “Craters” are cavities in a thin layer, but unlike dry spots, there is a residual layer of liquid in the centers of the “craters” [1]. When the pressure increased, single large bubbles formed instead of “craters”. In the pressure range P_v/P_{cr} = 2.758·10^{-3}–1.1·10^{-2} rapid bubble boiling was observed over the entire value of the change in the layer heights even at low heat fluxes. Ranges of the regime parameters of the formation of structures in liquid layers are presented in Table 1. The heat transfer coefficient was calculated by the following expression: $\alpha = q/(T_w - T_s)$, where q is the heat flux, W/m^2; T_w is the wall temperature, K; T_s is the saturation temperature. Shots of the processes observed in different layers of the liquid at different pressures are shown in figure 2.

Table 1. Ranges of the regime parameters of the formation of structures in liquid layers.

	h	h/l_σ	P_v, Pa	P_v/P_{cr}	q, W/m^2	α, W/(m^2·K)
“Funnels”	**1.7**	**0.99**	33-1200	1.822·10^{-5}-6.623·10^{-4}	512-13350	82-694
“Craters”	**2.5**	**1.4**	33-133	1.822·10^{-5}-7.358·10^{-5}	9628-70140	748-2345
	4.0	**2.32**	33-133	1.822·10^{-5}-7.358·10^{-5}	15580-74380	514-3260
Bubble systems	**1.7**	**0.99**	5000-20000	2.758·10^{-3}-1.1·10^{-2}	1740-22280	135-1765
	2.5	**1.4**	399-20000	2.203·10^{-4}-1.1·10^{-2}	2391-96330	252-8120
	4.0	**2.32**	399-20000	2.203·10^{-4}-1.1·10^{-2}	1515-194800	166-25699
	10.0	**5.56**	33-20000	1.822·10^{-5} - 1.1·10^{-2}	2409-181200	262-22970

“Macrolayer” zones were formed with an increase of the heat flux in the layers of (2.5–10.0) mm (figure 3). The presence of such “macrolayer” zones ensures the intensification of heat transfer during evaporation and boiling on a structured surface with a coating. A higher heat transfer in the zones of the “macrolayer” is caused by a significant decrease in the local temperature of the heat-releasing surface [6].

It is convenient to represent the data of table 1 in the form of maps of hydrodynamic regimes of evaporation and boiling. In [1], maps of hydrodynamic regimes were proposed, on which, for each height of the liquid layer, the regions of existence of structures are indicated, depending on heat fluxes and reduced pressure. The authors of [7] proposed maps on which, for a constant pressure above the layer, the regimes of liquid evaporation and boiling are shown depending on the heat fluxes and the layer height. Maps of the hydrodynamic regimes of evaporation and boiling on the modified surface are constructed in this paper. The abscissa axis was chosen as the dimensionless layer height h/l_σ to create these maps. The maps show the regimes from natural convection up to the boiling crisis. Maps of hydrodynamic regimes are shown in figure 4. Experimental data obtained at different heights of the liquid layer are also compared with the calculated dependences of Yagov [4] and Landau [5].

The maps show that the increase in the layer height from 1.7 mm ($h/l_\sigma = 0.99$) to 4.0 mm ($h/l_\sigma = 2.32$) increases the value of the CHF by an order of magnitude at all operating pressures. At the same time, a further increase in the layer height to 10 mm ($h/l_\sigma = 5.56$) does not significantly affect the increase in the CHF. In a layer of 1.7 mm at pressures up to 1000 Pa and heat fluxes, in which a transition from convection to "craters" or bubbles was observed in thicker layers, the surface was drained [8]. The CHF obtained on the modified surface for layers of 4.0 mm and 10.0 mm are higher than the calculated values according to the Yagov dependence. Deviations from the Yagov dependence are explained by the formation of "macrolayer" zones, which is characterized by a significant difference in the hydrodynamics of the movement of liquid and steam in the wall region on a surface with a capillary-porous coating in liquid layers of finite height compared to the movement of liquid in the wall region on a smooth surface. CHF at very low relative pressures on a capillary-porous coating is limited from above by the Landau dependence, which was obtained for processes without taking into account surface conditions.

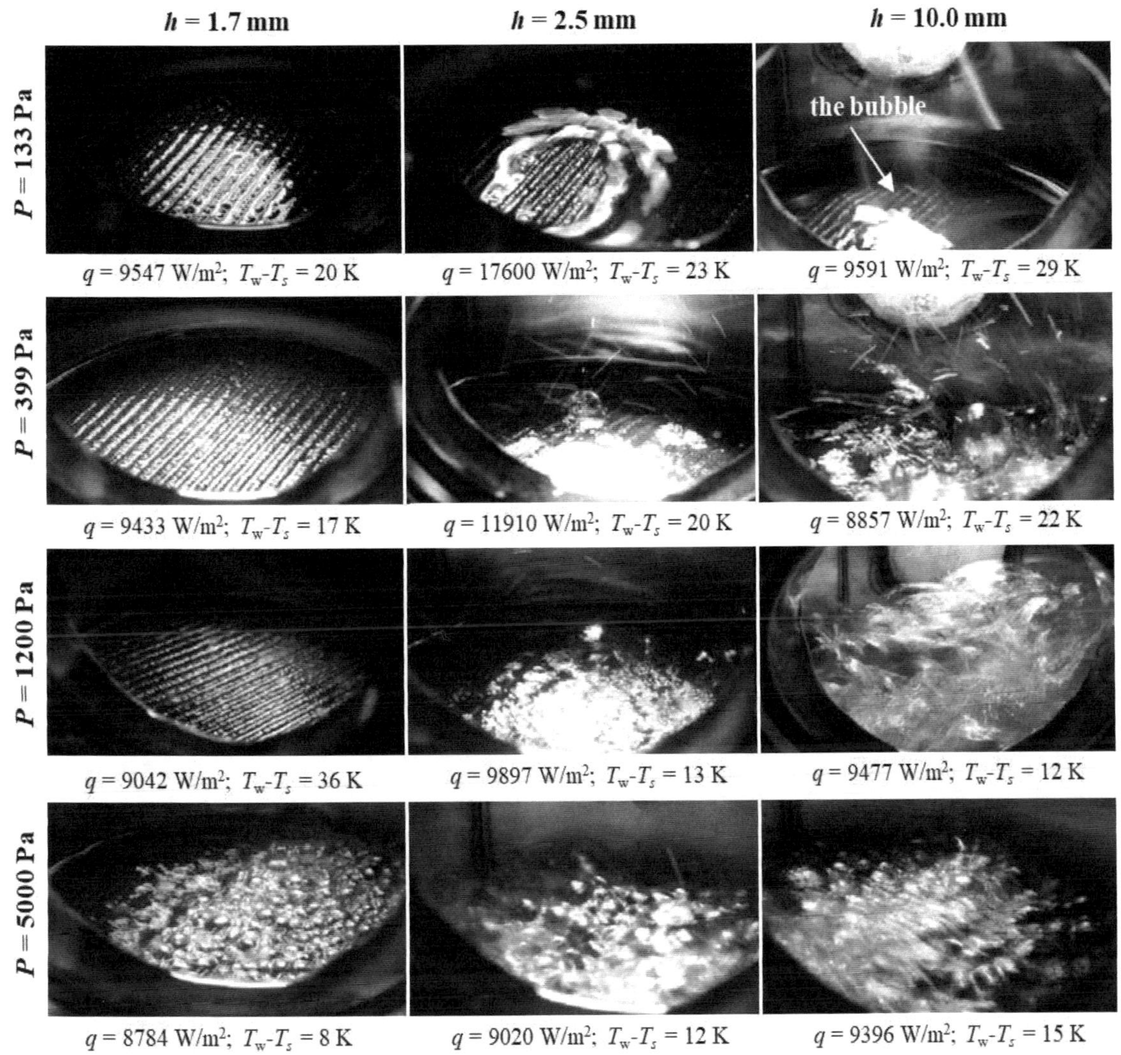

Figure 2. Shots of the structures observed in n-dodecane layers.

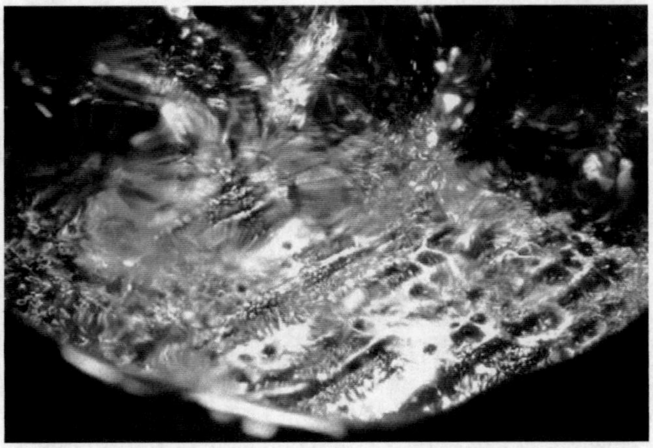

Figure 3. A shot of the "macrolayer":
h = 4.0 mm;
h/l_σ = 2.32;
P = 133 Pa;
P_v/P_{cr}=7.358·10^{-5};
q = 97230 W/m^2;
T_w-T_s = 33 K.

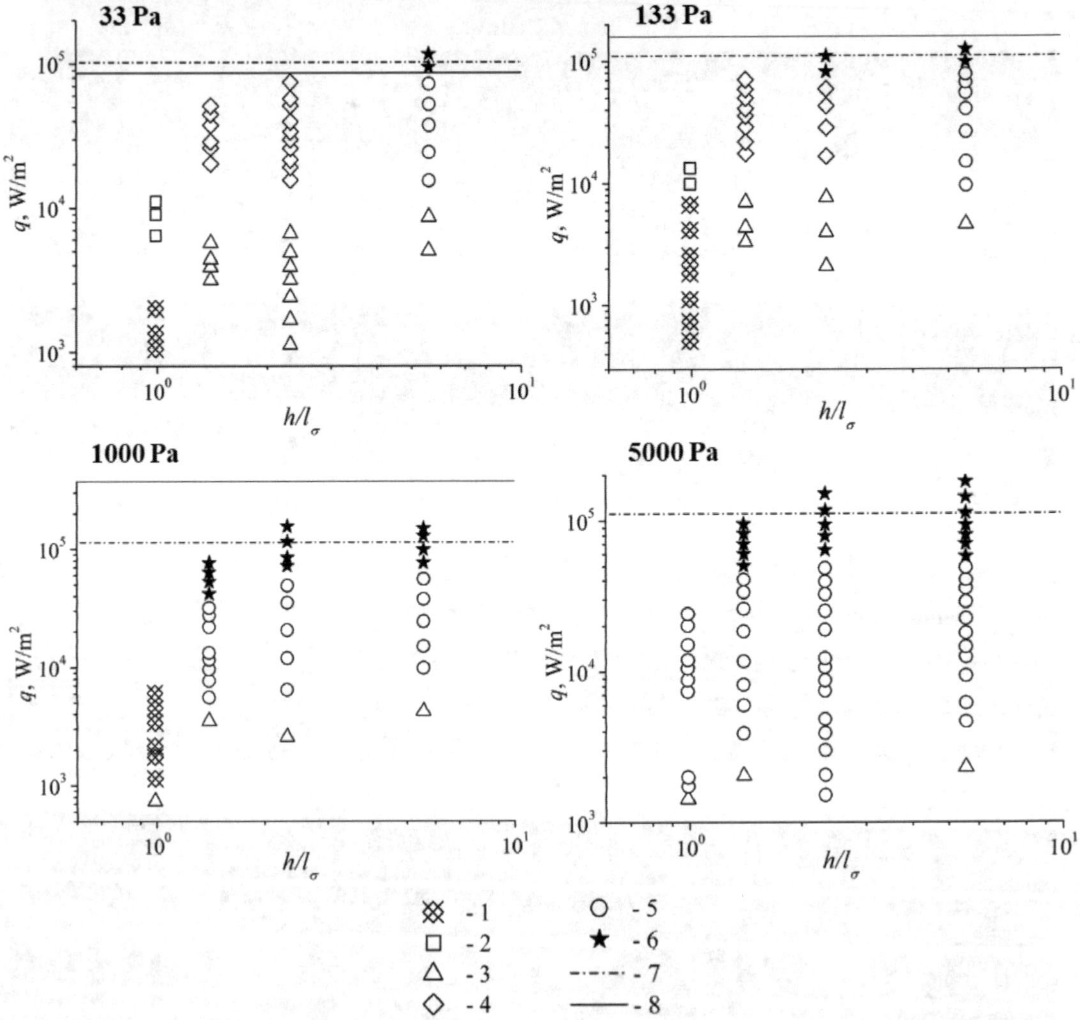

Figure 4. Maps of the hydrodynamic regimes of evaporation and boiling on the modified surface: 1 – "funnels"; 2 – the drying; 3 – convection; 4 – "craters"; 5 – bubble systems; 6 – zones of the "macrolayer"; 7 – CHF according to [4]; 8 – CHF according to [5].

4. Conclusion

The experimental study of heat transfer and crisis phenomena during evaporation and boiling of thin horizontal layers of liquid under reduced pressures on a modified heating surface is performed. In this paper, the dependences of CHF on the height of the liquid layer at different pressures are shown together with the designation of the regimes that were implemented in the heat transfer process. The resulting maps show the regimes of evaporation and boiling from the regime of natural convection up to the boiling crisis. The values of CHF increased by an order of magnitude with an increase in the layer height from 1.7 mm to 4.0 mm, but with a further increase in the layer height, the values of CHF increased slightly.

Acknowledgement

The study has been supported by the Russian Science Foundation grant (Project № 19-19-00180).

References

[1] Zhukov V I and Pavlenko A N 2018 *Int. J. Heat Mass Transfer* **117** 978
[2] Bessmeltsev V P, Pavlenko A N and Zhukov V I 2019 *Optoelectronics, Instrumentation and Data Processing* **55 (6)** 554
[3] Zhukov V I, Pavlenko A N and Shvetsov D A 2020 *Int. J. Heat Mass Transfer* **163** 120488
[4] Yagov V V 2014 *Int. J. Heat Mass Transfer* **73** 265
[5] Landau L D 1944 *ZhETF* **14(6)** 240 [in Russian]
[6] Surtaev A, Kuznetsov D, Serdyukov V, Pavlenko A, Kalitac V, Komlev D, Ivannikov A and Radyuk A 2018 *Applied Thermal Engineering* **133** 532
[7] Wojtasik K, Rullière R, Krolicki Z, Zajaczkowski B and Bonjour J 2020 *Experimental Thermal and Fluid Science* **118** 110150
[8] Shvetsov D A, Pavlenko A N, Brester A E and Zhukov V I 2020 *Journal of Physics: Conference Series* **1683** 022085

Method for calculating the critical heat flux at pool boiling on structured surfaces

A V Stupakova and A V Dedov

National Research University "MPEI"
Russia, 111250 Moscow, Krasnokazarmennaya, 14

StupakovaAV@mpei.ru

Abstract. A method of the critical heat flux enhancements at pool boiling using rough structures of various regular and irregular geometries is investigated. The experimental data are compared, and the critical heat flux is calculated since the V. V. Yagov model, which considers the additional inflow of liquid into the evaporation zone due to the action of capillary forces in the porous space. Based on the comparison of experimental and calculated data, it is concluded that the model is not universal for all microstructured surfaces. Recommendations for the calculation are given.

1. Introduction

When using nucleate boiling as a method of heat removal, there are restrictions on the heat flux removed, associated with the occurrence of a boiling crisis – the transition from nucleate boiling to film boiling, which leads to a sharp increase in the surface temperature, its burnout and destruction. When vaporization occurs on a solid surface, three phases interact: liquid, vapor, and solid. The heat transfer coefficient, critical heat fluxes (CHF), and boiling heat transfer depends both on the physical properties of the used fluids and on many geometric and physico-chemical characteristics of the heat-releasing wall. A promising way to increase the critical density of the heat flux at boiling is to produce a microstructured surface, which can increase the CHF by 2-3 times compared to a smooth surface. The result depends on a combination of many causes, and experimental verification of the proposed structures is necessary. Over the past decade, many works have been carried out, in which boiling and the crisis of heat transfer on surfaces structured by various technologies have been studied. An overview of the state of the problem of heat transfer intensification and critical heat loads during boiling in a large volume is presented in [1]. Comparison of the results [1] shows a large range of values, the lack of calculation methods. In this paper, we analyze the results of increasing the CHF obtained on various structured surfaces and calculate the CHF according to the model proposed in [2].

2. Review research

When using microstructured surfaces, the roughness is small enough to change the consistent pattern of single-phase heat transfer. Technologies for obtaining new variants of rough structures are currently being developed, which have known in advance the geometric characteristics and physico-chemical properties. The main goal in the development of structured surfaces is to create many vaporization centers on the surface, which will lead to boiling at lower temperature pressures or to an earlier onset of boiling.

In [3], the study of the deformation-cut surfaces is carried out. This technology is a mechanical processing of metal on standard metalworking apparatus. During the study, it was found that surfaces with multidimensional texture are characterized by a great heat transfer coefficients up to 9 times, and CHFs up to 4 times. For surfaces with a three-dimensional structure, the maximum values of heat transfer i enhancements are obtained.

In [4], horizontally oriented microstructural surfaces were used. Surfaces was with a circular cross-section pins with a surface development coefficient R in the range from 1.79 to 5.94. The CHF was increased up to 2 times.

In [5], a study of CHF on pool boiling of ethanol on horizontal microstructured surfaces with structures in the form of columns and cavities was performed. It is revealed that such surfaces can increase the number of centers of vaporization, thereby reducing the overheating of the wall and increasing the heat flux. The enhancements of heat transfer coefficient and CHF on microstructured surfaces was from 3 to 5 times in comparison with smooth surfaces.

In [6], the study of the maxima of the CHF enhancement with a change of the surface pin density was carried out. The effect of different distance between the pins on the pool boiling was studied. The CHF increases with a decrease in the step between the pins, and then decreases sharply after reaching a critical value.

In [7], an increase in heat transfer on pool boiling and CHF was studied with an increase in the surface development coefficient from 1 to 4. 12 samples were made with a structure in the form of microcylinders with different geometric dimensions (height, diameter, pitch). The result was that the heat transfer coefficient increased by 1.5 times, and the CHF by 1.5-2 times compared to a smooth surface. A critical gap size is found when analyzing the capillary flow velocity on a structured surface, which limits the CHF.

In [8], the study of the intensification of heat transfer during the boiling of acetone at atmospheric pressure on a horizontal heat-generating surface was carried out. Seven heated surfaces were made, which were divided into 3 classes. One class is a smooth surface, the next is a surface with open microchannels, and the last class is a surface with two-dimensional micropores. The optimal geometrical parameters of the micro-surfaces was found, which provides the maximum heat transfer and the CHF enhancements among the three two-dimensional test sections.

In [9], irregular microstructured surfaces with a FeCrAl layer on a metal substrate were produced by sputtering. For the first time, a significant increase in CHF was achieved using all types of surfaces. The greatest increase in CHF up to 1.6 times was observed for the deposition layer deposited for 1 hour at a substrate temperature of 150 degrees. The increase in CHF was explained by an increase in the roughness coefficient r at the micro-scale. This could influence reducing the re-wetting interval with FeCrAl-layered heaters.

In [10], pool boiling is experimentally investigated on thin homogeneous porous coatings using different diameters of copper particles, different manufacturing methods, and characteristics of particles (solid or porous) with a coating thickness of 3 to 5 particle diameters. The results obtained show that the critical heat flux for all coatings is approximately 1.8 times higher than on a smooth surface. It is assumed that the presence of a thin homogeneous porous coating affects the hydrodynamic (macroscale) stability in such a way that the area of the steam channels increases.

In [11], provides a fundamental understanding of the underlying mechanisms of boiling intensification using various nanomodifications of structures. Some nanostructures resulted in a 2.6-fold increase in CHF compared to smooth surfaces. Several methods of applying deformations to the surface have been studied. Some methods give an increase in CHF, and some a decrease. This analysis led to the conclusion that you need to use several methods at the same time to get even better results.

In [12], numerous studies of heat transfer during boiling of freon R-113 on specially created modified surfaces were carried out. It was found that the increase in the heat transfer coefficient and the critical heat flux is not proportional to the surface development coefficient. As an explanation for the growth of CHF, it was suggested that the possibility of evaporation of the micro-layer of liquid under the bubble remains due to the non-isothermal heating surface and the additional inflow of liquid into the evaporation zone. An increase in the CHF of up to 25% was obtained, for this purpose, a simple technology of surface modification by an electron beam was used.

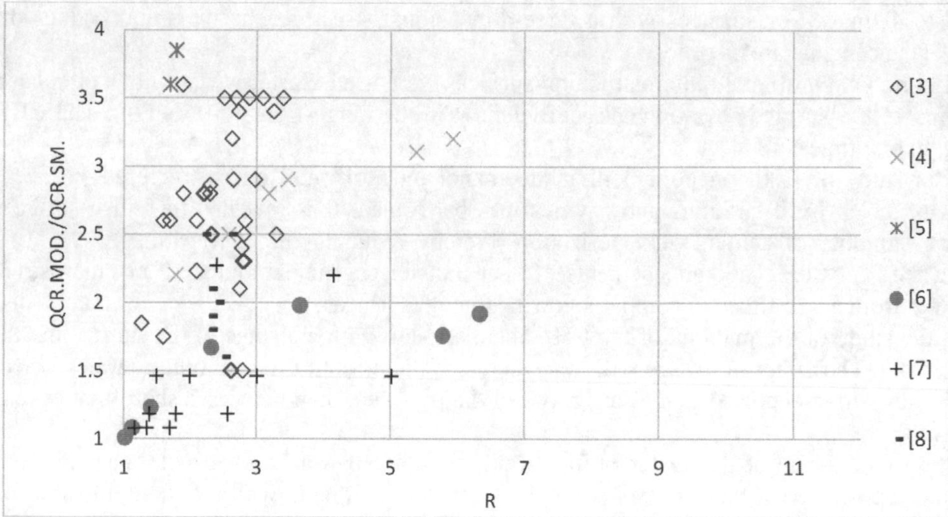

Figure 1. The CHF enhancements $q_{cr.mod.}/q_{cr.sm.}$ from the surface development coefficient for regular structures (experimental data from the above articles).

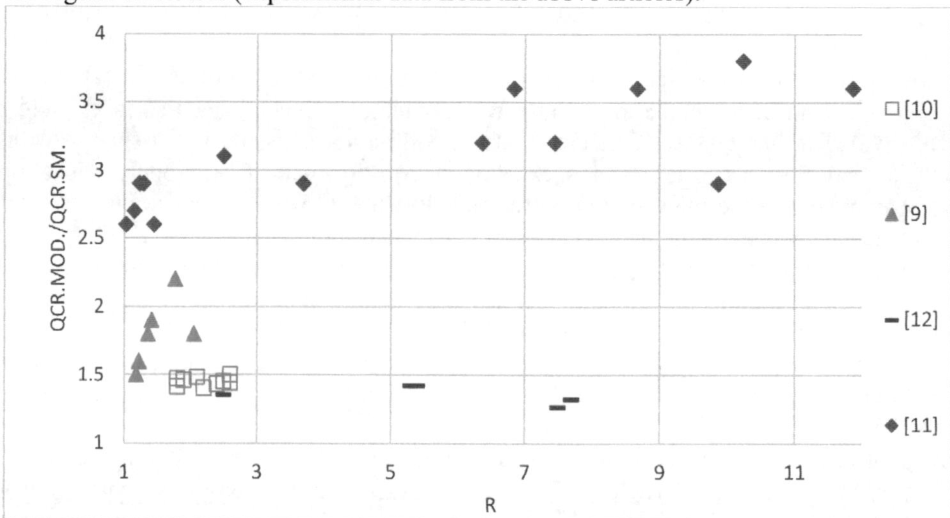

Figure 2. The CHF enhancements $q_{cr.mod.}/q_{cr.sm.}$ from the surface development coefficient for irregular structures (experimental data from the above articles).

Figure 1-2 shows the results of the studies [3–11] in the form of the CHF enhancements as the function on the surface development coefficient R. The CHF enhancement varies from 1.1 to 3.5 times. The maximum CHF enhancement up to 3-4 times was in [5] on the structure in the form of micro-columns and micro-cavities, in [12] on some surfaces, as well as in [11].

3. Calculation method

Most of the research on the effect of surface structuring on the value of the CHF in the vast majority of cases is experimental research. The theoretical consideration of pool boiling CHF enhancement on a structured surface with regular geometry is carried out in [2]. In [2] the contradictions of the hydrodynamic model of the CHF are analyzed. The different way to developed of the boiling crisis theory is proposed. The crisis conditioned by irreversible increase of the dry spots on the heating surface. The CHF occurs when the liquid inflow into the area of intense evaporation, caused by the gradient of the curvature of the liquid macrofilm surface, does not compensate for its evaporation at the boundary of the dry spot. In developing this model, the effect of micro-coating is introduced through a change in the capillary pressure gradient and the flow conditions of the liquid in the macro-film. At boiling on a structured surface, a capillary pressure gradient in the microchannels is added to the pressure gradient due to the curvature of the meniscus of the film. The final equation [2] for the CHF relative enhancement has the form:

$$\frac{q_{cr.mod}}{q_{cr}} = 1 + k_1 \left(\frac{\sigma}{\rho G}\right)^{0.4} \left(\frac{\mu}{g\Delta\rho}\right)^{0.8} (a+s)^{-2}, \quad (1)$$

where q_{cr} – CHF on a smooth surface; $q_{cr.mod}$ - CHF on a structured surface, k_1=7.5·10^{-4} the empirical coefficient, a - the characteristic size (for example, width or diameter), s - the step between the micro-roughnesses.

To calculate the projected increase in CHF, an analysis of the various structures was carried out separately. For structures with regular morphology, it was necessary to determine the characteristic size. It is concluded that for structures in the form of columns (in the cross section of which a square, rectangle, circle, or polygon, as well as for spherical recesses), the width or diameter is taken as the characteristic size. And for micro-edges or micro-channels, the height of the deformation is taken as the characteristic size. For structures with irregular morphology, some articles initially gave the average values of the necessary parameters, but for some it was necessary to calculate them according to the structure profile. As well as for micro-ribs with regular morphology, the average height of deformations was chosen as the characteristic size.

Using the geometric parameters of the surface structure, the surface development coefficient can be expressed:

$$r = \pi a h (a+s)^{-2}, \quad (2)$$

where h is the height of the structure.

For the above works, where the study of pool boiling at atmospheric pressure on surfaces with regular and irregular structures was performed, the relative increase in CHF was calculated using the formula (1) [13]. Figure 3-4 shows a graph of the dependence of the ratio of experimental data to those calculated by formula (1) on the surface development coefficient.

Based on figure 3-4, there are points showing the discrepancy between the experimental and calculated values. At the same time, most of the calculated values coincide with the empirical data or vary within the margin of error. For these structures, the formula (1) can be used for the theoretical prediction to assume the value of the critical heat flux on a certain surface. For example, for structures in the form of bayonet micro-ribs from work [3], for micro-cavities and micro-channels from work [5] with a moderate surface development coefficient, for micro-cylinders from work [11] with $4 < R < 5$, for work [6] with micro-columns.

The CHF date obtained in [8] is not calculated using formula (1). From the 34 CHF value considered in [8], only 4 microstructures have reasonable agreement with the calculation according to (1). The rest of the experimental data on CHF are several times higher than the calculated ones. Such structures require careful study and reflection.

Figure 5 shows a graph of the dependence of the ratio of CHF on a modified surface to CHF on a smooth surface from the surface development coefficient for both regular and irregular structures.

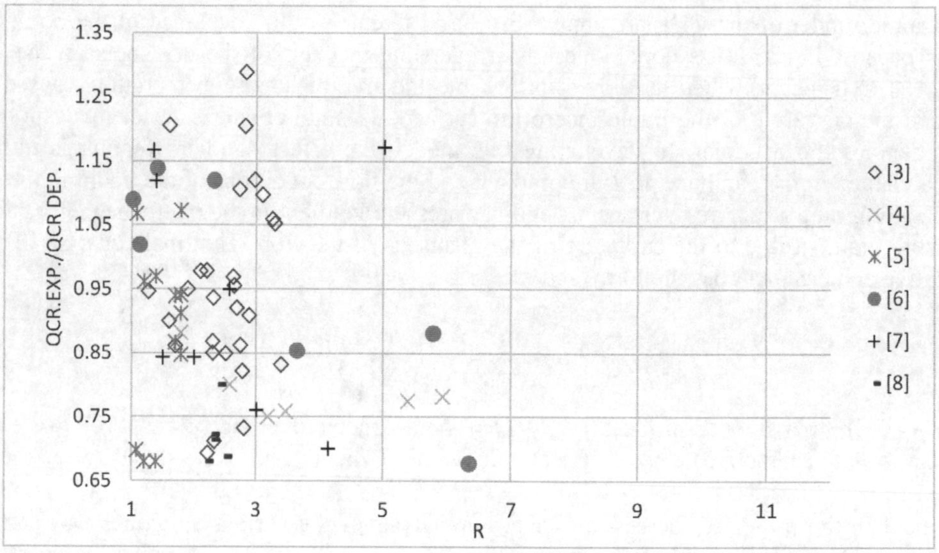

Figure 3. Dependence of $q_{cr.exp.}/q_{cr\ dep.}$ on the surface development coefficient for regular structures.

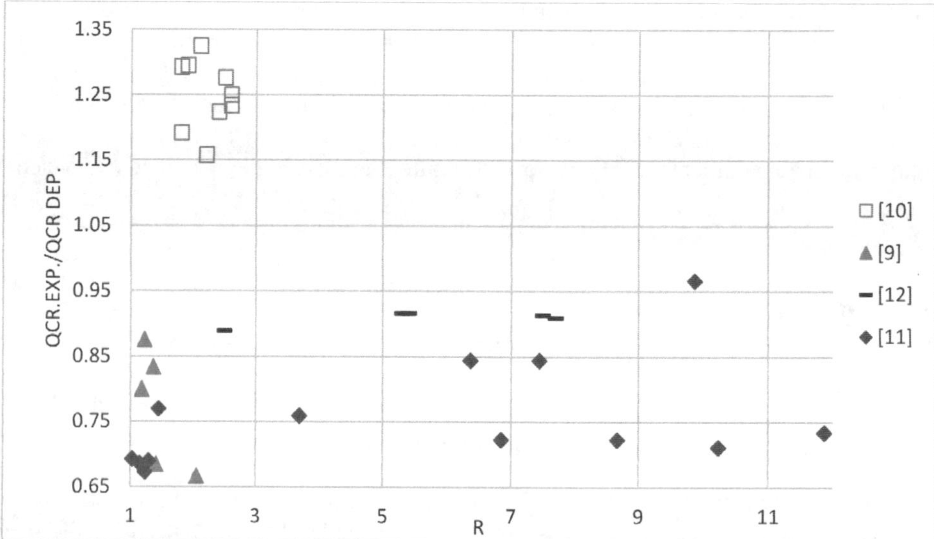

Figure 4. Dependence of $q_{cr.exp.}/q_{cr\ dep.}$ on the surface development coefficient for irregular structures.

The generalization of the data is made by approximating the polynomial function f(R) and its equation is obtained $f(R) = -0.03R^2+0.22R+0.81$. Thus, it is possible to predict the CHF value on the structured surface for some structures.

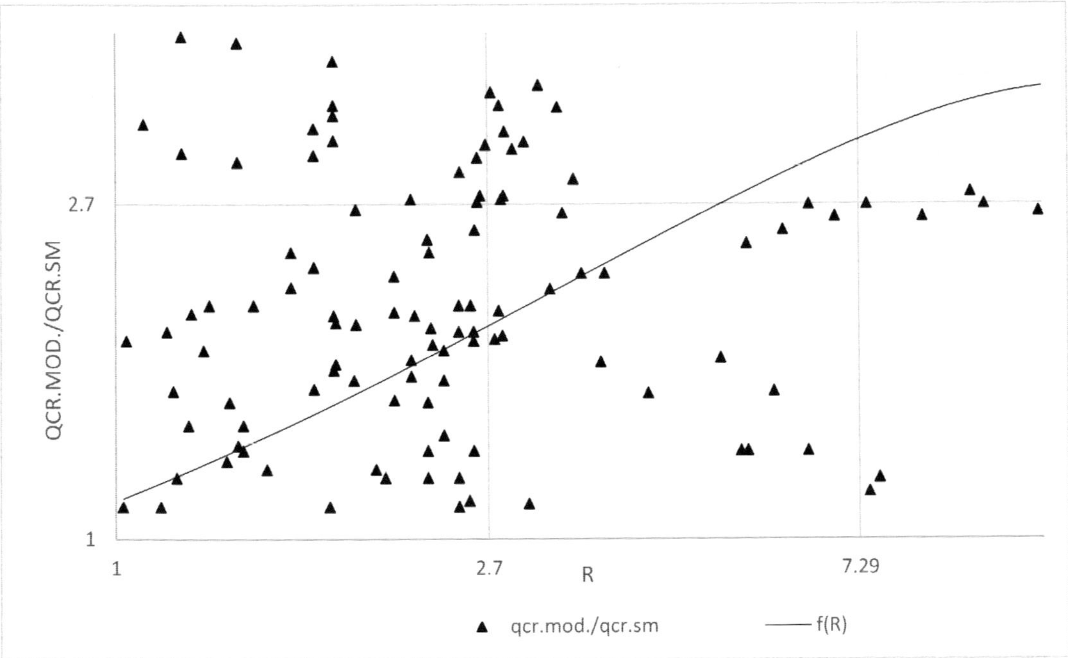

Figure 5. The approximating curve for predicting CHF.

4. Conclusion

The problem of increasing CHF due to the use of microstructured surfaces is considered. The conditions for the movement of liquid and vapor in a porous space are determined by the geometric parameters of microstructured coatings. The characteristic size and number of potential centers of vaporization play a key role in the intensification of heat transfer during boiling. Surfaces with a developed microstructure provide a greater number of potential centers of vaporization, and therefore have higher heat transfer coefficients in comparison with technically smooth surfaces. At the same time, structures that are too thick cause complex counter-flows of liquid and tend to "steaming". The resulting steam does not have time to leave the structure while the liquid does not have time to be brought to the centers of vaporization. Thus, the surface is burned. Therefore, such surfaces show a deterioration in heat transfer and CHF.

The calculation according to the formula (1) was performed for ten works, in which the values of CHF obtained on both regular and irregular microstructures with different geometric parameters were presented. Comparing the results of calculations with experimental data on CHF it can be concluded that it is possible to use formula (1) to generalize the results of most studies. For specific regular structures (microstructured surface with ribs, microcolumns with square and circular cross-sections, as well as structures in the form of microchannels and microcavities), the best fit of the data is observed at moderate surface development coefficients. Irregular structures also show a good correspondence between experimental and calculated data but require more careful study. Based on the comparison of experimental and calculated data, it is concluded that this model is not universal for all microstructured surfaces. During data comparison, it was suggested that it is possible to predict the value of CHF on various structured surfaces using the generalizing function f (R).

Acknowledgements

This research was supported by the Russian science Foundation (grant 19-19-00410).

References

[1] Dedov A V 2019 A Review of Modern Methods for Enhancing Nucleate Boiling Heat Transfer (Review) *Thermal Engineering* Volume 66, Issue 12, 1 p. 881-915

[2] Sukomel L A, Yagov V V 2017 Possibilities of increasing critical heat fluxes during boiling on surfaces with porous coatings (review) *Vestnik MPEI*, Pages 55-67

[3] Schelchkov A V, Khakimzyanov R R 2014 Heat transfer on microstructured surfaces during boiling water under conditions of free convection. Pages 263-266

[4] Kuang-Han Chu, Ryan Enright, Evelyn N Wang 2012 Structured surfaces for enhanced pool boiling heat transfer *American Institute of Physics* 5 p

[5] Lining Dong, Xiaojun Quan, Ping Cheng 2013 An experimental investigation of enhanced pool boiling heat transfer from surfaces with micro/nano-structures *International Journal of Heat and Mass Transfer*

[6] Navdeep Singh Dhillon, Jacopo Buongiorno, Kripa K Varanasi 2015 Critical heat flux maxima during boiling crisis on textured surfaces *Nature Communications* p 189-196

[7] Seol Ha Kima, Gi Cheol Leeb, Jun Young Kanga, Kiyofumi Moriyamaa, Moo Hwan Kima,c, Hyun Sun Park 2015 Boiling heat transfer and critical heat flux evaluation of the pool boiling on micro structured surface *International Journal of Heat and Mass Transfer* p 1140-1147

[8] Xianbing Ji, Jinliang Xu, Ziwei Zhao, Wolong Yang 2013 Pool boiling heat transfer on uniform and non-uniform porous coating surfaces *Experim ental Thermal and Fluid Science* p 198-212

[9] Gwang Hyeok Seo, Gyoodong Jeun, Sung Joong Kim 2016 Enhanced pool boiling critical heat flux with a FeCrAl layer fabricated by DC sputtering *International Communications in Heat and Mass Transfer* 102 p 1293-1307

[10] Hwang G-S, Kaviany M 2006 Critical heat flux in thin, uniform particle coatings International Journal of Heat and Mass Transfer 49 p 844–849

[11] Md Mahamudur Rahman, Emre Ölçeroğlu, Matthew McCarthy 2014 Role of Wickability on the Critical Heat Flux of Structured Superhydrophilic Surfaces *Langmuir* №30 p 11225-11234

[12] Dedov A V, Khaziev I A, Laharev D A & Fedorovich S D 2021 Study of Nucleate Pool Boiling Heat Transfer Enhancement on Surfaces Modified by Beam Technologies *Heat Transfer Engineering*, DOI: 10.1080/01457632.2021.1896834

[13] Stupakova A V, Dedov A V 2020 Calculation of critical heat fluxes at boiling on microstructured surfaces *Journal of Physics*: *Conference Series*, 1683(2), 022098

Modeling of vapor bubble growth at subatmospheric pressures

I V Vladyko[1,2], I P Malakhov[1,2], A S Surtaev[1,2], A A Pil'nik[1,2] and A A Chernov[1,2]

[1]Novosibirsk State University, Novosibirsk, 630090, Russia
[2]Kutateladze Institute of Thermophysics, Siberian Branch of the Russian Academy of Sciences, Russia, 630090 Novosibirsk

vsserd@gmail.com

Abstract. In this paper, the results of numerical calculations of a vapor bubble growth in superheated water at different pressures are presented. Modeling is based on a previously developed by the authors semi-analytical solution. The results are verified by experimental data obtained at atmospheric and subatmospheric pressures. The presented simulation results and experimental data are in good agreement. The advantage of the solution over the earlier ones (based on the thermal growth model) is shown.

1. Introduction

Boiling is a process wide spread both in common life and technology. It can be found not only in energy applications (various kinds of heat exchangers, where an efficient heat removal is the main problem), but also in modern medical techniques. For example, the medical technologies include the technique of minimally invasive laser surgery, where the explosive boiling occurs. The cumulative jet, which carry out a destructive effect on pathological formations, appears due to the boiling of a biological fluid heated by laser radiation [1]. The basic act of boiling is the nucleation and growth of a vapor / vapor-gas bubble in an overheated (or locally overheated under general subcooling) liquid. Without detailed knowledge about the mechanism of basic act a correct description of the boiling process remains impossible.

In addition to many experimental studies there are a number of theoretical works, in the literature, where the vapor bubble growth is described for various system conditions. It should be noted that the theoretical description of the vapor bubble growth mechanism is a complex nonlinear moving boundary problem. So, the complete solution of this problem is not yet available. Most of the known models, which became classical, and solutions obtained on their basis are either asymptotic or empirical. This fact does not allow us to call them to be a complete solution of the problem for a wide range of the system parameters.

Obviously, it is impossible to find general analytical solution. Therefore, most researchers use significant assumptions that limit the scope of the obtained solutions. This applies both to classical works [2-5] and to relatively modern studies [6-8], where the inertial and thermal stages of bubble growth are considered separately. Some attempts to take into account the dynamic and thermal effects simultaneously were undertaken in [9-13]. However, the transitional stage of bubble growth is still insufficiently described. This work is devoted to numerical modeling of the vapor bubble growth, based on the semi-analytical solution obtained earlier by the authors of the work [14, 15], for a wide range of Jacob numbers and all stages of the process, including the transition stage, which is extremely necessary to take into account in some cases.

2. Numerical simulation

Our calculations are based on the theoretical model presented in [14, 15] and the semi-analytical solution found under assumptions of this model. The assumptions are the following. The bubble is considered to be spherically symmetric during the entire process. Liquid around the bubble is incompressible. The bubble growth is determined only by external factors, which takes into account both the dynamic and thermal effects. To do this, the Rayleigh equation and the boundary value thermal problem with a moving boundary are jointly solved. Furthermore, the laws of conservation of mass, momentum, and energy at the bubble boundary are taken into account. For vapor in the bubble, the conditions of uniformity of temperature and pressure are met. The vapor is stationary and in a saturated state (the pressure and temperature of the vapor are functions of time). Heat released by viscous dissipation is assumed to be negligible. The liquid is initially uniformly superheated. The latter is generally not universal, but in the zero approximation, it is proposed to introduce a certain superheat averaged over the allocated volume of liquid and use it as a variable parameter of the problem, determining it from the experiment.

To perform numerical simulation of the system, the 4th order Runge-Kutta method and the implicit Krank-Nicholson scheme is used. To find the time functions of vapor parameters given implicitly, the Brent method is used. The calculations were performed using dimensionless variables with the following characteristic values of time, velocity, and size, respectively: $t_0 = a_l \rho_l / \Delta p_i$, $w_0 = (\Delta p_i / \rho_l)^{1/2}$, $R_0 = a_l (\rho_l / \Delta p_i)^{1/2}$, where a_l is the thermal conductivity coefficient, ρ_l liquid density, Δp_i the initial deviation of the liquid pressure from the saturated vapor pressure. The dimensionless phase transition criteria here is the Jacob number $\mathrm{Ja} = \dfrac{\rho_l}{\rho_v} \dfrac{c_l \Delta T_i}{L}$, where ρ_v is saturated vapor density; c_l heat capacity of liquid; L specific heat of the phase transition; ΔT_i initial overheating of the liquid. During comparing the calculation results with the experimental curves of bubble growth, the Jacob number was used as a fitting parameter.

3. Experimental setup

For verification and analysis of the modeling results, the experimental data obtained during heterogeneous boiling of saturated water under subatmospheric pressures in range of 8.8-103 kPa (the saturation temperature in this case varies in range 43.3–100.4 °C) were used [16,17].

The scheme and detailed description of the experimental setup can be found in [16]. The working volume was a sealed chamber connected to a vacuum pump to control pressure. The system pressure was controlled using the "Manometer" BO1227 vacuum gauge. To maintain the constant temperature during the experiment, the working volume was mounted in the isothermal bath.

To observe boiling process the experimental setup has two sealed windows on the side of the chamber. The main feature of the setup is a transparent heater, which is a thin ITO (indium-tin oxide) film with 1 μm thickness, vacuum deposited on a 3-millimeter sapphire substrate. This allows to carry out the high-speed visualization from the bottom side of the heater surface. In addition, opacity of the ITO film in IR spectrum makes it possible to measure integral value of the heater temperature using an IR camera, located under heater.

Figure 1 (a) shows a frame of side view high-speed visualization of the bubble growth at p_{sat} = 8.8 kPa. As shown in [16], the bubble has a flattened shape during its growth, so it is necessary to operate with the value of the equivalent diameter $D_{eq} = (D_h D_b^2)^{1/3}$. That is why the high-speed visualisation was performed from the side of the transparent heater. Frequency and spatial resolution of the high-speed visualization was 5 kHz and 0.072 mm/pix respectively.

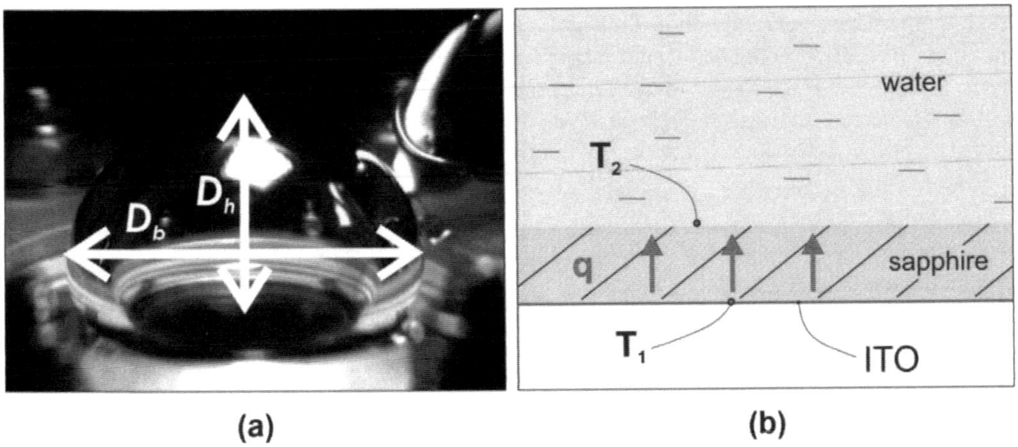

Figure 1. (a) A frame of high speed visualization and measurements of bubble diameter; (b) Scheme for recovering the temperature of a sapphire surface in direct contact with water.

To obtain the temperature of surface in direct contact with water, a stationary solution of the one-dimensional heat equation is used (see figure 1b): $T_1 - T_2 = q\delta/\lambda$, where q is the heat flux density, δ is the sapphire substrate thickness, λ is the thermal conductivity of sapphire, T_2 is the temperature on sapphire surface in contact with water, T_1 is the temperature of the ITO film.

4. Results and discussion

The results of numerical simulation are compared with experimental data on the growth of a vapor bubble at pressures 8.8, 42.1, 72.2 and 102.9 kPa (figure 2). For comparative analysis, the graphs show the Labuntsov-Yagov solution obtained in the framework of the thermal growth model (ignoring the inertial effects) [18] and Rayleigh model [2] developed within framework inertial scheme. As can be seen from the figure, the developed model works well at all stages of the process under consideration in contrast to the limiting laws of growth, which describe the inertial and thermal stages of growth separately and are characterized by power dependences of the form $R \sim t^n$, where $n = 1/2$ and 1 respectively. Both accordance to the laws of growth and to the experimental data confirm the success of the model. Thus, this model can be applied in a wide range of parameters of the process, in particular, in a wide range of initial superheats and pressures (Jacob numbers). It should be noted here that at large superheats, the kinetics of the phase transition can play a significant role (at least at the initial stage of the process). This can be expressed in limiting the mass flow due to the phase transition leading to the fact that the vapor in the bubble will be in an unsaturated state. However, this point requires further research.

Noteworthy, there is a deviation between the calculations and the experimental data (the figure shows that the experimental points lie slightly below the theoretical curve especially for low Ja numbers) over long periods of time This fact can be clearly explained. This deviation is observed at the thermal stage of growth when the growth of the bubble is mainly determined by heat transfer to the interphase boundary which directly depends on the overheating of the liquid away outside the bubble (at distances much greater than the characteristic thickness of the temperature boundary layer formed around the bubble during its growth at least). In the model used for the calculation, it is assumed that the overheating of the liquid around the bubble at the initial moment of time is uniform. In contrast, in the experiment at the time of nucleation of the bubble there is always a certain temperature gradient directed from the heating surface deep into the liquid (of course, this implies a negative gradient; the overheating of the liquid decreases away from the heated wall). Hence the bubble, as it grows, comes into contact with the colder layers of the liquid, and its rate of growth becomes less than in case of growing under the condition of uniform overheating. It should also be noted here that the deviation is less the lower pressure in the medium. This is explained by the fact that the lower pressure, the longer waiting time for an appearance of the vapor phase embryo (which is directly observed in the experiment), and with it the longer the heating time of the liquid adjacent to the heated surface. This is before the appearance of vapor bubble, a layer of liquid with a larger thickness warms up, and the bubble grows at the observed times under the conditions that are closest to our conditions.

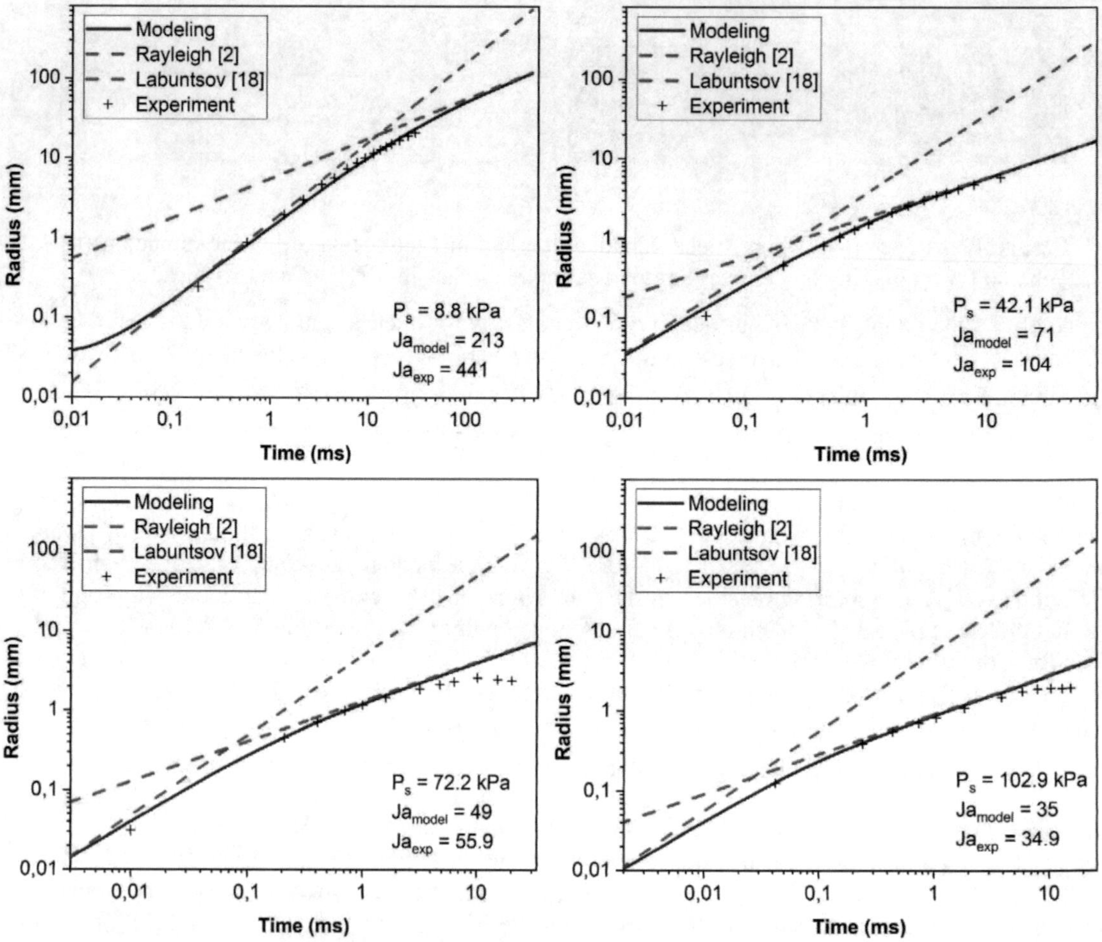

Figure 2. Dependence of bubble radius on time, calculated for different pressures: solid line - modeling carried out within the presented model; dashed lines - modeling carried out within the inertial model [2] and the thermal model [18]; points – the experimental data.

5. Conclusion

Thus, the present study shows a good agreement between the calculation results and experimental data on the dynamics vapor bubble growth during boiling of water in pressure range of 8.8-103 kPa. The calculations are carried out on the basis of the semi-analytical solution found earlier by the authors of the study, which simultaneously takes into account both inertial and thermal effects that determine the process of vapor bubble growth. The obtained solution is compared with the classical solutions found under certain assumptions (within the framework of inertial and thermal growth schemes) and representing power-low dependences of the form $R \sim t^n (n = 1/2, 1)$. It was shown that, in contrast to the classical solutions, which are valid only at a certain stage of the process, the found solution correctly describes the experimental data during the entire bubble growth time: from the initial inertial to "asymptotic" thermal one. Further development of the model implies taking into account kinetic effects, which can be significant at large Jacob numbers.

Acknowledgements
This work was supported by the Russian Foundation for Basic Research, project No. 19-48-540005.

References

[1] Chudnovskii V M, Levin A A, Yusupov V I, Guzev M A, Chernov A A 2020 The formation of a cumulative jet during the collapse of a vapor bubble in a subcooled liquid formed as a result of laser heating *International Journal of Heat and Mass Transfer* **150** 119286
[2] Lord Rayleigh 1917 On the pressure developed in a liquid during the collapse of a spherical cavity *The London, Edinburgh, and Dublin Philosophical Magazine and Journal of Science* **34**(200) pp 94–98
[3] Plesset M S and Zwick S A 1952 A nonsteady heat diff usion problem with spherical symmetry *Journal of Applied Physics* **23**(1) pp 95–98
[4] Forster H K and Zuber N 1954 Growth of a vapor bubble in a superheated liquid *Journal of Applied Physics* **25**(4) pp 474–478
[5] Scriven L E 1959 On the dynamics of phase growth *Chemical Engineering Science* **10** pp 1–13
[6] Robinson A J, Judd R L 2004 The dynamics of spherical bubble growth *International Journal of Heat and Mass Transfer* **47**(23) pp 5101–13
[7] Avdeev A A 2016 *Bubble Systems* p 446
[8] Prosperetti A 2017 Vapor Bubbles *Annual Review of Fluid Mechanics* **49**(1) pp 221–248
[9] Mikic B B, Rohsenow W M, Grith P 1970 On bubble growth rates *International Journal of Heat and Mass Transfer* **13** (4) pp 657-666
[10] Miyatake O, Tanaka I, Lior N 1997 A simple universal equation for bubble growth in pure liquids and binary solutions with a non-volatile solute *International Journal of Heat and Mass Transfer* **40**(7) pp 1577-1584
[11] Lesage F J, Siedel S, Cotton J S, Robinson A J 2014 A mathematical model for predicting bubble growth for low Bond and Jakob number nucleate boiling *Chemical Engineering Science* **112** pp 35–46
[12] Wang Q, Gu J, Li Z, Yao W 2017 Dynamic modelling of bubble growth in vapor-liquid phase change covering a wide range of superheats and pressures *Chemical Engineering Science* **172** pp 169–181
[13] Cai C, Liu H, Xi X, Jia M, Yin H 2018 Bubble growth model in uniformly superheated binary liquid mixture *International Journal of Heat and Mass Transfer* **127** pp 629-638
[14] Chernov A A, Pil'nik A A, Vladyko I V, Lezhnin SI 2020 New semi-analytical solution of the problem of vapor bubble growth in superheated liquid *Scientific Reports* **10** 16526
[15] Chernov A A, Guzev M A, Pilnik A A, Vladyko I V, Chudnovsky V M 2020 Analytical description of vapor bubble growth in a superheated liquid: a new approach *Doklady Physics* **65** pp 405-408
[16] Surtaev A, Serdyukov V, Malakhov I 2020 Effect of subatmospheric pressures on heat transfer, vapor bubbles and dry spots evolution during water boiling *Experimental Thermal and Fluid Science* **112** 109974
[17] Serdyukov V, Malakhov I, Surtaev A 2020 High-speed visualization and image processing of sub-atmospheric water boiling on a transparent heater *Journal of Visualization* **23** pp 873-884
[18] Labuntsov D, Kolchugin B, Golovin V, et al. 1964 Study of bubble growth during boiling of saturated water in a wide range of pressure changes by means of high-speed filming *High Temperature* **2** pp 446-453.

Development of water deaeration technologies with boiler exhaust gases

M M Zamaleev*, V N Kovalnogov, R I Kamalova and O V Pazushkina

Ulyanovsk State Technical University
Russia, 432027 Ulyanovsk, Severny Venets, 32

Mansur_zamaleev@mail.ru

Abstract. The calculation of the heat and mass transfer efficiency of the proposed technology for deaeration of water by the exhaust gases of a gas-proof boiler in deaerators of serial design has been performed. The conditions and criterial values of the media consumption in the apparatus have been studied. It is shown that the technology can be successfully applied in operating thermal power plants.

1. Introduction

Thermal deaeration of the make-up water of the heating system and the make-up of boiler water at thermal power plants is a traditional method of anticorrosive water treatment and is widely used in the Russian Federation and on the territory of the post-Soviet space. Over the decades of using this method, the modes of operation of deaerators have been studied, various designs have been proposed, measures have been developed to improve their effective operation, however, the use of unconventional media as a desorbent has recently been proposed.

Thermal deaeration is the most energy-consuming process of water preparation in thermal power plants. Foreign scientists dealt with the problems of optimizing this process. A. E. Kittredge and A. W. Kingsbury developed deaerator designs in the middle of the 20th century to improve their efficiency. R. E. Athey the 90s published the results of his study and tests [1]. The purpose of the study is to develop new effective methods for water degassing and substantiate the possibility of their application in modern energy systems.

2. Boiler exhaust gases degassing technology

The method of thermal deaeration of water with steam at thermal power plants has a number of disadvantages, such as the need for additional steam consumption for the deaeration process itself, unwanted heat losses with steam, as well as losses of steam and condensate during the preparation of the make-up water of the heating network in atmospheric deaerators. The owners of boiler houses operating on steam boilers in many medium and small cities of Russia face similar problems. In the case of transferring the boiler house to hot water mode in order to reduce fuel consumption and reduce heat losses, thermal deaeration becomes difficult to implement [2]. At the same time, most of the boiler houses do not have steam generators in the installed equipment, and the problem of thermal deaeration is quite acute and the anti-corrosion treatment of water should somehow be carried out in the absence of steam.

In the research laboratory "Heat power systems and installations" of the Ulyanovsk State Technical University, the researchers proposed the use of a gas- proof boiler in the deaerator as a desorbing agent of flue gases, operating under pressurization and burning natural gas [3].

The developed technology is aimed not only at increasing the efficiency of the water degassing process, but also at the beneficial use of the heat of the boiler exhaust gases in the deaeration process. The possibility of organizing deaeration of the heating network make-up water on hot water boilers without steam sources is the main advantage of the proposed solution [4]. The authors believe that the highest efficiency can be achieved with non-stoichiometric combustion of fuel - natural gas [5], as well as in the operation of the boiler under pressure, which avoids an increase in the oxygen content in the exhaust gases due to possible air leaks into the furnace and gas ducts.

3. Mass transfer efficiency of flue gas deaeration

The determination of the mass transfer efficiency is presented in the form of a calculation based on the solution of the balance equations for the processes of mass transfer and heat transfer.

The area of application of the developed technology is proposed to be determined after calculating the residual flow rate of the desorbing agent for the reaction. The operating parameters of the boiler and atmospheric deaerator with the required capacity of 25 t/h are taken.

The heat balance equation to determine the flue gas flow rate sufficient to ensure the deaeration process with the boiler flue gases:

$$Q_{sw} + Q_{ex.gas} = Q_{dw} + Q_{vent} . \quad (1)$$

Also, the equation for the heat balance of deaeration can be expanded as

$$G_{sw}h_{sw} + D_{ex.gas}h_{ex.gas} = G_{dw}h_{dw} + D_{vent}h_{vent} , \quad (2)$$

where G_{sw} and G_{dw} - the quantity of source and water, that was deaerated, kg/h; $D_{ex.gas}$- rate of the desorbing agent, kg/h; D_{vent} - vapor consumption, that consist of mixture of corrosive gases released from water, kg / h; h_{sw} - the enthalpy of water at the entrance to the deaerator, kj; h_{dw} - the enthalpy of water at the exit from deaerator, kj; $h_{ex.gas}$ - the enthalpy of flue gases at the entrance to the deaerator, kj; h_{vent} - the enthalpy of flue gases at the exit from the deaerator, kj.

The required consumption of flue gases for the process will be determined as

$$D_{ex.gas} = \frac{G_{dw}h_{dw} + D_{vent}h_{vent} - G_{sw}h_{sw}}{h_{ex.gas}} . \quad (3)$$

The calculations carried out showed that the required flue gas flow rate will be 71.5 m^3/h to ensure the deaeration process of 25 t/h of water at a standard value of the vapor consumption of 50 kg/t.

The equation of material balance for deaeration with exhaust boiler gases for determining the oxygen concentration can be written as

$$G_{sw}X_{sw}^{O_2} + D_{ex.gas}Y_{ex.gas}^{O_2} = G_{dw}X_{dw}^{O_2} + D_{vent}Y_{vent}^{O_2} , \quad (4)$$

where G_{sw} and G_{dw} - the amount of source and deaerated water, kg / h; $D_{ex.gas}$- consumption of flue gases supplied to the deaerator, kg / h; D_{vent}- deaerator vapor consumption (mixture of corrosive gases and flue gases released from water), kg / h; $X_{sw}^{O_2}$, $X_{dw}^{O_2}$ - oxygen concentration in water at the inlet to the deaerator and at the outlet from it, mg/dm^3; $Y_{ex.gas}^{O_2}$, $Y_{vent}^{O_2}$ are the oxygen content in flue gases at the inlet to the deaerator and in the vapor at the outlet from the deaerator.

The oxygen content in the flue gas at the water inlet and in the steam can be expressed in terms of the gas concentration in the water. The total pressure of a gas or vapor-gas mixture is equal to the sum of the partial pressures of gases and vapors that make up the mixture (Dalton's law). At the same time, the concentration of a gas dissolved in water is proportional to the partial pressure of this gas above the water surface.

At the inlet of the deaerator, the concentration of oxygen in the flue gases will be:

$$Y_{ex.gas}^{O_2} = K_H^{O_2} X_{dw}^{O_2} / p, \tag{5}$$

where $K_H^{O_2}$ is Henry's coefficient (constant of phase equilibrium for oxygen [6]), Pa; p - pressure in the deaerator, Pa.

With countercurrent flow of water and gases in the deaerator, the oxygen concentration in the steam-gas mixture is

$$Y_{vent}^{O_2} = K_{H_1}^{O_2} X_{sw}^{O_2} / p, \tag{6}$$

where $K_{H_1}^{O_2}$ is Henry's coefficient (constant of phase equilibrium for oxygen), Pa; p - pressure in the deaerator, Pa.

With certain assumptions, the material balance equation can be written in the form:

$$G_{sw} X_{sw}^{O_2} + D_{ex.gas} K_H^{O_2} X_{dw}^{O_2} / p = G_{dw} X_{dw}^{O_2} + D_{vent} K_{H_1}^{O_2} X_{sw}^{O_2} / p. \tag{7}$$

We find the oxygen concentration in deaerated water, for which we take the values of the initial parameters $X_{sw}^{O_2}$ = 8000 mg/dm³, D_{vent} = 50 kg/h, $K_{H_1}^{O_2}$ = 7104·10⁶ Pa, $K_H^{O_2}$ = 6372·10⁶ Pa, p = 1.2 ·10⁶ Pa, $G_{sw} = G_{dw}$ = 25·10³ kg/h:

$$X_{dw}^{O_2} = \frac{X_{sw}^{O_2}\left(D_{vent} K_{H_1}^{O_2} / p - G_{sw}\right)}{\left(D_{ex.gas} K_H^{O_2} / p - G_{dw}\right)} = 48.5 \text{ mg/dm}^3. \tag{8}$$

For makeup water, the permissible oxygen content is up to 50 mg/dm³ [7]. The proposed technical solution with the use of boiler flue gases as a desorbing agent in atmospheric deaerators makes it possible to provide the standard oxygen content. According to the above calculations, the theoretically required specific gas consumption for deaeration will be 2.86 m³ per 1 ton of deaerated water.

The results of calculating the deaeration process performed according to formulas (3) and (8) at a concentration of dissolved oxygen in water at the inlet to the atmospheric deaerator of 8 mg/dm³ are shown in figure 1. Thus, the calculations performed not only show the possibility of using the technology of deaeration of water by the flue gases of the boiler, but also prove its effectiveness in the case of use with deaerators of a common design.

Due to the high content of carbon monoxide in flue gases, the question of neutralizing carbon dioxide to the minimum standard values for the implementation of the possibility of using this method for deaeration of feed water without damage to boiler equipment arises. The authors propose to dose an alkaline agent into the feed water pipeline after the deaerator. This can be sodium sulfite or hydrazine.

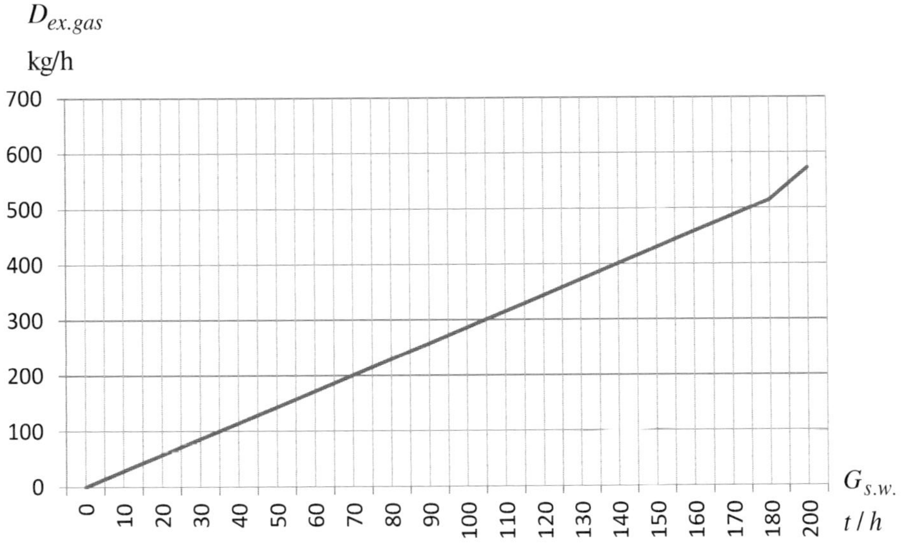

Figure 1. Consumption of flue gases for degassing dissolved oxygen in a countercurrent flow of water and flue gases in the deaerator.

4. Conclusions

A new technology for deaeration of water has been developed due to the use of a gas-proof boiler as a desorbent of flue gases. The estimation of the residual oxygen concentration in the deaerated water was made, as well as the theoretically required flow rate of the boiler flue gases for the ralization of the make-up water deaeration process was calculated.

The suitability of the new technology for deaeration of make-up water using boiler flue gases as a desorbing agent has been proven by calculation calculations based on the equations of heat and material balance.

References

[1] Athey R E 1990 *Condensate oxygen control in a combined ccycle system without a conventional deaerator* (Elec. Power Research Institute, Condencer Tech. Conf.) 1-8
[2] Semenov V G 2006 *Heat supply news* **1** 55-63
[3] Zamaleev M M, Pazushkina O V and Kamalova R I 2020 J. Phys. Conf. Series **1683** 042062
[4] Sharapov V I, Kamalova R I, Kudryavtseva E V and Rogachev S S 2017 Byul. Izobr. **5** Russian Federation Patent 2629321
[5] Roslyakov P V and Zakirov I A 2001 *Non-stoichiometric combustion of natural gas and fuel oil in heat power plants* (Moscow: MEU)
[6] Pavlov K F, Romankov P G and Noskov A A 1981 *Examples and tasks on the course of processes and devices of chemical technology. Textbook for high schools* (Leningrad: Chemistry)
[7] *Rules for the technical operation of power plants and networks of the Russian Federation* 2003 (Ministry of Energy of Russia) **229**

AUTHOR INDEX